T0261214

C O S M O S

VOLUME 1

FOUNDATIONS OF NATURAL HISTORY

Foundations of Natural History is a series from the Johns Hopkins University Press for the republication of classic scientific writings that are of enduring importance for the study of origins, properties, and relationships in the natural world.

Published in the Series:

C O S M O S

A SKETCH

OF

A PHYSICAL DESCRIPTION OF THE UNIVERSE

BY

ALEXANDER VON HUMBOLDT

TRANSLATED FROM THE GERMAN
BY E. C. OTTÉ

Naturae vero rerum vis atque majestas in omnibus momentis fides caret, si quis modo
partes ejus ac non totam complectatur animo.—Plin., *Hist. Nat.*, lib. vii, c. 1.

VOLUME I

WITH AN INTRODUCTION
BY NICOLAAS A. RUPKE

THE JOHNS HOPKINS UNIVERSITY PRESS
Baltimore and London

Introduction ©1997 The Johns Hopkins University Press
All rights reserved. Published 1997
Printed in the United States of America on acid-free paper
Reprinted from the English-language edition published by
Harper & Brothers, Publishers, New York, 1858
Johns Hopkins Paperbacks edition, 1997
06 05 04 03 02 01 00 99 98 97   5 4 3 2 1

The Johns Hopkins University Press
2715 North Charles Street
Baltimore, Maryland 21218-4319
The Johns Hopkins Press Ltd., London

Library of Congress Cataloging-in-Publication Data will be found at the
end of this book.

A catalog record for this book is available from the British Library.

ISBN 0-8018-5502-0

Frontispiece: Alexander von Humboldt, etching, Jacobs Pinx, Prudhomme.
From the 1858 Harper & Brothers edition of *Cosmos,* volume 1.

INTRODUCTION
TO THE 1997 EDITION

Nicolaas A. Rupke

THE LIBERAL STANDARD OF SCIENCE LITERACY
OF THE MID-NINETEENTH CENTURY

Humboldt's *Cosmos* was an immensely popular book, a great success, both for its author and its publishers, appearing in five volumes under the title *Kosmos: Entwurf einer physischen Weltbeschreibung* (vol. 1, 1845; vol. 2, 1847; vol. 3, 1850–51; vol. 4, 1858; and a posthumous vol. 5, 1862). There were five nineteenth-century authorized German editions, including a German-American one, and a string of translations (see below). Of the original edition alone, published by J. G. Cotta in Stuttgart and Tübingen, in response to popular demand, no fewer than 22,000 copies of volume 1 were printed, 20,000 of volume 2, and 15,000 each of volumes 3, 4, and 5. As far as sales are concerned, *Cosmos* made Humboldt the most successful author of his generation.[1]

Friedrich Wilhelm Heinrich Alexander von Humboldt, born into a family of recent, minor nobility,[2] published *Cosmos* toward the end of his long life (1769–1859), when he was in his late seventies and eighties. He is best known for this book, although he was internationally celebrated long before the year in which the first volume of *Cosmos* appeared. Humboldt wrote many books and articles, which can be grouped into four categories.

First, there were his early works, the most famous being a book on organic electricity, *Versuche über die gereizte Muskel-*

und Nervenfaser (2 vols., 1797, [1798]). During the period of
these *Jugendarbeiten*, Humboldt studied at various German
universities in succession, was employed in the Prussian mining
service, and in 1795 conducted an extensive tour of the Swiss,
French, and Italian Alps.

Second, there were the publications that resulted from his
renowned journey of exploration in equatorial America (1799–
1804), on which he was accompanied by the French botanist
Aimé Bonpland (1773–1858). Most of these publications were
written during Humboldt's Parisian period (1804–27), when he
worked as an independently wealthy private scholar. The production of the American oeuvre (Humboldt's *amerikanisches
Reisewerk*) was a major undertaking, which exhausted his personal fortune. The 30-volume work carried the collective title
*Voyages aux régions équinoxiales du Nouveau Continent fait en
1799, 1800, 1801, 1802, 1803, et 1804, par Al. de Humboldt et
A. Bonpland.* It dealt with botany, plant geography, zoology,
physical geography, political economy, and included such clas-
sics as the *Essai politique sur le royaume de la Nouvelle-
Espagne* (2 vols., 1808–11) and the *Relation historique du
voyage aux régions équinoxiales du Nouveau Continent* (3 vols.,
1814–17, 1819–21, 1825–31).

Third, there were the publications that resulted from his
relatively short Russian journey of exploration (1829)—a journey that lasted approximately seven months and during which
Humboldt turned sixty years of age—for example, *Asie centrale.
Recherches sur les chaines de montagnes et la climatologie
comparée* (3 vols, 1843). These belonged to Humboldt's Berlin
period (1827–59), during which he was employed as a royal
chamberlain at the Prussian Court.

The fourth and last category includes the publications of
Humboldt's old age, the most famous being *Cosmos*, a work
that constituted both the summary of many of Humboldt's
lifelong interests and a holistic digest of the scientific study of
celestial and terrestrial phenomena. On the coattails of its
success, a new, third German edition of *Ansichten der Natur*

was produced (1849; first and second editions, 1808, 1826, respectively), followed by new translations into English (1849), French (1851), and other languages.[3]

THE LONG GESTATION OF COSMOS

Humboldt became infamous for dragging out the writing of several of his major books. The results of his American journey, for example, which were to be published in just two years, took some three and a half decades to appear in print (1805–39). The first, introductory part, his *Relation historique* (*Personal Narrative*) began appearing nearly a decade after Humboldt's return and was never completed: the travel account comes to an end with Humboldt's first visit to Cuba, in the spring of 1801, and the planned fourth volume failed to materialize (see below). Part of the problem was that while working on his *Reisewerk*, Humboldt more than once changed its overall plan and significantly expanded some of the parts.[4]

The same fate of delay, expansion, and ultimate incompletion befell *Cosmos*. The book began to take shape with a series of Parisian salon lectures, delivered from the end of 1825 until the beginning of 1827. Upon his return from Paris to Berlin, Humboldt delivered similar lectures (November 1827–April 1828): 61 lectures at the University of Berlin and 16 at the Music Academy ("Sing-Akademie").[5] No sooner had Humboldt completed his lecture series than he planned to get an *Entwurf einer physischen Weltbeschreibung* in print.[6] Yet it took nearly two decades for this plan to be realized.

Apparently, the *Cosmos* concept antedated the Paris-Berlin lectures by many years. As Humboldt explained in an often-cited letter, written 27 October 1834 to the diplomat-writer Karl August Varnhagen von Ense (1785–1858), he began the project fifteen years earlier (i.e., 1819), in French, entitling it "Essai sur la physique du monde."[7] As Hanno Beck points out, much earlier yet, in 1793, Humboldt had already conceived the idea of a "physique du monde," but at this time "monde" still meant to Humboldt "Erde" ("the earth"), and his vision of the study of the physical world was holistic only to the extent that it combined physical and human geography. The additional concept of covering in a single book both terrestrial and celestial phenomena, changing the meaning of "monde" or "Welt" to "universe," was not born until the Paris and Berlin lectures and—as Beck believes—did in fact not assume mature clarity until 1834.[8]

In this year, too, Humboldt chose the title *Cosmos* for his planned book, in order to emphasize that he was not writing a conventional physical geography (*physische Erdbeschreibung*), but that both heaven and earth were an integral part of his conception. Also, this title, in addition to indicating the vast scope of his book, gave expression to Humboldt's aesthetic-holistic epistemology, since the word *cosmos* in Homeric times had meant "ornament" and "elegance" and later had come to denote the order or harmonious arrangement of the world.[9]

ELISE C. OTTÉ AND THE ENGLISH TRANSLATIONS

International interest in *Cosmos* was so strong and the commercial potential of translations so considerable that renditions into other languages began appearing as soon as volume 1 had been completed. By the time the fourth volume had come out, *Cosmos* had been translated into no fewer than eleven different languages,[10] and in some languages more than one translation had been produced (there were, for example three in English).

The first of these translations was by the Bristol eye surgeon Augustin Prichard (1818–98), who had studied in Berlin and Vienna and held an M.D. degree from the University of Berlin. His was a pirated edition and the quality of the translation was poor; moreover, he only translated the first two volumes of *Cosmos* (1845, 1848). Humboldt feared that the Prichard edition would do serious damage to his reputation in England.[11] The second translation (4 vols., 1846–58)—an authorized one— was done by Elizabeth Juliana Sabine, née Leeves (1807–79), wife of the Humboldtian geophysicist Edward Sabine (1788– 1883). Elizabeth was an accomplished translator, who also rendered Humboldt's *Ansichten der Natur* into English (2 vols., 1849). She received help from Humboldt's friend, the Prussian diplomat and evangelical theologian Christian Karl Josias Bunsen (1791–1860), at the time an envoy in London. Edward Sabine, who was much interested in terrestrial magnetism and promoted the establishment of a global network for the measurement of magnetic variations, provided footnote commentary to his wife's *Cosmos* translation, aided in this by the London-based Scottish geologist Roderick Impey Murchison (1792–1871) and by other eminent metropolitan colleagues.

At the time, the Sabine translation ranked as the most authoritative English *Cosmos*, and it formed the basis for virtually all of the reviews that appeared in leading British periodicals (see below). Although the translation was "said to be singularly accurate and elegant," the book was criticized for being expensive and for not being entirely complete in that some passages that might be unpalatable to "national prejudices" were left out;[12] Humboldt himself commented on the omission of a passage in which uncertainty was expressed about the geographic location of the cradle of mankind: "das sonderbare Eiland" ("that strange island"), he wrote to Bunsen in reference to such British, religious traditionalism.[13] Moreover, only the first part of the fourth volume ever appeared.

The third, and in the long run most often reprinted and widely sold translation, was the work of Elise C. Otté (1818– 1903). It covered in five volumes (1848–65) Humboldt's original four and was reprinted in parts and as a whole a number of times, both in Britain and in the United States. In working on volumes 1 and 2, Otté had the benefit of two existing translations, the Sabine one and the French translation by the astronomer Hervé-Auguste-Étienne-Albans Faye (1814–1902).

For her fourth volume she had as co-translator the chemist Benjamin Horatio Paul (dates unknown), and for her fifth the surgeon-naturalist William Sweetland Dallas (1824–90). The 1858 American edition of Otté's translation is reprinted here (without the posthumous index volume of 1862). The text of the 1858 American edition remained unchanged, but volume 1 of the edition carried a frontispiece picture of Humboldt different from the one in the original London edition, and on the title page it incorrectly stated "in four volumes" (there were five).

Otté had not been the first choice for the job of *Cosmos* translator. The bookseller and publisher Henry George Bohn (1796–1884), who, from the moment volume 1 of *Cosmos* had seen the light, had wanted to produce a popular edition (*Volksbuch*) in English, was delayed in carrying out his plan because of "the procrastination and debauchery" of the person to whom initially he had entrusted the task of translation.[14] With Otté, however, the publisher had engaged an exceptional woman, a scholar in her own right, who had acquired from her philologist stepfather a taste for Anglo-Saxon and Icelandic. During a period in the United States, Otté studied physiology at Harvard University, and back in Britain she became known for her scientific translations and for essays that she published in scientific periodicals. Born in Copenhagen of a Danish father and an English mother, Otté continued to show much interest in Scandinavia, publishing a Danish and a Swedish grammar and (her best-known book) *Scandinavian History* (1875). As with Elise Sabine, she translated not only *Cosmos* but also *Ansichten der Natur* (*Views of Nature* [1850]).[15]

THE PLAN OF *COSMOS*

Cosmos grew larger than originally projected, as the writing progressed and Humboldt attempted to keep up with the latest scientific developments and literature.[16] Originally, Humboldt envisaged a two-volume book that would be "the work of my life," encompassing the entire natural world, from galaxies, at one extreme, to moss growing on granitic rocks, at the other. The material was to be divided into a general part, the introductory prolegomena, which Humboldt regarded as "die Hauptsache" ("the main thing"), followed by a general part, consisting of discussions in detail. The range of the latter would be as follows:

> Outer space—all of astronomical physics—the solid body of our earth, interior, exterior, electro-magnetism of the interior. Vulcanism, i.e., the reaction of the interior of a planet at its surface. Arrangement of [rock] masses. A brief geognosy—sea—atmosphere—climate—organic phenomena—plant geography. Geography of animals—human races and language—of which the subsequent organisation (articulation of sounds) is governed by intelligence (whose product, manifestation is language).[17]

In the end, the four volumes that Humboldt actually wrote covered only the introductory parts (vols. 1 and 2), the details of astronomy or the "uranologic" phenomena (vol. 3), and the detailed discussions of geology or the "telluric" phenomena (vol. 4). A further part or volume on the treatment in detail of organic life, its global distribution, especially of human races and human culture, never saw the light. The posthumous, fifth volume contained a further fragment of Humboldt's writing (adding to the telluric phenomena of vol. 4), some addenda and corrigenda, and a very elaborate index of no fewer than 1,146 pages, prepared at Humboldt's request and directions by the philologist-librarian Johann Karl Eduard Buschmann (1805–80).

The incompletion of the final product adds significance to volume 1, in which Humboldt presents a complete survey of the range of natural phenomena that he intended to discuss in the detailed volumes, including plant and animal distribution and humankind. This first volume contains two of four prolegomenal parts, one defining the very concept and boundaries of the discipline of a "physical description of the universe," another describing the factual content of the visible heaven and earth.

Humboldt argues that it is a far superior accomplishment to perceive connections than to study isolated facts. To help readers grasp his holistic concept of the universe, Humboldt speaks of "Naturgemälde" ("painting of nature," inadequately translated by Otté as "delineation of nature"), adding the element of an aesthetic appreciation of nature to its scientific study. Use of the fine arts metaphor and of a poetic presentation forms the vehicle for bringing the results of specialized scientific research to the public at large, integrating the study of nature with its human, societal context.[18]

The scope of the *Naturgemälde* of volume 1, consisting of a substantial, first part on celestial phenomena, was wider than that of Humboldt's previous publications record. The astronomical part of his American journey, the *Receuil d'observations astronomique* (2 vols., 1808, 1811) had in fact been edited by the Berlin University astronomer Jabbo Oltmanns (1783–1833). The celestial section of volume 1 of *Cosmos* included not only stars and planets, but also meteors, meteorites, and the physics of light. For the description of these, Humboldt had to rely almost exclusively on the secondary literature and on the assistance of his circle of scientific friends, such as, in the case of comets, another Berlin University astronomer, namely, Johann Franz Encke (1791–1865). With respect to the terrestrial part, which covered phenomena ranging from the shape, density, and internal heat of the Earth to climate and weather, Humboldt was on more familiar ground. His particular expertise concerned volcanoes, stratigraphy, and physical geography—on all of which, primarily during the 1810s and 1820s, he had published books and articles, such as his *Vues des Cordillères* (1810–13), "Des lignes isothermes et de la distribution de la chaleur sur le globe,"[19] and *Essai géognostique sur le gisement des roches dans les deux hémisphères* (1823).

Whereas the celestial part represented an expansion of the

scope of Humboldt's expertise, the concluding section of volume 1, dealing with plants, animals, and humans, was oddly narrowed in comparison with his earlier writings, Humboldt having previously devoted entire volumes to human geography and political economy. Examples of such books range from the already cited *Essai politique sur le royaume de la Nouvelle-Espagne* (Humboldt's most popular pre-*Cosmos* book) to the *Essai politique sur l'ile de Cuba* (2 vols., 1826) (a favorite of Humboldt himself).[20] Possible reasons for this narrowing are discussed below.

<hr />

THE BERGHAUS ATLAS

Humboldt made an effort to enhance the readability of *Cosmos* by gathering many details into extensive footnotes. Yet there were no illustrations to facilitate access to his story. Why should this have been? He had, after all, pioneered some remarkable representational forms for the purpose of illustrating the global variations of a range of natural and cultural phenomena. Humboldt's political essay on the kingdom of New Spain (Mexico), for example, had been accompanied by a splendid *Atlas géographique et physique du royaume de la Nouvelle-Espagne* (1808–12), which featured improved topographic maps of Central America, a novel cross section of the Mexican plateau, diagrams of the gold and silver production of Mexican mines compared with mines elsewhere, a world map of transport routes of "metal riches," and block diagrams showing the ratios of territorial expanse and population for various colonies compared with the colonizing motherlands.[21]

Most famously, Humboldt added to his *Essai sur la géographie des plantes* (1805–07) the "Tableau physique des Andes et des pays voisins," the iconic cross-sectional profile of South America, from the Pacific to the Atlantic at the latitude of Chimborazo, showing the zoned occurrence of different plants at different altitudes. Later, Humboldt joined to his *Nova genera et species plantarum* (vol. 1, 1815) a table titled "Geographiae plantarum lineamenta" that depicted the vertical, zoned distribution of plants on three mountains, one in the tropics, another in the temperate region, and a third in a polar region—demonstrating Humboldt's famous "law" that the changes in plant distribution by altitude match the ones by latitude (Figure 1).

Shortly after, Humboldt devised the enormously successful representational technique of isolines, proposing in an 1817 paper to depict the distribution of heat across the Northern Hemisphere by means of isotherms.[22] Moreover, his American oeuvre included such atlases as the beautiful *Atlas géographique et physique des régions équinoxiales du Nouveau Continent* (1814–34) with thirty-nine plates. The many botanical volumes, executed together with Bonpland, were also lavishly and exquisitely illustrated.

There is less of a contradiction here than one might think, in that from the outset Humboldt did plan a volume of illustrations to accompany *Cosmos* and such a volume was in fact produced. Already in 1827, he expressed the wish to his cartographic collaborator Heinrich Berghaus (1797–1884) that a volume of illustrations to what was to become *Cosmos* be begun.[23] Book and atlas became separated, however, when Humboldt signed a contract for the publication of *Cosmos* with J. G. Cotta in Stuttgart and Berghaus ended up publishing the atlas with the then-leading cartographic publishing house Justus Perthes in Gotha. The maps of Berghaus's *Physikalischer Atlas* appeared over a period of eleven years (1838–48) and were published in two volumes in 1845 and 1848, respectively. During the preparations, Humboldt functioned as an advisory editor. The second edition of 1852 explicitly mentioned Humboldt in the title.[24] According to Hanno Beck, Humboldt's connection with the *Physikalischer Atlas* was so close that a

Figure 1. *Opposite page*. Examples of Humboldtian visual representation, integrating physiographic data from across the globe, such as average and maximum heights of mountains (top) and distribution of vegetation by altitude versus latitude (bottom). From Heinrich Berghaus, *Physikalischer Schul-Atlas* (Gotha: Justus Perthes, 1850), plate 20.

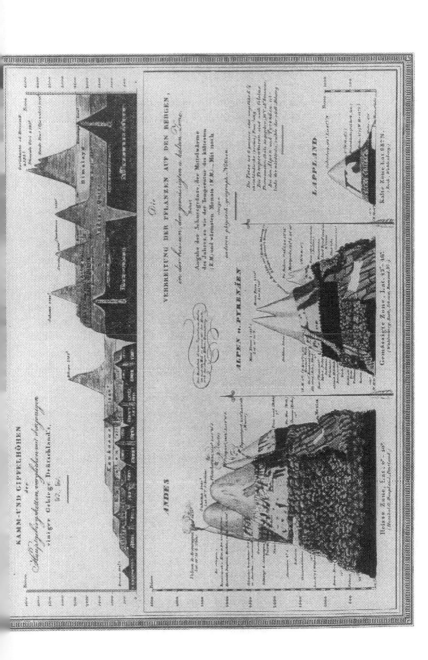

KAMM- UND GIPFELHÖHEN
der
Hauptgebirgsketten, verglichen mit Hohengruppe einiger Gebirge Deutschland's,

Die
VERBREITUNG DER PFLANZEN AUF DEN BERGEN,
in der Linea der gemässigten halben Zone.

ANDES

ALPEN u. PYRENÄEN

LAPPLAND

Heisse Zone, Lat. 0°–10°

Gemässigte Zone, Lat. 42°–46°

Kalte Zone, Lat. ERTN.

proper bibliographical reference should cite Humboldt as se-
nior author.[25]

A similar atlas, Traugott Bromme's *Atlas zur Physik der Welt*
(1851), was also presented as an illustrated commentary to
Cosmos and carried a second title page, *Atlas zu Alex. v.
Humboldt's Kosmos.* The degree to which this atlas was a
plagiarized version of Berghaus's atlas is a matter of controversy.[26] An authorized English version of the Berghaus atlas,
The Physical Atlas (1848; 2d ed. 1856; reduced version, 1850),
was produced by Alexander Keith Johnston (1804–71), Edinburgh geographer in ordinary to the queen.

The second edition of the *Physikalischer Atlas* contained
ninety-three distribution maps, divided into eight groups: meteorology and climatography, hydrology and hydrography, geology, earth magnetism, plant geography, zoogeography, anthro-
pology, and ethnography. This 1852 edition represented a
complete delineation of the Humboldtian program; and because
Cosmos itself remained unfinished, truncated by Humboldt's
death before the completion of the fifth volume, the Berghaus
atlas is a significant document in solving the problem of what
was on the Humboldtian agenda.

These Humboldtian visual representations were tremendously successful and popular—copied, plagiarized, and imitated in a wide range of other publications. In an essay on
physical geography, discussing three publications, one of which
was Johnston's atlas, the *Quarterly Review* (1848) stated enthu-
siastically:

> Such works are as essential to the study of physical geography as
> are experiments to the chemical student, or models and dia-
> grams to instruction in the mechanical sciences—or what is
> more pertinent in this case, as common maps to common
> geography. Their linear delineations to the eye are an admirable
> shorthand-writing, conveying impressions to the mind far more
> explicit and forcible than any mere descriptions can afford;
> suggesting comparisons and relations, and giving facilities of
> reference, which can in no other way be equally attained.[27]

RAVE REVIEWS

If we want to appreciate Humboldt's *Cosmos* for the contribution it made to the dynamics of historical change, it is less important to know what Humboldt actually wrote and what precisely he meant by it than it is to explore the impact of *Cosmos* on the readership. A crucial question is, What did *Cosmos* mean to Humboldt's contemporaries, and what did they read in it? One approach to answering this question is to survey the reviews of Humboldt's last book.[28]

Initial reactions showed a great, international interest. Across Europe and in the United States, substantial reviews in leading periodicals appeared, written by some of the best-known scientists or literati. There was nothing in print comparable to *Cosmos*. In Britain, the scientific expositor Mary Fairfax Greig Somerville (1780–1872) had written *On the Connexion of the Physical Sciences* (1834) and, more specifically, *Physical Geography* (1848), but neither of these books possessed the authority or the magnitude of Humboldt's endeavor. Somerville's *Physical Geography* was, however, translated into German but was put in its place by being entitled *Kosmos für gebildete Frauen* (*Cosmos for Educated Women*) (1851).[29] This German title, given to Somerville's treatise on physical geography, illustrates that at least six years after the appearance of volume 1 of *Cosmos*, the word *cosmos* had become a generic description of a Humboldtian study of the physical world.

The most important factor in generating widespread interest in *Cosmos* was Humboldt's fame rather than the contents of his work. Humboldt was regarded not merely as a German scientist but as the consummate European, "Mr. European scientist," who formed the hub of an international network of correspondence, knew a variety of languages, had lived in both Paris and Berlin, wrote with equal facility in French and German, possessed a genial temperament, commanded an encyclopeadic knowledge, and was a nobleman and a statesman-patron of the sciences—and therefore deemed the person most qualified to sum up and weigh the natural sciences at midcentury. As one reviewer exclaimed:

> Were the republic of letters to alter its constitution, and choose a sovereign, the intellectual sceptre would be offered to Alexander von Humboldt. The New World would send deputies across the Atlantic to assist at his installation, and the princes and philosophers of every clime,—the autocrats of the East and the democrats of the West,—would hail the enterprising traveller who trod the mountain crests of Europe, ascended the American Cordilleras, and explored the auriferous beds of the Uralian chain.[30]

Reactions to *Cosmos* differed from country to country. In Germany, almost without exception, superlatives of praise were added to expressions of national pride. The *Heidelberger Jahrbücher für Literatur* (1845), for example, in the person of the aged Heidelberg physicist Georg Wilhelm Muncke (1772–1847), raised *Cosmos* on a high pedestal by stating that the work was beyond the critique of a single learned reviewer: the grand task of providing a description of the universe in the coherence of its parts had previously been partially attempted by the physicists/astronomers Johann Heinrich Lambert (1728–77), Joseph Johann von Littrow (1781–1840), and Pierre-Simon, Marquis de Laplace (1749–1827), yet its completion had been left to Humboldt.[31] The five-volume *Briefe über Alexander von Humboldt's Kosmos* (1848–60) was more substantive than the reviews in denoting *Cosmos* as an exceptional accomplishment; this multivolume work was a commentary for "educated laymen," which appeared nearly in tandem with the successive *Cosmos* volumes and to which the Freiberg geologist Carl Bernhard von Cotta (1808–79) made a major contribution. The supplementary fifth volume, appearing in the year following Humboldt's death, was in fact a Humboldt biography, written by the Munich scholar Wilhelm Constantin Wittwer (1822–?).[32] The Cotta commentary was a notable instance of a burgeoning *Cosmos* spin-off industry.

In France, praise for Humboldt and his book was strong,

too, although not as unexampled. It was gently pointed out that
Cosmos contained no new, penetrating insights nor novel contributions to any scientific speciality but was a condensation of
principal discoveries and theories. In a forty-page essay for the
Revue des deux mondes (1846), it was pointed out by Jean-
Louis-Armaud de Quatrefages de Bréau (1810–92), the Parisian
zoologist and later holder of the Chair of the Natural History of
Man at the Muséum d'Histoire Naturelle, that Humboldt in
the individual branches of science on which he had worked
(chemistry, botany, geology, and zoology) had his superiors in
Antoine-Laurent Lavoisier (1743–94), Antoine-Laurent de
Jussieu (1748–1836), Leopold von Buch (1774–1853), and
Georges Cuvier (1769–1832), but that as "physicien du globe"
("physicist of the globe") Humboldt deserved to be ranked
alongside these "rois d'intelligence" ("kings of the mind").[33]

By far the longest and largest number of reviews were
written in Britain. One reason for this type of output was that
reviews by periodicals, in which there appeared major essays
often with substantial extracts from the discussed texts, were
more numerous in the British Isles than on the Continent. The
Scottish physicist and writer David Brewster (1781–1868), in
the *North British Review,* and his fellow Scotsman, James
David Forbes (1809–68), the physicist-geologist and Edinburgh
professor of natural philosophy, in the *Quarterly Review,* used
the occasion to write a brief biography of Humboldt, which
Otté used for the biographical sketch of Humboldt that she
added to her *Cosmos* translation.[34] Brewster's biography of
Humboldt was particularly good, and his essay was translated
into French,[35] attracting praise also in the German *Magazin für
die Literatur des Auslandes.* Nevertheless, Brewster was accused of having plagiarized a German source, namely, the
Illustrirte Zeitung of 1844.[36] Brewster's claim was patently
untrue that no biography of Humboldt existed in the English
language when he wrote his biography, because *The Life,
Travels, and Researches of Alexander von Humboldt,* by the
Edinburgh naturalist William MacGillivray (1796–1852), had

been in print since 1832 and had enjoyed a second printing, an American edition, and translations.

BRITISH DETRACTIONS

Unlike the German and French reviews of *Cosmos,* the British reviews, while containing highly favorable passages about both author and book, pronounced explicit and, in places, irreverent criticisms. Although the Tory *Quarterly Review* devoted more pages to *Cosmos* than any other British periodical did, twice publishing a long essay—first a thirty-eight-page one by Forbes and later a thirty-two-page one by the London society physician Henry Holland (1788–1873)—it was at the same time the most critical of the periodicals.[37] Only the astronomer John Frederick William Herschel (1792–1871), himself of German descent, in his sixty-page essay for the Whig *Edinburgh Review,* withheld any but minor, technical censure.[38]

Cosmos was criticized for being too wordy and too voluminous: more could have been attained if Humboldt had written less. Moreover, the book was far too ambitious a scheme. Humboldt's sketch of a physical description of the universe was "premature" and "partially failed."[39] Holland observed:

> At the risk of appearing presumptious we must express our doubt whether he has ever entirely defined the term of Cosmos to his own mind. A grand and spacious idea was before him; congenial to the temperament of German thought, and according well with his own vast and various knowledge, and his desire to concentrate the labours of a life in one great closing work. He sought to mark by the name the magnitude of the conception. But the conception itself is beyond the power of adequate fulfilment, even by one possessing the resources of our author. The Universe, as expressing all the material phenomena of nature (and we shall see presently that Humboldt has superadded other topics having relation to the human faculties and progress), is too vast a theme for a single man or a single work.[40]

Added to this British dislike of the encyclopeadic, system-bound style of German scholarship was a sense that Humboldt, in his role as judge of European science, had sold the British short. Humboldt's account, it was felt, was not fair and balanced but slanted away from Britain toward the countries where he had resided: France and Germany. Forbes in particular accused Humboldt of having lavished an unfair share of praise on his compatriot, Leopold von Buch: "We miss the recognition of the place which our geologists are entitled to hold in the history of science." Paris and Berlin had been put on the map, but not London:

> Neither France nor Germany has any right to complain of the share which Humboldt has assigned to them in the great struggle for physical discovery. But we cannot rise from the careful perusal of this elaborate work without feeling that our own country has come off second, or rather *third,* best. The physics have (it seems to us) been written for the longitude of Paris, and the geology for that of Berlin; and no one, we think, who is conversant with the scientific circles of those capitals, can fail to see that the selection of topics and of authors is tinged with the unconscious prejudice of local opinion.[41]

NO MENTION OF GOD

The most severe strictures upon *Cosmos* did not concern Humboldt's silence with respect to the accomplishments of London's geologists but his omission of God. As the gentleman-scholar John Crosse (1810–?), writing in the Benthamite *Westminster Review*, remarked: "A sketch of the universe in which the *word* 'God' appears nowhere, but the *spirit of God* is supposed everywhere, will perhaps be regarded as dangerously Atheistical by the stickler for *the word*."[42] To most of the British reviewers, Humboldt's demonstrations of the harmonies and beauty of the physical world required the mention of a Supreme Harmonizer. The absence of "proofs of divine design" was noted with dismay. Whereas in France the positivistically inclined orientalist-theologian Ernest Renan (1823–92), in his thirteen-page *Cosmos* review for *La liberté de penser,* explicitly praised Humboldt for having avoided the language of natural theology "as it is understood in England,"[43] the British reviewers sorely missed references to "the power, wisdom and goodness of God as manifested in the creation," as the multiauthor *Cosmos* of a sort, the Bridgewater Treatises, had put it. Forbes rebuked:

> We conceive it to be impossible for any well-constituted mind to contemplate the sum and totality of creation, to generalize its principles, to mark the curious relations of its parts, and especially the subtle chain of connexion and unity between beings and events apparently the most remote in space, time, and constitution, without referring more or less to the doctrine of final causes, and to the *design* of a superintending Providence. We call it the highest pedantry of intellect to put to silence suggestions which arise spontaneously in every mind, whether cultivated or not, when engaged in such contemplations; and we are sorry to observe in the work before us a silence on such topics so pointed as must attract the attention of at least every English reader.[44]

In his thirty-eight-page essay for the Congregational *British Quarterly Review,* the clergyman-astronomer Thomas John Hussey (1797–1866/67?) went further, chiding Humboldt not only for the omission of proper references to God but also for surreptitiously introducing Hegelian pantheism. This reproach echoed a wider concern that Humboldt might be making common cause with Berlin's Hegelian radicals. Humboldt had shown "the very height of affectation, or something worse," by completing a treatise on the harmony of the natural world "without one reference to Him whom Faith recognises as the Source and the Life of all things." Hussey did not feel that his review was the place for a conventional natural theology, continuing:

> But if Baron Humboldt is at liberty to refer us on almost every page to the eternal order and the eternal laws of Nature, it were surely hard that we should be denied all right of reference to Him, the alone eternal, without whose preordination we assert

that this order had never been, and without whose co-ordina-
tion, these laws had been powerless as the infant's whisper, to
direct or control the worlds which hang upon them.[45]

LIBERAL SCIENCE LITERACY

The British press did not stand alone when rebuking
Humboldt for failing to raise the banner of orthodox Christianity. On the Continent, especially in Catholic-conservative circles,
accusations were made of Hegelian radicalism or heterodoxy
with respect to the biblical creation story;[46] and the question
was disquietly posed: Does the author of Cosmos ever talk
about God? The *Journal historique et littéraire* (1858) answered in the affirmative; yet the passages that were cited as
proof from *Cosmos* demonstrated that Humboldt in fact avoided
using the name "God" and merely referred to "a first impulse."
The writer could show, however, that Humboldt, in his correspondence, also wrote about the "Almighty."[47]

For Humboldt to keep natural theology out of *Cosmos* was
unlikely a matter of unbelief or atheistical sympathies but
reflected a conviction that an expository scientific treatise was
not the proper place for the argument of design. To many of his
British critics, by contrast, it was inconceivable that a popular
exposition of science should be without the stated aim that the
study of nature leads up to nature's God; and a noticeable
feature of the British reviews was that they added the argument
of design to *Cosmos*. Thus, Humboldt's book was "domesticated" for an English readership by making it consonant with
natural theology.

Humboldt read the British reviews with great interest, but
pointed out that the accusations of atheism were unfounded.
Already in January 1845, when personally translating for the
French edition of *Cosmos* the introductory "Considérations"
("Reflections"), Humboldt added a self-defense in the form of
a reference to "the very Christian Immanuel Kant": all things
that are beyond the material world do not belong to the

physical description of the universe but to a higher kind of speculation.[48]

Historians of Victorian science have pointed out that, at an unspoken level, the language of natural theology played a mediating role in bringing people of different Christian creeds together in a latitudinarian pursuit of science.[49] Humboldt's concern, however, was not primarily with the literally parochial one of denominational strife. Religious conservatives in Prussia were in fact deeply disappointed that the author of *Cosmos* did not join "the victorious battle for the Christian revelation as foundation of German unity in church and state."[50] The mediating mission of *Cosmos* was first and foremost at the level of sociopolitical divisions, nationally and internationally. Humboldt was "the instructor of mankind," as the liberal *Fraser's Magazine* remarked.[51]

The same sentiment was expressed by an American commentator, James Davenport Welpley (1817–72). In an essay on the *Cosmos* prolegomena for the Whig *American Review*, Welpley commented that those who studied the world and contemplated the connections of its multifarious phenomena were knitted together in a cosmopolitan network, "making common cause against ignorance and prejudice." He continued: "if the world is ever to be harmonized, it must be through a community of knowledge, for there is no other universal or non-exclusive principle in the nature of man."[52] London's *Athenaeum* used similar language in a review of volume 2 of *Cosmos*:

> Never since the world began has there been any epoch so marked as the present by the wonderful application of the powers of nature to the want of man. We hold the key by which we may lock in one common brotherhood all the nations of Europe—and finally the world; making peace the universal desire and the interchange of thought the universal instinct of every people.[53]

People who were scientifically literate and knowledgeable about the orderly arrangement of the world would rise above party division. *Cosmos* could thus become an argument from nature in the promotion of liberal causes, social and political. Those who disapproved of revolutionary chaos, on the one hand, or reactionary despotism, on the other, in the sequence of events from the French Revolution to the Revolution of 1848, could take up the *Cosmos* banner. As Humboldt stated at the beginning of volume 1, nature is the realm of freedom. By implication, the proper study of nature would lead to liberty, away from religious and political absolutism and oppression. Thus, *Cosmos* could become a metonymy of the politics of liberalization.

After the initial excitement, press interest precipitously dropped. No major reviews in leading reviewing periodicals appeared at the completion of volume 4, nor at the completion of volume 5, published posthumously; and the *Athenaeum*, which did publish short notices on volumes 3 and 4 regarded any discussions in detail as essentially superfluous.[54] Whereas some reviewers had hailed volume 1 as a major work of science, the five-volume completed *Cosmos* no longer was given this label but was seen as an enclopeadia of all the things an educated person would need to know about the physical world— a work of science literacy. Its encyclopaedic nature was underscored by the fifth volume, with its enormously long and detailed index to the entire work. Some reviewers had from the very beginning characterized *Cosmos* as a distillation of European scientific culture. Welpley observed the following about the first few chapters of *Cosmos*:

> A careful examination discovers them to be an exposition of the very spirit of liberal culture. They show the tendency of the most enlightened minds in Europe and in the civilized world. They seem to give an impression of the age itself, in its best features, and might serve, almost, as a preface to its intellectual history.[55]

Cosmos was the supremely respectable book of science literacy, defining both the scope and the form of scientific discourse. Humboldt combined an astonishing memory and

breadth of grasp with an uncanny talent for avoiding anything
even vaguely disreputable, whether religious or sociopolitical.
Perceptions that had overtly divisive implications he tended to
circumnavigate.

A fine example of this was the then-frequently depicted
division of the surface of the earth into a continental and an
oceanic hemisphere: the first contained the maximum of dry
land, and the second continued the greatest concentration of
water. A map of such hemispheres had first appeared in
Handatlas über alle Theile der Erde by the amateur cartographer Adolf Stieler (1775–1836) and formed the first map of the
geological section of Berghaus's atlas.[56] This very popular map
was reproduced in simplified form in a wide range of texts. The
reason for its popularity was its Eurocentricity: the center of
the continental hemisphere was northwestern Europe. Humboldt's friend and fellow geographer Carl Ritter (1779–1859),
together with many followers, used this map to argue that
Europe formed the geographical hub of the world, which
produced the superiority of European culture, thus adding up
to a mandate for Europe's imperial sway of the globe.[57]

The precise center of the terrestrial hemisphere was a
matter of nationalistic controversy among Europeans: some
centered the map on Paris, others on Berlin, and yet others on
southern England. Humboldt, in his *Cosmos* discussion of the
global distribution of sea and land, remained aloof from this
jingoistically and deterministically tainted issue, merely observing that the Northern Hemisphere contains nearly three times
as much land as the Southern Hemisphere and that the Eastern
Hemisphere has far more land than the Western Hemisphere.[58]

The English, however, did not let Humboldt get away with
this and claimed the centrality of their island. Herschel, for
example, in his *Edinburgh Review* essay, commented that
Humboldt's discussion made for "a much less lively and distinct
impression of the law of distribution of the globe into two
hemispheres, a terrene and an aqueous one, the former having
Great Britain, the latter her antipodes, for its vertex." He

continued: "In fact, if we endeavour to include the maximum of land in one hemisphere, and that of water in the other, according to our present knowledge of the globe, we shall find as the centre of the terrene hemisphere a point in the south of England somewhat eastward of Falmouth."[59]

NARROWED AGENDA

Margarita Bowen has looked at Humboldt's *Cosmos* in the broad context of European political developments of the second quarter of the nineteenth century, pointing to the pressures exerted on Humboldt by the reactionary forces of absolute monarchism, both in France and in Prussia. In the secondary literature, Humboldt's politics have in places been characterized as liberal to radical/revolutionary. His support for Latin American revolutions and his encounter in 1804 with the "liberator of Spanish America" Simon Bolivar (1783–1830) have been cited in support. Moreover, Humboldt's participation in the 22 March 1848 funeral procession of revolutionaries killed in the Berlin uprising has been given prominence.[60]

Bowen sees the agenda of Humboldt's *Cosmos* narrowed, and in fact impoverished, under the pressures of reactionary politics. She locates Humboldt on the sociopolitical map of the period by contrasting him with several prominent contemporaries, among whom was the socialist Karl Heinrich Marx (1818–83), who was deprived of his citizenship by the Prussian government in 1845, the year of publication of volume 1 of *Cosmos*, which was dedicated by Humboldt "in tiefer Ehrfurcht und mit herzlichem Dankgefühl" ("in deep reverence and with sincere feelings of gratitude") to the Prussian King Friedrich Wilhelm IV (1795–1861). In 1845, too, Marx, who had been living in Paris, was expelled from the French capital. Marx's allies in Paris and the radical Hegelians in Berlin accused Humboldt of complicity.[61]

This accusation was probably unjust. But though Humboldt, like Marx, was concerned with social reform, Humboldt's position—in Bowen's words—was more that of a liberal conservative, advocating freedom of communication and an improved constitution through democratic processes; and Humboldt evidently hoped to achieve this reform initially with the support of an enlightened monarchy. Given his sense of historical continuity of social patterns, Humboldt was likely to be "wary of radical change through revolution."[62] Bowen also believes that "some form of conservative intervention" suppressed the publication of the planned fourth volume of the *Relation historique*, which was to have dealt with the period from May 1801 to August 1804, when Humboldt traveled in the modern states of Colombia, Ecuador, Peru, Mexico, Cuba (for a second time), and the United States, and which would have been of great interest in Europe, especially to antimonarchists.[63] Such intervention, or suppression, is speculation; but an explanation needs to be found for the fact that in 1810, when the wars of independence from the Spanish Crown in South and Central America were about to erupt, the manuscript, though it was virtually finished and the printing of it had already begun, was destroyed, obliging Humboldt to pay the publisher the substantial compensation of 9,500 francs.[64] The surmise that Humboldt was less liberal than widely thought is not without foundation.

A pinpointing of the precise sociopolitical anchoring ground of *Cosmos* is impossible, because the missing part of the never-completed book would have been the most sensitive and informative part, dealing with human geography. One can put forward the Bowenite argument that the ultimate nonappearance of the detailed discussions in the planned human section of *Cosmos* indicated that Humboldt had yielded to reactionary restoration pressures, in that he had put off the writing of this part, reluctant to deal with divisive sociopolitical issues. Given Humboldt's tendency to mediate and seek the middle ground, it is equally possible to argue that Humboldt's delay was motivated by a concern not to have to deal with the traditional racist and nationalistic prejudices of his associates.

Beck firmly maintains that the *Cosmos* program was not

narrowed but simply was cut short by Humboldt's death and that the program's full intended extent can be read from the scope and content of Berghaus's atlas.[65] Humboldt's active participation in Berghaus's *Hertha. Zeitschrift für Erd-, Völker- und Staatenkunde* and in its successor magazine *Annalen der Erd-, Völker- und Staatenkunde* would support the contention that Humboldt had all along stood for an inclusion of humans and human society in the study of physical geography. If the Berghaus atlas can indeed be used as an index of Humboldt's agenda, then not only its politically liberal features but also its racist Eurocentricity should be taken into account. The four anthropological maps of Berghaus's atlas, which were separately published under the title *Allgemeiner anthropographischer Atlas* (1852), depicted the global distribution of the human races and their diet, of human diseases, of dress, and, on the last map, of occupation, religion, government, and mental development. Berghaus, who completed the text to the maps on 10 March 1848, that is, shortly after the Paris uprising and just days before the revolts of Vienna and Berlin, expected in the long term "segensreiche Wirkungen" ("blessed results") from the 1848 revolutions. He condemned the absolute monarchy as a primitive institution that would disappear as *"Humanität und Bildung"* ("humanity and education") would spread.[66]

Simultaneously, however, Berghaus expressed doubts about democratic rule and republicanism, for example, as found in the United States, regarding the American political system as unstable and liable to give way to government by a privileged minority and limited monarchy. Next to showing opposition to conservative monarchism, the Berghaus maps exhibited features that were emphatically Christian, anti-Catholic and pro-Protestant, racist and strongly Eurocentric.[67]

Such views, expressed in illustrations, captions, and accompanying text were more blunt and partisan than is likely ever to have been replicated by Humboldt. Yet Humboldt's close association with Berghaus and his atlas, and an absence of indications that he distanced himself from the contents of the atlas, should be taken into account. And the conclusion is difficult to counter that Humboldt, in his late *Cosmos* period, was no more revolutionary than any liberal who preferred a constitutional over an absolute monarchy and that Humboldt's liberalism was tainted with various common and bourgeois prejudices of his time. He skillfully managed to keep most of these prejudices out of his writings, making it possible for different and opposing political parties to lay claim to the *Cosmos* heritage.

COSMOS AFTER HUMBOLDT'S DEATH

Among academics, it has been the geographers who have claimed Humboldt the founder of physical geography, as one of theirs.[68] Yet *Cosmos* should not be seen too exclusively as a source book for Humboldtian ideas about the physical world and humans' study of it. The enormous popularity of the book, among the educated of the entire Western world, marked it out, far more broadly, as the standard of science literacy. In fact, as pointed out above, *Cosmos* made no original contributions to any branch of modern science, not even to geography; and its holistic-aesthetic philosophy had no impact on any scientific discipline during the mid- and late-nineteenth century. On the contrary, even in geography, the person who left an organized school behind was not Humboldt but Ritter, who did not pursue the epistemological issue, nor did his several famous disciples, such as the Swiss-American geographer-geologist Arnold Henry Guyot (1807–84) in the United States.[69] *Cosmos* marked the end of a period.

Like Darwin, Humboldt has become a field of historical scholarship by himself. The literature on Humboldt is vast, and its bibliography alone fills several volumes. Entire catalogs have been published, separately listing Humboldt studies in the German language, in Polish, in Russian, and in Spanish. Dozens of Humboldt biographies have been written, and many volumes of collected essays on Humboldt and his legacy have been edited. The "Humboldt industry" curve, tracing the output of books and papers over time, has not followed an even course, but shows distinct peaks. Humboldt biographies in the form of books or articles already appeared during the last one-third of his life. In the course of the two decades following his death, a first modest publications peak took shape, centered on the 1869 centenary of his birth. This period was followed by a long flat stretch. New peaks emerged, first around 1959, the centenary of Humboldt's death, and, a decade later, around 1969, the bicentenary of his birth.[70]

Humboldt's name became synonymous with the title of his last book, *Cosmos*, and came to denote the totality of the sciences as well as their popular treatment (Figure 2). The *Monatsschrift für die gesamten Naturwissenschaften* (*Monthly for all the Sciences*), edited by the physics lecturer Georg Krebs (1833–1907), which intended to bring to the educated public the results of modern science in a popular form, was simply called *Humboldt*.[71] The ideological use that could be made of Humboldt by identifying him with the *Cosmos* ideal was two-pronged. Already Welpley, in his essay on *Cosmos* for the *American Review*, formulated the two prongs with clarity. The harmonious relations that Humboldt sought to portray, connecting the different parts of the physical universe, were also present among all those who shared in the knowledge of the cosmos. Thus, as the author of *Cosmos*, Humboldt, represented the cause of liberalism and internationalism.

The *Cosmos* ideal not only removed the lateral barriers of nationalism (first prong) but it also removed vertical class barriers in any society (second prong). Humboldt spoke "to the free intelligence of man," and his *Cosmos* made science "accessible to the people."[72] The fusion of science and poetic presentation in *Cosmos* was aimed, it has recently been argued, at the emancipation of all classes of society through scientific education.[73]

Interestingly, during the jingoistic exchanges between the German academics, on the one hand, and the Anglo-French academics, on the other, that accompanied World War I and

Figure 2. In the case of *Cosmos*, author and book title became virtually synonymous, as shown by this title page of a monthly magazine for all the sciences, which simply was called "Humboldt."

that concerned the question of which of the warring camps had produced the superior culture, the German side made little, if any use of Humboldt's name and reputation. It is significant, that, during the Third Reich, Humboldt was no Nazi favorite and appears to have been a difficult subject to put to work for fascism. One reason may have been that his philo-Semitism was a matter of public record. Yet, an attempt was made to incorporate Humboldt into the conception of a Greater Europe of German National Socialism.[74]

In stark contrast to its insignificant role in the ideological battles that accompanied the two world wars, Humboldt scholarship became a major instrument of German cultural politics in the post-1945 period. The name of Humboldt, through *Cosmos*, came to stand for what was, and had all along been, good about German culture: not militarism and nationalism, but humanism, democracy, and cosmopolitan liberalism. Humboldt was put forward—in particular by German emigrés to the English-speaking world[75]—as "the good European," defining a Germany with which one could coexist and cooperate.

During the 1945–90 period of the division of Germany, both East and West claimed Humboldt for itself, portraying him in colors that matched those of their respective post-Nazi political banners. In the West, the universality and cosmopolitanism of Humboldt as manifested as *Cosmos* was emphasized and turned into an instrument of postwar reconciliation and the re-establishment of international contacts, especially with the enemies and victims of the Nazi period.[76] In the East, the inclusion of the common man in the *Cosmos* ideal was emphasized, and Humboldt was depicted as a typical democrat who had shown concern for repressed peoples: thus, *Cosmos* was linked to socialism and international communism.[77] More recently, in the wake of the collapse of the Berlin Wall, the *Cosmos* ideal has served the cause of German reunification.[78] The recent and present-day use that is made of Humboldt's *Cosmos*, combined with the book's historical significance as a major document of mid-nineteenth-century science, amply justifies the reprint presented here by the Johns Hopkins University Press.

Notes

I warmly thank Wolfgang Böker, Robert J. Brugger, Michael Dettelbach, and David N. Livingstone for their help and instructive comments.

1. These and many other details can be found in Hanno Beck, ed., "Zu dieser Ausgabe des *Kosmos*," in Alexander von Humboldt, *"Kosmos": Entwurf einer physischen Weltbeschreibung* in Hanno Beck, ed., *Alexander von Humboldt: Studienausgabe* (Darmstadt: Wissenschaftliche Buchgesellschaft, 1993), vol. 7(2) p. 355.

2. Kurt-R. Biermann relates that the brothers Alexander and Wilhelm von Humboldt believed in an oral family legend that they were barons (*Freiherren*) but that the formal conferment of the title did not occur until 1875, on Wilhelm's progeny, sixteen years after Alexander's death (see Biermann, *Alexander von Humboldt* [Leipzig: B. G. Teubner Verlag, 1983], p. 11).

3. For Humboldt's bibliography, see Julius Löwenberg, "Alexander von Humboldt: Bibliographische Übersicht seiner Werke, Schriften und zerstreuten Abhandlungen," in Karl Bruhns, ed., *Alexander von Humboldt. Eine wissenschaftliche Biographie* 1872, reprint ed. (Osnabrück: Otto Zeller, 1969), vol. 2, pp. 485–552. See also Wolfgang-Hagen Hein, "Verzeichnis der Schriften Alexander von Humboldts," in Wolfgang-Hagen Hein, ed., *Alexander von Humboldt: Leben und Werk* (Frankfurt: Weisbecker, 1985), pp. 310–14. The difficulties and pitfalls of the Humboldt bibliography are discussed by Ulrike Leitner, *Alexander von Humboldts Werk: Probleme damaliger Publikation und heutiger Bibliographie* (Berlin: Alexander-von-Humboldt-Forschungsstelle, 1992).

4. Leitner, *Alexander von Humboldts Werk*.

5. Humboldt objected to the idea that these lectures be published from notes taken by listeners. Lecture notes (*Kollegnachschriften*) were published, however, although much later than 1828, when announcements of the publication of a two-volume edition of Humboldt's lectures appeared in several papers (see "Ouvrage de M. de Humboldt sur la géographie physique," *Nouvelles Annales des Voyages* 39 [1828]: 387–88). See *Alexander von Humboldts Vorlesungen über physikalische Geographie nebst Prolegomenen über die Stellung der Gestirne* (Berlin: Miron Goldstein, 1934). The sixteen lectures delivered at the Berlin Music Academy were edited by Jürgen Hamel and Klaus-Harro Tiemann, *Alexander von Humboldt über das Universum* (Frankfurt am Main and Leipzig: Insel, 1993).

6. Humboldt to Varnhagen, 15 April 1828, in *Briefe von Alexander von Humboldt an Varnhagen von Ense aus den Jahren 1827 bis 1858,* 4th ed. (Leipzig: Brockhaus, 1860), pp. 4–5.

7. Humboldt to Varnhagen, 24 October 1834, ibid., pp. 20–23.

8. Beck, *Alexander von Humboldt,* pp. 345–50.

9. Humboldt himself discussed the meaning of the word *cosmos* in some detail; see this volume, pp. 68–70.

10. Humboldt to Cotta, 12 December 1857, cited by Leitner, *Alexander von Humboldts Werk,* p. 25.

11. Humboldt to Bunsen, 18 July 1845, in *Briefe von Alexander von Humboldt an Christian Carl Josias Freiherr von Bunsen* (Leipzig: F. A. Brockhaus, 1869), p. 70.

12. Elise Otté, this volume, p. 6.

13. Humboldt to Bunsen, 28 September 1846, *Briefe von Humboldt an Freiherr von Bunsen,* p. 82.

14. Bohn to Cotta, 30 January 1849, Cotta-Archiv (Stiftung der Stuttgarter Zeitung), Schiller-Nationalmuseum/Deutsches Literaturarchiv, Marbach am Neckar.

15. Otté has an entry in the *Dictionary of National Biography.* Elisabeth Sabine is included in her husband's entry.

16. Humboldt in a public letter, 22 September 1847, in *Beilage zur Allgemeinen Zeitung,* 1 October 1847.

17. *Briefe von Humboldt an Varnhagen von Ense,* p. 20.

18. Ibid., p. 21.

19. Alexander von Humboldt, "Sur des lignes isothermes," *Annales de chimie et de physique* 5 (1817): 102–112. See note 22.

20. On the international reception of Humboldt's major publications, see Nicolaas A. Rupke, "Enlightened imperialism: The European appreciation of Humboldt's American journey of exploration," in Charles W. J. Withers and David N. Livingstone, eds., *Geography and Enlightenment* (Chicago: University of Chicago Press, in press).

21. See Anne Godlewska, "Humboldt's visual thinking: From Enlightenment vision to modern science," ibid.

22. Alexander von Humboldt, "Sur des lignes isothermes," *Annales de chimie et de physique* 5 (1817): 102–112. This article was accompanied by an illustration of isotherms across the Northern Hemisphere. Humboldt's isotherm paper was multiply published in French, appearing also in various British and German periodicals.

23. Humboldt to Berghaus, 20 December 1827, in *Briefwechsel Alexander von Humboldt's mit Heinrich Berghaus aus den Jahren 1825 bis 1858* (Leipzig: Hermann Costenoble, 1863), vol. 1, pp. 117–19.

24. Heinrich Berghaus, *Physikalischer Atlas: Eine unter der fördernden Anregung Alexander von Humboldt's verfasste Sammlung von 93 Karten, auf denen die hauptsächlichsten Erscheinungen der anorganischen und organischen Natur nach ihrer geographischen Verbreitung und Vertheilung bildlich dargestellt sind* (Gotha: Justus Perthes, 1852). See also Jane Camerini, "The *Physical Atlas* of Heinrich Berghaus: Distribution maps as scientific knowledge," in Renato G. Mazzolini, ed., *Non-Verbal Communication in Science Prior to 1900* (Florence: Leo S. Olschki, 1993), pp. 479–512.

25. Beck, "Zu dieser Ausgabe des *Kosmos,*" p. 363.

26. The plagiarism thesis is put forward by Gerhard Engelmann, *Heinrich Berghaus: Der Kartograph von Potsdam* (Halle/Saale: Deutsche Akademie der Naturforscher Leopoldina, 1977), pp. 69–71. Beck disagrees with Engelmann (see Beck, "Zu dieser Ausgabe des *Kosmos*," pp. 385–87).

27. [Henry Holland], "Physical geography," *Quarterly Review* 83 (1848), p. 308.

28. See Rupke, "Enlightened imperialism."

29. *Kosmos für gebildete Frauen: Nach der dritten Auflage des englischen Werks der Mstrs. Maria Somerville, unter Bezugnahme auf den Berghaus'schen physikalischen Atlas, bearbeitet von Dr. Carl Hartmann* (Grimma and Leipzig, 1851).

30. [David Brewster], "Physical phenomena of the universe," *North British Review* 4 (1845), p. 202. A similar sentiment was expressed in a review of the first volume of the Dutch translation of *Cosmos* (see the review of *Cosmos* in the Dutch journal *Nieuwe Algemene Vaderlandsche Letter Oefeningen*, 1848, p. 241).

31. [Georg Wilhelm] Muncke, "v. Humboldt's *Kosmos*," *Heidelberger Jahrbücher für Literatur* 38 (1845), pp. 810–11. From among the many other German reviews, see, for example, "Alexander von Humboldt's 'Kosmos,'" *Literarische Zeitung*, 31 May 1845, pp. 677–83; and "A. von Humboldt's 'Kosmos,'" *Blätter für literarische Unterhaltung*, 7 February 1848, pp. 149–51.

32. *Briefe über Alexander von Humboldt's "Kosmos": Ein Commentar zu diesem Werke für gebildete Laien*, 5 vols. (Leipzig: T. O. Weigel, 1848–60).

33. Armand de Quatrefages, "Illustrations scientifiques: Alexandre de Humboldt (*Cosmos*)," *Revue des deux mondes* 14 (1846): 755. A discussion of the somewhat muted reception of *Cosmos* in France appeared in the *Deutsche Allgemeine Zeitung*, 28 May 1845, p. 1393.

34. Otté, this volume, p. 5.

35. "F. H. Alexandre de Humboldt," *Revue Brittanique*, ser. 5, vol. 30 (1845), pp. 241–86.

36. "Alex. von Humboldt: Die englische und die deutsche Wissenschaft," *Magazin für die Literatur des Auslandes*, 17 January 1846, pp. 29–30. The allegedly plagiarized article was "Friedrich Heinrich Alexander Freiherr von Humboldt," *Illustrirte Zeitung* 2 (January–June 1844), pp. 38–41.

37. [J. D. Forbes], "Humboldt's *Cosmos*," *Quarterly Review* 77 (1845–46), pp. 154–91. [Henry Holland], "Humboldt's *Cosmos*: Sidereal astronomy," ibid. 94 (1853), pp. 49–79. For a discussion of Humboldt's appeal to the British, see W. H. Brock, "Humboldt and the British: A note on the character of British science," *Annals of Science* 50 (1993): 365–72.

38. [John Herschel], "Humboldt's *Kosmos*," *Edinburgh Review* 87 (1848), pp. 170–229.

39. [T. Hussey], "*Kosmos*: Astronomical science," *British Quarterly Review* 3 (1846), p. 322. In addition, Humboldt's "discursiveness" was spoken of in a review of volume 3 of *Cosmos* in the *Athenaeum*, 12 April 1851, p. 401.

40. [Holland], "Physical geography," p. 50.

41. [Forbes], "Humboldt's *Cosmos*," pp. 189–90. In addition, the section that involves the "progress of geological investigation" was regarded as unsatisfactory in a review of volume 2 of *Cosmos* in the *Athenaeum*, 8 July 1848, p. 675.

42. [John Crosse], "The Vestiges, etc.," *Westminster Review* 44 (1845), p. 154. This essay combined a review of *Cosmos* with that of the anonymously published *Vestiges of the Natural History of Creation*.

43. Ernest Renan, "*Cosmos* de M. de Humboldt," *La liberté de penser: Revue philosophique et littéraire* 2 (1848), p. 571.

44. [Forbes], "Humboldt's *Cosmos*," pp. 163–64.

45. [Hussey], "*Kosmos*," p. 354.

46. See Beck, "Zu dieser Ausgabe des *Kosmos*," pp. 411–12. A Catholic reaction to *Cosmos*, concerned with Humboldt's omission of the "Genesis and geology" issue, appeared in the *Historisch-politische Blätter für das katholische Deutschland* 16 (1845), pp. 453–56. See also "Humboldt un die Augsburger Postzeitung," *Die Grenzboten: Zeitschrift für Politik und Literatur* 4 (1845), pp. 591–92.

47. *Journal historique et littéraire* 24 (1857–58), pp. 493–97.

48. Humboldt to Bunsen, 4 January 1846, *Briefe von Humboldt an Freiherr von Bunsen*, p. 72. See also Humboldt to Encke, 26 February 1845, in Julius Löwenberg, *Alexander von Humboldt's Werk*, p. 39.

49. See, for example, J. H. Brooke, *Science and Religion: Some Historical Perspectives* (Cambridge: Cambridge University Press, 1991); and J. B. Morrell and Arnold Thackray, *Gentlemen of Science: The Early Years of the British Association for the Advancement of Science* (Oxford: Clarendon Press, 1981).

50. See Michael Dettelbach's introduction to volume 2 of this edition of *Cosmos*, pp. xii–xiii.

51. "Alexander von Humboldt's *Kosmos*," *Fraser's Magazine* 37 (1848), p. 208.

52. J. D. Welpley, "Humboldt's *Cosmos*," *American Review* 3 (1846), p. 603.

53. *Athenaeum,* 12 February 1848, p. 162.

54. Ibid., 12 April 1851, p. 401.

55. Welpley, "Humboldt's *Cosmos*," p. 598. See also *Athenaeum*, 12 April 1851, p. 401; and *Athenaeum*, 6 November 1858, pp. 589–90.

56. "Erdkarte zur Übersicht der Vertheilung des Starren und Flüssigen" (separately printed 1839), in Heinrich Berghaus, *Physikalischer Atlas*, 2d ed. (Gotha: Justus Perthes, 1852), 3d section, "Geologie," no. 1.

57. See, for example, Carl Ritter, "Über geographische Stellung und horizontale Ausbreitung der Erdtheile," in *Abhandlungen der historisch-*

xl INTRODUCTION TO THE 1997 EDITION

philologischen Klasse der Königlichen Akademie der Wissenschaften zu Berlin, 1826 (Berlin: Königl. Akademie der Wissenschaften, 1829), pp. 106–107. See also Nicolaas A. Rupke, "Eurocentric ideology of continental drift," *History of Science* 34 (1996), pp. 251–72.

58. Humboldt, this volume, p. 292.

59. [Herschel], "Humboldt's *Kosmos,*" p. 213. See also [Holland], "Physical geography," p. 318.

60. The secondary literature on Humboldt's advocacy of liberal and revolutionary points of view is fairly extensive and mainly concerns his antislavery, philo-Semitism and support for Latin American revolutions and wars of independence. See, for example, Philip S. Foner, ed., *Alexander von Humboldt on Slavery in the United States* (Berlin: Humboldt University, 1984); and J. Fred Rippy and E. R. Brann, "Alexander von Humboldt and Simon Bolivar," *American Historical Review* 52 (1947), pp. 697–703. See also Lutz Raphael, "Freiheit und Wohlstand der Nationen: Alexander von Humboldts Analysen der politischen Zustände Amerikas und das politische Denken seiner Zeit," *Historische Zeitschrift* 260 (1995), pp. 749–76.

61. Margarita Bowen, *Empiricism and Geographical Thought: From Francis Bacon to Alexander von Humboldt* (Cambridge: Cambridge University Press, 1981), pp. 240–55.

62. Ibid., p. 252.

63. Ibid., p. 251.

64. See Beck, *Alexander von Humboldt* (Wiesbaden: Frank Steiner Verlag, 1961), vol. 2, p. 69.

65. Beck, "Zu dieser Ausgabe des *Kosmos,*" pp. 363–84, in particular p. 379.

66. Berghaus, *Physikalischer Atlas* (Gotha: Justus Perthes, 1848), vol. 2, p. 233.

67. Berghaus, *Allgemeiner Anthropographischer Atlas* (Gotha: Justus Perthes, 1852), p. 4.

68. See, for example, Hanno Beck, *Grosse Geographen* (Berlin: Dietrich Reimer, 1982), pp. 83–102. See also David N. Livingstone, *The Geographical Tradition: Episodes in the History of a Contested Enterprise* (Oxford: Blackwell, 1992), pp. 134–38; and Anne Buttimer, *Geography and the Human Spirit* (Baltimore: Johns Hopkins University Press, 1993), p. 59.

69. Bowen, *Empiricism and Geographical Thought,* p. 239.

70. See Nicolaas A. Rupke, "Alexander von Humboldt," in Arne Hessenbruch, ed., *Reader's Guide to the History of Science* (London: Fitzroy Dearborn, in press).

71. The monthly was edited by Krebs from 1882 until 1887.

72. Welpley, "Humboldt's *Cosmos,*" p. 603.

73. This point is made by, among others, Gisela Brude-Firnau, "Alexander von Humboldt's socio-political intentions: Science and poet-

ics," in Nancy Kaiser and D. E. Wellbery, eds., *Traditions of Experiment from the Enlightenment to the Present* (Ann Arbor: University of Michigan Press, 1992), pp. 45–61.

74. Walther Linden, *Alexander von Humboldt: Weltbild der Naturwissenschaft* (Hamburg: Hoffmann and Campe, 1940).

75. Examples are Helmut de Terra, *Humboldt: The Life and Times of Alexander von Humboldt, 1769–1859* (New York: Knopf, 1955); and Lotte Kellner, *Alexander von Humboldt* (London: Oxford University Press, 1963).

76. Examples are Joachim H. Schultze, *Alexander von Humboldt: Studien zu seiner universalen Geisteshaltung* (Berlin: de Gruyter, 1959); and Heinrich Pfeiffer, ed., *Alexander von Humboldt: Werk und Weltgeltung* (München: Piper, 1969). See also Hanno Beck's authoritative *Alexander von Humboldt*, 2 vols. (Wiesbaden: Steiner, 1959–61), which presents Humboldt as the good, and in some ways superior, German scientist whose memory could help reestablish Germany's international contacts, especially with the French, the Americans, the Russians, and the Jews.

77. Examples are Hans Ertel, ed., *Alexander von Humboldt, 14.9.1769–6.5.1859. Gedenkschrift zur 100. Wiederkehr seines Todestages. Herausgegeben von der Alexander von Humboldt-Kommission der Deutschen Akademie der Wissenschaften zu Berlin* (Berlin: Akademie-Verlag, 1959); and, another product of the Deutsche Akademie der Wissenschaften zu Berlin, *Alexander von Humboldt: Wirkendes Vorbild für Fortschritt und Befreiung der Menschheit* (Berlin: Akademie-Verlag, 1969). The larger part of the latter volume was written by Kurt-R. Biermann, East Germany's answer to West Germany's Hanno Beck.

78. See, for example, Martin Guntau, Peter Hardetert, and Martin Pape, eds., *Alejandro de Humboldt: La Naturaleza, idea y aventura* (Essen: Projekt Agentur, 1993). This volume accompanied an exhibition of Humboldtiana, which was staged first in what used to be West Germany, then in the former East Berlin, and finally in Venezuela.

In the meantime, English-speaking historians of science have become engaged in a scholarly discourse that was initiated by Susan Cannon in the 1970s. Cannon coined the phrase "Humboldtian science" to describe the Humboldt-inspired activities of many of Humboldt's scientific contemporaries. These scientists not only made Baconian observations and measurements in the study of nature but also brought these observations and measurements together into global patterns, looking for connections and laws and making use of isomaps and other representational devices. Elaborating on this notion, a number of interesting studies have followed in Cannon's wake. See Susan F. Cannon, *Science in Culture: The Early Victorian Period* (New York, Dawson and Science History Publications, 1978), pp. 73–110; Malcolm Nicolson, "Alexander von Humboldt, Humboldtian science and the origins of the study of vegetation," *History of Science* 25 (1987): 167–94; Roderick W. Home, "Humboldtian science revisited: An Australian case study," *History of Science* 33 (1995): 1–22; Michael Dettelbach, "Humboldtian science," in N. Jardine, J. A. Secord, and E. C. Spary, eds., *Cultures of Natural History* (Cambridge: Cambridge University Press, 1996), pp. 287–304; Nicolaas Rupke, "Humboldtian medicine," *Medical History* 40 (1996): 293–310.

C O S M O S

VOLUME 1

TRANSLATOR'S PREFACE.

I can not more appropriately introduce the Cosmos than by presenting a brief sketch of the life of its illustrious author.* While the name of Alexander von Humboldt is familiar to every one, few, perhaps, are aware of the peculiar circumstances of his scientific career and of the extent of his labors in almost every department of physical knowledge. He was born on the 14th of September, 1769, and is, therefore, now in his 80th year. After going through the ordinary course of education at Göttingen, and having made a rapid tour through Holland, England, and France, he became a pupil of Werner at the mining school of Freyburg, and in his 21st year published an "Essay on the Basalts of the Rhine." Though he soon became officially connected with the mining corps, he was enabled to continue his excursions in foreign countries, for, during the six or seven years succeeding the publication of his first essay, he seems to have visited Austria, Switzerland, Italy, and France. His attention to mining did not, however, prevent him from devoting his attention to other scientific pursuits, among which botany and the then recent discovery of galvanism may be especially noticed. Botany, indeed, we know from his own authority, occupied him almost exclusively for some years; but even at this time he was practicing the use of those astronomical and physical instruments which he afterward turned to so singularly excellent an account.

The political disturbances of the civilized world at the close

* For the following remarks I am mainly indebted to the articles on the Cosmos in the two leading Quarterly Reviews.

of the last century prevented our author from carrying out
various plans of foreign travel which he had contemplated,
and detained him an unwilling prisoner in Europe. In the
year 1799 he went to Spain, with the hope of entering Africa
from Cadiz, but the unexpected patronage which he received
at the court of Madrid led to a great alteration in his plans,
and decided him to proceed directly to the Spanish posses-
sions in America, " and there gratify the longings for foreign
adventure, and the scenery of the tropics, which had haunted
him from boyhood, but had all along been turned in the diametrically opposite direction of Asia." After encountering
various risks of capture, he succeeded in reaching America,
and from 1799 to 1804 prosecuted there extensive researches
in the physical geography of the New World, which have indelibly stamped his name in the undying records of science.

Excepting an excursion to Naples with Gay-Lussac and
Von Buch in 1805 (the year after his return from America),
the succeeding twenty years of his life were spent in Paris, and
were almost exclusively employed in editing the results of his
American journey. In order to bring these results before the
world in a manner worthy of their importance, he commenced
a series of gigantic publications in almost every branch of
science on which he had instituted observations. In 1817,
after twelve years of incessant toil, four fifths were completed,
and an ordinary copy of the part then in print cost considerably more than one hundred pounds sterling. Since that time
the publication has gone on more slowly, and even now, after
the lapse of nearly half a century, it remains, and probably
ever will remain, incomplete.

In the year 1828, when the greatest portion of his literary
labor had been accomplished, he undertook a scientific journey
to Siberia, under the special protection of the Russian government. In this journey—a journey for which he had prepared
himself by a course of study unparalleled in the history of
travel—he was accompanied by two companions hardly less
distinguished than himself, Ehrenberg and Gustav Rose, and

the results obtained during their expedition are recorded by
our author in his *Fragments Asiatiques*, and in his *Asie
Centrale*, and by Rose in his *Reise nach dem Oural*.   If the
Asie Centrale had been his only work, constituting, as it
does, an epitome of all the knowledge acquired by himself and
by former travelers on the physical geography of Northern
and Central Asia, that work alone would have sufficed to
form a reputation of the highest order.

I proceed to offer a few remarks on the work of which I
now present a new translation to the English public, a work
intended by its author "to embrace a summary of physical
knowledge, as connected with a delineation of the material
universe."

The idea of such a physical description of the universe had,
it appears, been present to his mind from a very early epoch.
It was a work which he felt he must accomplish, and he devoted almost a lifetime to the accumulation of materials for
it.   For almost half a century it had occupied his thoughts;
and at length, in the evening of life, he felt himself rich
enough in the accumulation of thought, travel, reading, and
experimental research, to reduce into form and reality the
undefined vision that has so long floated before him.   The
work, when completed, will form three volumes.   The *first*
volume comprises a sketch of all that is at present known of
the physical phenomena of the universe; the *second* comprehends two distinct parts, the first of which treats of the in-
citements to the study of nature, afforded in descriptive poetry, landscape painting, and the cultivation of exotic plants;
while the second and larger part enters into the consideration
of the different epochs in the progress of discovery and of the
corresponding stages of advance in human civilization.   The
third volume, the publication of which, as M. Humboldt himself informs me in a letter addressed to my learned friend and
publisher, Mr. H. G. Bohn, "has been somewhat delayed,
owing to the present state of public affairs, will comprise the
special and scientific development of the great Picture of Na-

ture." Each of the three parts of the *Cosmos* is therefore, to
a certain extent, distinct in its object, and may be considered
complete in itself. We can not better terminate this brief
notice than in the words of one of the most eminent philos
ophers of our own country, that, "should the conclusion correspond (as we doubt not) with these beginnings, a work will
have been accomplished every way worthy of the author's
fame, and a crowning laurel added to that wreath with which
Europe will always delight to surround the name of Alexan
der von Humboldt."

In venturing to appear before the English public as the in-
terpreter of "*the great work of our age*,"* I have been en-
couraged by the assistance of many kind literary and scientific
friends, and I gladly avail myself of this opportunity of ex-
pressing my deep obligations to Mr. Brooke, Dr. Day, Pro
fessor Edward Forbes, Mr. Hind, Mr. Glaisher, Dr. Percy, and
Mr. Ronalds, for the valuable aid they have afforded me.

It would be scarcely right to conclude these remarks without a reference to the translations that have preceded mine.
The translation executed by Mrs. Sabine is singularly accurate and elegant. The other translation is remarkable for
the opposite qualities, and may therefore be passed over in silence. The present volumes differ from those of Mrs. Sabine
in having all the foreign measures converted into corresponding English terms, in being published at considerably less
than one third of the price, and in being a translation of the
entire work, for I have not conceived myself justified in omit-
ting passages, sometimes amounting to pages, simply because
they might be deemed slightly obnoxious to our national prejudices.

* The expression applied to the Cosmos by the learned Bunsen, in
his late Report on Ethnology, in *the Report of the British Association
for* 1847, p. 265.

AUTHOR'S PREFACE.

In the late evening of an active life I offer to the German public a work, whose undefined image has floated before my mind for almost half a century. I have frequently looked upon its completion as impracticable, but as often as I have been disposed to relinquish the undertaking, I have again—although perhaps imprudently—resumed the task. This work I now present to my cotemporaries with a diffidence inspired by a just mistrust of my own powers, while I would willingly forget that writings long expected are usually received with less indulgence.

Although the outward relations of life, and an irresistible impulse toward knowledge of various kinds, have led me to occupy myself for many years—and apparently exclusively—with separate branches of science, as, for instance, with descriptive botany, geognosy, chemistry, astronomical determinations of position, and terrestrial magnetism, in order that I might the better prepare myself for the extensive travels in which I was desirous of engaging, the actual object of my studies has nevertheless been of a higher character. The principal impulse by which I was directed was the earnest endeavor to comprehend the phenomena of physical objects in their general connection, and to represent nature as one great whole, moved and animated by internal forces. My inter course with highly-gifted men early led me to discover that, without an earnest striving to attain to a knowledge of special branches of study, all attempts to give a grand and general view of the universe would be nothing more than a vain illusion. These special departments in the great domain of nat-

ural science are, moreover, capable of being reciprocally fruc-
tified by means of the appropriative forces by which they are
endowed. Descriptive botany, no longer confined to the narrow circle of the determination of genera and species, leads
the observer who traverses distant lands and lofty mountains
to the study of the geographical distribution of plants over the
earth's surface, according to distance from the equator and vertical elevation above the sea. It is further necessary to investigate the laws which regulate the differences of temperature and climate, and the meteorological processes of the atmosphere, before we can hope to explain the involved causes
of vegetable distribution ; and it is thus that the observer who
earnestly pursues the path of knowledge is led from one class
of phenomena to another, by means of the mutual dependence
and connection existing between them.

I have enjoyed an advantage which few scientific travelers
have shared to an equal extent, viz., that of having seen not
only littoral districts, such as are alone visited by the majority
of those who take part in voyages of circumnavigation, but
also those portions of the interior of two vast continents which
present the most striking contrasts manifested in the Alpine
tropical landscapes of South America, and the dreary wastes
of the steppes in Northern Asia. Travels, undertaken in districts such as these, could not fail to encourage the natural
tendency of my mind toward a generalization of views, and to
encourage me to attempt, in a special work, to treat of the
knowledge which we at present possess, regarding the sidereal
and terrestrial phenomena of the Cosmos in their empirical
relations. The hitherto undefined idea of a physical geography has thus, by an extended and perhaps too boldly imagined a plan, been comprehended under the idea of a physical
description of the universe, embracing all created things in the
regions of space and in the earth.

The very abundance of the materials which are presented
to the mind for arrangement and definition, necessarily impart
no inconsiderable difficulties in the choice of the form under

which such a work must be presented, if it would aspire to the honor of being regarded as a literary composition. Descriptions of nature ought not to be deficient in a tone of life-like truthfulness, while the mere enumeration of a series of general results is productive of a no less wearying impression than the elaborate accumulation of the individual data of observation. I scarcely venture to hope that I have succeeded in satisfying these various requirements of composition, or that I have myself avoided the shoals and breakers which I have known how to indicate to others. My faint hope of success rests upon the special indulgence which the German public have bestowed upon a small work bearing the title of *Ansichten der Natur*, which I published soon after my return from Mexico. This work treats, under general points of view, of separate branches of physical geography (such as the forms of vegetation, grassy plains, and deserts). The effect produced by this small volume has doubtlessly been more powerfully manifested in the influence it has exercised on the sensitive minds of the young, whose imaginative faculties are so strongly manifested, than by means of any thing which it could itself impart. In the work on the Cosmos on which I am now engaged, I have endeavored to show, as in that entitled *Ansichten der Natur*, that a certain degree of scientific completeness in the treatment of individual facts is not wholly incompatible with a picturesque animation of style.

Since public lectures seemed to me to present an easy and efficient means of testing the more or less successful manner of connecting together the detached branches of any one science, I undertook, for many months consecutively, first in the French language, at Paris, and afterward in my own native German, at Berlin (almost simultaneously at two different places of assembly), to deliver a course of lectures on the physical description of the universe, according to my conception of the science. My lectures were given extemporaneously, both in French and German, and without the aid of written notes, nor have I, in any way, made use, in the present work.

of those portions of my discourses which have been preserved
by the industry of certain attentive auditors.   With the exception of the first forty pages, the whole of the present work
was written, for the first time, in the years 1843 and 1844.

A character of unity, freshness, and animation must, I
think, be derived from an association with some definite
epoch, where the object of the writer is to delineate the present condition of knowledge and opinions.   Since the additions constantly made to the latter give rise to fundamental
changes in pre-existing views, my lectures and the Cosmos
have nothing in common beyond the succession in which the
various facts are treated.   The first portion of my work con
tains introductory considerations regarding the diversity in
the degrees of enjoyment to be derived from nature, and the
knowledge of the laws by which the universe is governed ; it
also considers the limitation and scientific mode of treating a
physical description of the universe, and gives a general picture of nature which contains a view of all the phenomena
comprised in the Cosmos.

This general picture of nature, which embraces within its
wide scope the remotest nebulous spots, and the revolving
double stars in the regions of space, no less than the telluric
phenomena included under the department of the geography
of organic forms (such as plants, animals, and races of men),
comprises all that I deem most specially important with regard to the connection existing between generalities and specialities, while it moreover exemplifies, by the form and style
of the composition, the mode of treatment pursued in the selection of the results obtained from experimental knowledge.
The two succeeding volumes will contain a consideration of
the particular means of incitement toward the study of nature (consisting in animated delineations, landscape painting,
and the arrangement and cultivation of exotic vegetable
forms), of the history of the contemplation of the universe, or
the gradual development of the reciprocal action of natural
forces constituting one natural whole ; and, lastly, of the spe-

cial branches of the several departments of science, whose mutual connection is indicated in the beginning of the work. Wherever it has been possible to do so, I have adduced the authorities from whence I derived my facts, with a view of affording testimony both to the accuracy of my statements and to the value of the observations to which reference was made.   In those instances where I have quoted from my own writings (the facts contained in which being, from their very nature, scattered through different portions of my works), I have always referred to the original editions, owing to the importance of accuracy with regard to numerical relations, and to my own distrust of the care and correctness of translators.   In the few cases where I have extracted short passages from the works of my friends, I have indicated them by marks of quotation ; and, in imitation of the practice of the ancients, I have invariably preferred the repetition of the same words to any arbitrary substitution of my own paraphrases.   The much-contested question of priority of claim to a first discovery, which it is so dangerous to treat of in a work of this uncontroversial kind, has rarely been touched upon.   Where I have occasionally referred to classical antiquity, and to that happy period of transition which has rendered the sixteenth and seventeenth centuries so celebrated, owing to the great geographical discoveries by which the age was characterized, I have been simply led to adopt this mode of treatment, from the desire we experience from time to time, when considering the general views of nature, to escape from the circle of more strictly dogmatical modern opinions, and enter the free and fanciful domain of earlier presentiments.

It has frequently been regarded as a subject of discouraging consideration, that while purely literary products of intellectual activity are rooted in the depths of feeling, and interwoven with the creative force of imagination, all works treating of empirical knowledge, and of the connection of natural phenomena and physical laws, are subject to the most marked modifications of form in the lapse of short periods of time, both

by the improvement in the instruments used, and by the consequent expansion of the field of view opened to rational ob
servation, and that those scientific works which have, to use
a common expression, become *antiquated* by the acquisition
of new funds of knowledge, are thus continually being con-
signed to oblivion as unreadable.   However discouraging such
a prospect must be, no one who is animated by a genuine love
of nature, and by a sense of the dignity attached to its study,
can view with regret any thing which promises future additions and a greater degree of perfection to general knowledge.
Many important branches of knowledge have been based upon
a solid foundation which will not easily be shaken, both as regards the phenomena in the regions of space and on the earth ;
while there are other portions of science in which general
views will undoubtedly take the place of merely special ;
where new forces will be discovered and new substances will
be made known, and where those which are now considered
as simple will be decomposed.   I would, therefore, venture to
hope that an attempt to delineate nature in all its vivid animation and exalted grandeur, and to trace the *stable* amid the
vacillating, ever-recurring alternation of physical metamorphoses, will not be wholly disregarded even at a future age.

Potsdam, Nov., 1844.

CONTENTS OF VOL. I.

GENERAL REVIEW OF NATURAL PHENOMENA.

SUMMARY.

Insight into the connection of phenomena as the aim of all natural
investigation. Nature presents itself to meditative contemplation as a
unity in diversity. Differences in the grades of enjoyment yielded by
nature. Effect of contact with free nature; enjoyment derived from
nature independently of a knowledge of the action of natural forces, or
of the effect produced by the individual character of a locality. Effect
of the physiognomy and configuration of the surface, or of the character
of vegetation. Reminiscences of the woody valleys of the Cordilleras
and of the Peak of Teneriffe. Advantages of the mountainous region
near the equator, where the multiplicity of natural impressions attains
its maximum within the most circumscribed limits, and where it is
permitted to man simultaneously to behold all the stars of the firmament and all the forms of vegetation—p. 23-33.

Tendency toward the investigation of the causes of physical phenom
ena. Erroneous views of the character of natural forces arising from
an imperfect mode of observation or of induction. The crude accu
mulation of physical dogmas transmitted from one century to another.
Their diffusion among the higher classes. Scientific physics are asso
ciated with another and a deep-rooted system of untried and misunder
stood experimental positions. Investigation of natural laws. Appre
hension that nature may lose a portion of its secret charm by an inquiry
into the internal character of its forces, and that the enjoyment of na
ture must necessarily be weakened by a study of its domain. Advant
ages of general views which impart an exalted and solemn character
to natural science. The possibility of separating generalities from
specialities. Examples drawn from astronomy, recent optical discov
eries, physical geognosy, and the geography of plants. Practicabil
ity of the study of physical cosmography—p. 33-54. Misunderstood
popular knowledge, confounding cosmography with a mere encyclopedic enumeration of natural sciences. Necessity for a simultaneous regard for all branches of natural science. Influence of this study on
national prosperity and the welfare of nations; its more earnest and
characteristic aim is an inner one, arising from exalted mental activity.
Mode of treatment with regard to the object and presentation; reciprocal connection existing between thought and speech—p. 54-5C

The notes to p. 28-33. Comparative hypsometrical data of the elevations of the Dhawalagiri, Jawahir, Chimborazo, Ætna (according to the
measurement of Sir John Herschel), the Swiss Alps, &c.—p. 28. Rarity

of palms and ferns in the Himalaya Mountains—p. 29. European v~g. etable forms in the Indian Mountains—p. 30. Northern and southern limits of perpetual snow on the Himalaya; influence of the elevated plateau of Thibet—p. 30–33. Fishes of an earlier world—p. 46.

Limits and Method of Exposition of the Physical Description of the Universe. .... Page 56–78

Subjects embraced by the study of the Cosmos or of physical cosmog raphy. Separation of other kindred studies—p. 56–62. The urano logical portion of the Cosmos is more simple than the telluric; the im possibility of ascertaining the diversity of matter simplifies the study of the mechanism of the heavens. Origin of the word *Cosmos*, its sig nification of adornment and order of the universe. The *existing* can not be absolutely separated in our contemplation of nature from the *future.* History of the world and description of the world—p. 62–73. Attempts to embrace the multiplicity of the phenomena of the Cos mos in the unity of thought and under the form of a purely rational combination. Natural philosophy, which preceded all exact observa tion in antiquity, is a natural, but not unfrequently ill-directed, effort of reason. Two forms of abstraction rule the whole mass of knowl edge, viz.: the *quantitative*, relative determinations according to num ber and magnitude, and *qualitative*, material characters. Means of submitting phenomena to calculation. Atoms, mechanical methods of construction. Figurative representations; mythical conception of im ponderable matters, and the peculiar vital forces in every organism. That which is attained by observation and experiment (calling forth phenomena) leads, by analogy and induction, to a knowledge of *empir ical laws;* their gradual simplification and generalization. Arrange ment of the facts discovered in accordance with leading ideas. The treasure of empirical contemplation, collected through ages, is in no dan ger of experiencing any hostile agency from philosophy—p. 73–78. [In the notes appended to p. 66–70 are considerations of the general and comparative geography of Varenius. Philological investigation into the meaning of the words κοσμος and *mundus.*]

Delineation of Nature. General Review of Natural Phenomena
p. 79–359

Introduction—p. 79–83. A descriptive delineation of the world embraces the whole universe (τὸ πᾶν) in the celestial and terrestrial spheres. Form and course of the representation. It begins with the depths of space, of which we know little beyond the existence of laws of gravitation, and with the region of the remotest nebulous spots and double stars, and then, gradually descending through the starry stratum to which our solar system belongs, it contemplates this terres trial spheroid, surrounded by air and water, and, finally, proceeds to the consideration of the form of our planet, its temperature and mag netic tension, and the fullness of organic vitality which is unfolded on its surface under the action of light. Partial insight into the relative dependence existing among all phenomena. Amid all the mobile and unstable elements in space, *mean numerical values* are the ultimate aim of investigation, being the expression of the physical laws, or forces of the Cosmos. The delineation of the universe does not begin with the earth, from which a merely subjective point of view might have led us to start, but rather with the objects comprised in the regions of space. Distribution of matter, which is partially conglomerated into rotating and circling heavenly bodies of very different density and magnitude, and partly scattered as self-luminous vapor. Review of the separate portions of the picture of nature, for the purpose of explaining the reciprocal connection of all phenomena.

a. Form of the earth, its mean density, quantity of heat, electro-magnetic activity, process of light—p. 154–202.

b. Vital activity of the earth toward its external surface. Reaction of the interior of a planet on its crust and surface. Subterranean noise without waves of concussion. Earthquakes dynamic phenomena— p. 202–217.

c. Material products which frequently accompany earthquakes. Gaseous and aqueous springs. Salses and mud volcanoes Upheavals of the soil by elastic forces—p. 217–228.

d. Fire-emitting mountains. Craters of elevation. Distribution of volcanoes on the earth—p. 228–247.

e. Volcanic forces form new kinds of rock, and metamorphose those already existing. Geognostical classification of rocks into four groups. Phenomena of contact. Fossiliferous strata; their vertical arrangement. The faunas and floras of an earlier world. Distribution of masses of rock—p. 247–284.

f. Geognostical epochs, which are indicated by the mineralogical difference of rocks, have determined the distribution of solids and fluids into continents and seas. Individual configuration of solids into horizontal expansion and vertical elevation. Relations of area. Articulation. Probability of the continued elevation of the earth's crust in ridges—p. 284–301.

g. Liquid and aëriform envelopes of the solid surface of our planet. Distribution of heat in both. The sea. The tides. Currents and their effects—p. 301–311.

h. The atmosphere. Its chemical composition. Fluctuations in its density. Law of the direction of the winds. Mean temperature. Enumeration of the causes which tend to raise and lower the temperature. Continental and insular climates. East and west coasts. Cause of the curvature of the isothermal lines. Limits of perpetual snow. Quantity of vapor. Electricity in the atmosphere. Forms of the clouds—p. 311–339.

i. Separation of inorganic terrestrial life from the geography of vital organisms; the geography of vegetables and animals. Physical gradations of the human race—p. 339–359.

Special Analysis of the Delineation of Nature, including References to the Subjects treated of in the Notes.

The universe and all that it comprises—multiform nebulous spots, planetary vapor, and nebulous stars. The picturesque charm of a southern sky—note, p. 85. Conjectures on the position in space of the world. Our stellar masses. A cosmical island. Gauging stars. Double stars revolving round a common center. Distance of the star 61 Cygni—p. 88 and note. Our solar system more complicated than was conjectured at the close of the last century. Primary planets with Neptune, Astrea, Hebe, Iris, and Flora, now constitute 16; secondary planets 18; myriads of comets of which many of the inner ones are inclosed in the orbits of the planets; a rotating ring (the zodiacal light) and meteoric stones, probably to be regarded as small cosmical bodies. The telescopic planets, Vesta, Juno, Ceres, Pallas, Astrea, Hebe, Iris, and Flora, with their frequently intersecting, strongly inclined, and more eccentric orbits, constitute a central group of separation between the inner planetary group (Mercury, Venus, the Earth, and Mars) and the outer group (Jupiter, Saturn, Uranus, and Neptune). Contrasts of these planetary groups. Relations of distance from one central body. Differences of absolute magnitude, density, period of revolution, eccentricity, and inclination of the orbits. The so-called law of the distances of the planets from their central sun. The planets which have the largest number of moons—p. 96 and note. Relations in space, both absolute and relative, of the secondary planets. Largest and smallest of the moons. Greatest approximation to a primary planet. Retrogressive movement of the moons of Uranus. Libration of the Earth's satellite—p. 98 and note. Comets; the nucleus and tail; various forms and directions of the emanations in conoidal envelopes, with more or less dense walls. Several tails inclined toward the sun; change of form of the tail; its conjectured rotation. Nature of light. Occultations of the fixed stars by the nuclei of comets. Eccentricity of their orbits and periods of revolution. Greatest distance and greatest approximation of comets. Passage through the system of Jupiter's satellites. Comets of short periods of revolution, more correctly termed inner comets (Encke, Biela, Faye)—p. 107 and note. Revolving aërolites (meteoric stones, fire-balls, falling stars). Their planetary velocity. magnitude, form, observed height. Periodic return in streams; the November stream and the stream of St. Lawrence. Chemical composition of meteoric asteroids—p. 130 and notes. Ring of zodiacal light. Limitation of the present solar atmosphere—p. 141 and note. Translatory motion of the whole solar system—p. 145–149 and note. The existence of the law of gravitation beyond our solar system. The milky way of stars and its conjectured breaking up. Milky way of nebulous spots, at right angles with that of the stars. Periods of revolutions of bi-colored double stars. Canopy of stars; openings in the stellar stratum. Events in the universe; the apparition of new stars. Propagation of light, the aspect of the starry vault of the heavens conveys to the mind an idea of inequality of time—p. 149–154 and notes.

a. Figure of the earth. Density, quantity of heat, electro-magnetic tension, and terrestrial light—p. 154–202 and note. Knowledge of the compression and curvature of the earth's surface acquired by measurements of degrees, pendulum oscillations, and certain inequalities in the moon's orbit. Mean density of the earth. The earth's crust, and the depth to which we are able to penetrate—p. 159, 160, note. Threefold movement of the heat of the earth; its thermic condition. Law of the increase of heat with the increase of depth—p. 160, 161 and note. Magnetism electricity in motion. Periodical variation of terrestrial magnetism. Disturbance of the regular course of the magnetic needle. Magnetic storms; extension of their action. Manifestations of magnetic force on the earth's surface presented under three classes of phenomena, namely, lines of equal force (isodynamic), equal inclination (isoclinic), and equal deviation (isogonic). Position of the magnetic pole. Its probable connection with the poles of cold. Change of all the magnetic phenomena of the earth. Erection of magnetic observatories since 1828; a far-extending net-work of magnetic stations—p. 190 and note.   Development of light at the magnetic poles; terrestrial light as a consequence of the electro-magnetic activity of our planet. Elevation of polar light.   Whether magnetic storms are accompanied by noise.   Connection of polar light (an electro-magnetic development of light) with the formation of cirrus clouds.   Other examples of the generation of terrestrial light—p. 202 and note.

b. The vital activity of a planet manifested from within outward, the principal source of geognostic phenomena.   Connection between merely dynamic concussions or the upheaval of whole portions of the earth's crust, accompanied by the effusion of matter, and the generation of gaseous and liquid fluids, of hot mud and fused earths, which solidify into rocks.   Volcanic action, in the most general conception of the idea, is the reaction of the interior of a planet on its outer surface.   Earthquakes.   Extent of the circles of commotion and their gradual increase. Whether there exists any connection between the changes in terrestrial magnetism and the processes of the atmosphere.   Noises, subterranean thunder without any perceptible concussion.   The rocks which modify the propagation of the waves of concussion.   Upheavals; eruption of water, hot steam, mud mofettes, smoke, and flame during an earthquake—p. 202–218 and notes.

c. Closer consideration of material products as a consequence of internal planetary activity.   There rise from the depths of the earth, through fissures and cones of eruption, various gases, liquid fluids (pure or acidulated), mud, and molten earths.   Volcanoes are a species of intermittent spring.   Temperature of thermal springs; their constancy and change.   Depth of the foci—p. 219–224 and notes.   Salses, mud volcanoes.   While fire-emitting mountains, being sources of molten earths, produce volcanic rocks, spring water forms, by precipitation, strata of limestone.   Continued generation of sedimentary rocks—p 228 and note.

d. Diversity of volcanic elevations.   Dome-like closed trachytic mountains.   Actual volcanoes which are formed from craters of elevation or among the detritus of their original structure.   Permanent connection of the interior of our earth with the atmosphere.   Relation to certain rocks.   Influence of the relations of height on the frequency of the eruptions.   Height of the cone of cinders.   Characteristics of those volcanoes which rise above the snow-line.   Columns of ashes and fire. Volcanic storm during the eruption.   Mineral composition of lavas— p. 236 and notes.   Distribution of volcanoes on the earth's surface; central and linear volcanoes; insular and littoral volcanoes.   Distance of volcanoes from the sea-coast.   Extinction of volcanic forces—p. 246 and notes.

e. Relation of volcanoes to the character of rocks.   Volcanic forces form new rocks, and metamorphose the more ancient ones.   The study of these relations leads, by a double course, to the mineral portion of geognosy (the study of the textures and of the position of the earth's strata), and to the configuration of continents and insular groups elevated above the level of the sea (the study of the geographical form and outlines of the different parts of the earth).   Classification of rocks according to the scale of the phenomena of structure and metamorphosis, which are still passing before our eyes.   Rocks of eruption, sedimentary rocks, changed (metamorphosed) rocks, conglomerates—compound rocks are definite associations of oryctognostically simple fossils There are four phases in the formative condition: rocks of eruption, endogenous (granite, sienite, porphyry, greenstone, hypersthene, rock, euphotide, melaphyre, basalt, and phonolithe); sedimentary rocks (silurian schist, coal measures, limestone, travertino, infusorial deposit); metamorphosed rock, which contains also, together with the detritus of the rocks of eruption and sedimentary rocks, the remains of gneiss, mica schist, and more ancient metamorphic masses. Aggregate and sandstone formations. The phenomenon of contact explained by the artificial imitation of minerals. Effects of pressure and the various rapidity of cooling. Origin of granular or saccharoidal marble, silicification of schist into ribbon jasper. Metamorphosis of calcareous marl into micaceous schist through granite. Conversion of dolomite and granite into argillaceous schist, by contact with basaltic and doleritic rocks. Filling up of the veins from below. Processes of cementation in agglomerate structures. Friction conglomerates—p. 269 and note. Relative age of rocks, chronometry of the earth's crust. Fossiliferous strata. Relative age of organisms. Simplicity of the first vital forms. Dependence of physiological gradations on the age of the formations. Geognostic horizon, whose careful investigation may yield certain data regarding the identity or the relative age of formations, the periodic recurrence of certain strata, their parallelism, or their total suppression. Types of the sedimentary structures considered in their most simple and general characters; silurian and devonian formations (formerly known as rocks of transition); the lower trias (mountain limestone, coal measures, together with *todlliegende* and zechstein); the upper trias (bunter sandstone, muschelkalk, and keuper); Jura limestone (lias and oolite); freestone, lower and upper chalk, as the last of the flötz strata, which begin with mountain limestone; tertiary formations in three divisions, which are designated by granular limestone, lignite, and south Apennine gravel—p. 269-278.

The faunas and floras of an earlier world, and their relations to existing organisms. Colossal bones of antediluvian mammalia in the upper alluvium. Vegetation of an earlier world; monuments of the history of its vegetation. The points at which certain vegetable groups attain their maximum; cycadeæ in the keuper and lias, and coniferæ in the bunter sandstone. Lignite and coal measures (amber-tree). Deposition of large masses of rock; doubts regarding their origin—p. 285 and note

f. The knowledge of geognostic epochs—of the upheaval of mountain chains and elevated plateaux, by which lands are both formed and destroyed, leads, by an internal causal connection, to the distribution into solids and fluids, and to the peculiarities in the natural configuration of the earth's surface. Existing areal relations of the solid to the fluid differ considerably from those presented by the maps of the physical portion of a more ancient geography. Importance of the eruption of quartzose porphyry with reference to the then existing configuration of continental masses. Individual conformation in horizontal extension (relations of articulation) and in vertical elevation (hypsometrical views). Influence of the relations of the area of land and sea on the temperature, direction of the winds, abundance or scarcity of organic products, and on all meteorological processes collectively. Direction of the major axes of continental masses. Articulation and pyramidal termination toward the south. Series of peninsulas. Valley-like formation of the Atlantic Ocean. Forms which frequently recur—p. 285-293 and notes. Ramifications and systems of mountain chains, and the means of determining their relative ages. Attempts to determine the center of gravity of the volume of the lands upheaved above the level of the sea. The elevation of continents is still progressing slowly, and is being compensated for at some definite points by a perceptible sinking. All geognostic phenomena indicate a periodical alternation of activity in the interior of our planet. Probability of new elevations of ridges—p. 293–301 and notes.

g. The solid surface of the earth has two envelopes, one liquid, and the other aeriform. Contrasts and analogies which these envelopes—the sea and the atmosphere—present in their conditions of aggregation and electricity, and in their relations of currents and temperature. Depths of the ocean and of the atmosphere, the shoals of which constitute our highlands and mountain chains. The degree of heat at the surface of the sea in different latitudes and in the lower strata. Tendency of the sea to maintain the temperature of the surface in the strata nearest to the atmosphere, in consequence of the mobility of its particles and the alteration in its density. Maximum of the density of salt water. Position of the zones of the hottest water, and of those having the greatest saline contents. Thermic influence of the lower polar current and the counter currents in the straits of the sea—p. 302–304 and notes. General level of the sea, and permanent local disturbances of equilibrium; the periodic disturbances manifested as tides. Oceanic currents; the equatorial or rotation current, the Atlantic warm Gulf Stream, and the further impulse which it receives; the cold Peruvian stream in the eastern portion of the Pacific Ocean of the southern zone. Temperature of shoals. The universal diffusion of life in the ocean. Influence of the small submarine sylvan region at the bottom of beds of rooted algæ, or on far-extending floating layers of fucus—p. 302–311 and notes.

h. The gaseous envelope of our planet, the atmosphere. Chemical composition of the atmosphere, its transparency, its polarization, pressure, temperature, humidity, and electric tension. Relation of oxygen to nitrogen; amount of carbonic acid; carbureted hydrogen; ammoniacal vapors. Miasmata. Regular (horary) changes in the pressure of the atmosphere. Mean barometrical height at the level of the sea in different zones of the earth. Isobarometrical curves. Barometrical windroses. Law of rotation of the winds, and its importance with reference to the knowledge of many meteorological processes. Land and sea winds, trade winds and monsoons—p. 311–317. Climatic distribution of heat in the atmosphere, as the effect of the relative position of transparent and opaque masses (fluid and solid superficial area), and of the hypsometrical configuration of continents. Curvature of the isothermal lines in a horizontal and vertical direction, on the earth's surface and in the superimposed strata of air. Convexity and concavity of the isothermal lines. Mean heat of the year, seasons, months, and days. Enumeration of the causes which produce disturbances in the form of the isothermal lines, *i. e.*, their deviation from the position of the geographical parallels. Isochimenal and isotheral lines are the lines of equal winter and summer heat. Causes which raise or lower the temperature. Radiation of the earth's surface, according to its inclination, color, density, dryness, and chemical composition. The form of the cloud which announces what is passing in the upper strata of the atmosphere is the image of the strongly radiating ground projected on a hot summer sky. Contrast between an insular or littoral climate, such as is experienced by all deeply-articulated continents, and the climate of the interior of large tracts of land. East and west coasts. Difference between the southern and northern hemispheres. Thermal scales of cultivated plants, going down from the vanilla, cacoa, and musaceæ, to citrous and olives, and to vines yielding potable wines. The influence which these scales exercise on the geographical distribution of cultivated plants. The favorable ripening and the immaturity of fruits are essentially influenced by the difference in the action of direct or scattered light in a clear sky or in one overcast with mist. General summary of the causes which yield a more genial climate to the greater portion of Europe considered as the western peninsula of Asia—p. 326. Determination of the changes in the mean annual and summer temperature, which correspond to one degree of geographical latitude. Equality of the mean temperature of a mountain station, and of the polar distance of any point lying at the level of the sea. Decrease of temperature with the decrease in elevation. Limits of perpetual snow, and the fluctuations in these limits. Causes of disturbance in the regularity of the phenomenon. Northern and southern chains of the Himalaya; habitability of the elevated plateaux of Thibet—p. 331. Quantity of moisture in the atmosphere, according to the hours of the day, the seasons of the year, degrees of latitude, and elevation. Greatest dryness of the atmosphere observed in Northern Asia, between the river districts of the Irtysch and the Obi. Dew, a consequence of radiation. Quantity of rain—p. 335. Electricity of the atmosphere, and disturbance of the electric tension. Geographical distribution of storms. Predetermination of atmospheric changes. The most important climatic disturbances can not be traced, at the place of observation, to any local cause, but are rather the consequence of some occurrence by which the equilibrium in the atmospheric currents has been destroyed at some considerable distance—p. 335–339.

i. Physical geography is not limited to elementary inorganic terrestrial life, but, elevated to a higher point of view, it embraces the sphere of organic life, and the numerous gradations of its typical development. Animal and vegetable life. General diffusion of life in the sea and on the land; microscopic vital forms discovered in the polar ice no less than in the depths of the ocean within the tropics. Extension imparted to the horizon of life by Ehrenberg's discoveries. Estimation of the mass (volume) of animal and vegetable organisms—p. 339–346. Geography of plants and animals. Migrations of organisms in the ovum, or by means of organs capable of spontaneous motion. Spheres of distribution depending on climatic relations. Regions of vegetation, and classification of the genera of animals. Isolated and social living plants and animals. The character of floras and faunas is not determined so much by the predominance of separate families, in certain parallels of latitude, as by the highly complicated relations of the association of many families, and the relative numerical value of their species. The forms of natural families which increase or decrease from the equator to the poles. Investigations into the numerical relation existing in different districts of the earth between each one of the large families to the whole mass of phanerogamia—p. 346–351. The human race considered according to its physical gradations, and the geographical distribution of its simultaneously occurring types. Races and varieties. All races of men are forms of one single species. Unity of the human race. Languages considered as the intellectual creations of mankind, or as portions of the history of mental activity, manifest a character of nationality, although certain historical occurrences have been the means of diffusing idioms of the same family of languages among nations of wholly different descent—p. 351–359.

INTRODUCTION.

In attempting, after a long absence from my native country, to develop the physical phenomena of the globe, and the
simultaneous action of the forces that pervade the regions of
space, I experience a two-fold cause of anxiety.   The subject
before me is so inexhaustible and so varied, that I fear either
to fall into the superficiality of the encyclopedist, or to weary
the mind of my reader by aphorisms consisting of mere generalities clothed in dry and dogmatical forms.   Undue conciseness often checks the flow of expression, while diffuseness is
alike detrimental to a clear and precise exposition of our ideas.
Nature is a free domain, and the profound conceptions and
enjoyments she awakens within us can only be vividly deline
ated by thought clothed in exalted forms of speech, worthy of
bearing witness to the majesty and greatness of the creation.

In considering the study of physical phenomena, not merely in its bearings on the material wants of life, but in its general influence on the intellectual advancement of mankind,
we find its noblest and most important result to be a knowledge of the chain of connection, by which all natural forces
are linked together, and made mutually dependent upon each
other ; and it is the perception of these relations that exalts
our views and ennobles our enjoyments.   Such a result can,
however, only be reaped as the fruit of observation and intellect, combined with the spirit of the age, in which are reflected all the varied phases of thought.   He who can trace,
through by-gone times, the stream of our knowledge to its
primitive source, will learn from history how, for thousands
of years, man has labored, amid the ever-recurring changes
of form, to recognize the invariability of natural laws, and
has thus, by the force of mind, gradually subdued a great portion of the physical world to his dominion.   In interrogating
the history of the past, we trace the mysterious course of ideas
yielding the first glimmering perception of the same image of

a Cosmos, or harmoniously ordered whole, which, dimly shad-
owed forth to the human mind in the primitive ages of the
world, is now fully revealed to the maturer intellect of man
kind as the result of long and laborious observation.

Each of these epochs of the contemplation of the external
world—the earliest dawn of thought and the advanced stage
of civilization—has its own source of enjoyment. In the
former, this enjoyment, in accordance with the simplicity of
the primitive ages, flowed from an intuitive feeling of the or
der that was proclaimed by the invariable and successive reappearance of the heavenly bodies, and by the progressive development of organized beings; while in the latter, this sense
of enjoyment springs from a definite knowledge of the phenomena of nature. When man began to interrogate nature,
and, not content with observing, learned to evoke phenomena
under definite conditions; when once he sought to collect and
record facts, in order that the fruit of his labors might aid investigation after his own brief existence had passed away, the
philosophy of Nature cast aside the vague and poetic garb
in which she had been enveloped from her origin, and, having
assumed a severer aspect, she now weighs the value of observations, and substitutes induction and reasoning for conjecture and assumption. The dogmas of former ages survive
now only in the superstitions of the people and the prejudices
of the ignorant, or are perpetuated in a few systems, which,
conscious of their weakness, shroud themselves in a vail of
mystery. We may also trace the same primitive intuitions
in languages exuberant in figurative expressions; and a few
of the best chosen symbols engendered by the happy inspiration of the earliest ages, having by degrees lost their vagueness through a better mode of interpretation, are still preserved
among our scientific terms.

Nature considered *rationally*, that is to say, submitted to
the process of thought, is a unity in diversity of phenomena;
a harmony, blending together all created things, however dis-
similar in form and attributes; one great whole ($\tau\grave{o}$ $\pi\tilde{a}\nu$) animated by the breath of life. The most important result of
a rational inquiry into nature is, therefore, to establish the
unity and harmony of this stupendous mass of force and matter, to determine with impartial justice what is due to the
discoveries of the past and to those of the present, and to analyze the individual parts of natural phenomena without suc-
cumbing beneath the weight of the whole. Thus, and thus
alone, is it permitted to man, while mindful of the high des-

tiny of his race, to comprehend nature, to lift the vail that shrouds her phenomena, and, as it were, submit the results of observation to the test of reason and of intellect.

In reflecting upon the different degrees of enjoyment presented to us in the contemplation of nature, we find that the first place must be assigned to a sensation, which is wholly independent of an intimate acquaintance with the physical phenomena presented to our view, or of the peculiar character of the region surrounding us. In the uniform plain bounded only by a distant horizon, where the lowly heather, the cistus, or waving grasses, deck the soil; on the ocean shore, where the waves, softly rippling over the beach, leave a track, green with the weeds of the sea; every where, the mind is penetrated by the same sense of the grandeur and vast expanse of nature, revealing to the soul, by a mysterious inspiration, the existence of laws that regulate the forces of the universe. Mere communion with nature, mere contact with the free air, exercise a soothing yet strengthening influence on the wearied spirit, calm the storm of passion, and soften the heart when shaken by sorrow to its inmost depths. Every where, in every region of the globe, in every stage of intellectual culture, the same sources of enjoyment are alike vouchsafed to man. The earnest and solemn thoughts awakened by a communion with nature intuitively arise from a presentiment of the order and harmony pervading the whole universe, and from the contrast we draw between the narrow limits of our own existence and the image of infinity revealed on every side, whether we look upward to the starry vault of heaven, scan the far-stretching plain before us, or seek to trace the dim horizon across the vast expanse of ocean.

The contemplation of the individual characteristics of the landscape, and of the conformation of the land in any definite region of the earth, gives rise to a different source of enjoyment, awakening impressions that are more vivid, better defined, and more congenial to certain phases of the mind, than those of which we have already spoken. At one time the heart is stirred by a sense of the grandeur of the face of nature, by the strife of the elements, or, as in Northern Asia, by the aspect of the dreary barrenness of the far-stretching steppes; at another time, softer emotions are excited by the contemplation of rich harvests wrested by the hand of man from the wild fertility of nature, or by the sight of human habitations raised beside some wild and foaming torrent. Here I regard less the degree of intensity than the difference existing in the various sensations that derive their charm and permanence from the peculiar character of the scene.

If I might be allowed to abandon myself to the recollections of my own distant travels, I would instance, among the most striking scenes of nature, the calm sublimity of a tropical night, when the stars, not sparkling, as in our northern skies, shed their soft and planetary light over the gently-heaving ocean ; or I would recall the deep valleys of the Cordilleras, where the tall and slender palms pierce the leafy vail around them, and waving on high their feathery and arrow-like branches, form, as it were, " a forest above a forest ;"* or I would describe the summit of the Peak of Teneriffe, when a horizontal layer of clouds, dazzling in whiteness, has separated the cone of cinders from the plain below, and suddenly the ascending current pierces the cloudy vail, so that the eye of the traveler may range from the brink of the crater, along the vine-clad slopes of Orotava, to the orange gardens and banana groves that skirt the shore. In scenes like these, it is not the peaceful charm uniformly spread over the face of nature that moves the heart, but rather the peculiar physiognomy and conformation of the land, the features of the landscape, the ever-varying outline of the clouds, and their blending with the horizon of the sea, whether it lies spread before us like a smooth and shining mirror, or is dimly seen through the morning mist. All that the senses can but imperfectly comprehend, all that is most awful in such romantic scenes of nature, may become a source of enjoyment to man, by opening a wide field to the creative powers of his imagination. Impressions change with the varying movements of the mind, and we are led by a happy illusion to believe that we receive from the external world that with which we have ourselves invested it.

When far from our native country, after a long voyage, we tread for the first time the soil of a tropical land, we experience a certain feeling of surprise and gratification in recognizing, in the rocks that surround us, the same inclined schistose strata, and the same columnar basalt covered with cellular amygdaloids, that we had left in Europe, and whose identity of character, in latitudes so widely different, reminds us that the solidification of the earth's crust is altogether independent of climatic influences. But these rocky masses of schist and of basalt are covered with vegetation of a character with which we are unacquainted, and of a physiognomy wholly

* This expression is taken from a beautiful description of tropical forest scenery in *Paul and Virginia*, by Bernardin de Saint Pierre.

unknown to us ; and it is then, amid the colossal and majestic
forms of an exotic flora, that we feel how wonderfully the flex·
ibility of our nature fits us to receive new impressions, linked
together by a certain secret analogy.   We so readily perceive
the affinity existing among all the forms of organic life, that
although the sight of a vegetation similar to that of our native
country might at first be most welcome to the eye, as the sweet
familiar sounds of our mother tongue are to the ear, we nev·
ertheless, by degrees, and almost imperceptibly, become famil
iarized with a new home and a new climate.   As a true citi
zen of the world, man every where habituates himself to tha'
which surrounds him ; yet fearful, as it were, of breaking tl
links of association that bind him to the home of his childhood,
the colonist applies to some few plants in a far-distant clime the
names he had been familiar with in his native land ; and by
the mysterious relations existing among all types of organiza·
tion, the forms of exotic vegetation present themselves to his
mind as nobler and more perfect developments of those he had
loved in earlier days.   Thus do the spontaneous impressions
of the untutored mind lead, like the laborious deductions of
cultivated intellect, to the same intimate persuasion, that one
sole and indissoluble chain binds together all nature.

It may seem a rash attempt to endeavor to separate, into
its different elements, the magic power exercised upon our
minds by the physical world, since the character of the landscape, and of every imposing scene in nature, depends so materially upon the mutual relation of the ideas and sentiments
simultaneously excited in the mind of the observer.

The powerful effect exercised by nature springs, as it were,
from the connection and unity of the impressions and emotions produced ; and we can only trace their different sources
by analyzing the individuality of objects and the diversity of
forces.

The richest and most varied elements for pursuing an analysis of this nature present themselves to the eyes of the traveler in the scenery of Southern Asia, in the Great Indian
Archipelago, and more especially, too, in the New Continent,
where the summits of the lofty Cordilleras penetrate the con-·
fines of the aërial ocean surrounding our globe, and where the
same subterranean forces that once raised these mountain
chains still shake them to their foundation and threaten their
downfall.

Graphic delineations of nature, arranged according to sys·
tematic views, are not only suited to please the imagination,

but may also, when properly considered, indicate the grades of the impressions of which I have spoken, from the uniformity of the sea-shore, or the barren steppes of Siberia, to the inexhaustible fertility of the torrid zone. If we were even to picture to ourselves Mount Pilatus placed on the Schreckhorn,* or the Schneekoppe of Silesia on Mont Blanc, we should

---

* These comparisons are only approximative. The several elevations above the level of the sea are, in accurate numbers, as follows:

The Schneekoppe or Riesenkoppe, in Silesia, about 5270 feet, according to Hallaschka. The Righi, 5902 feet, taking the height of the Lake of Lucerne at 1426 feet, according to Eschman. (See *Compte Rendu des Mesures Trigonométriques en Suisse*, 1840, p. 230.) Mount Athos, 6775 feet, according to Captain Gaultier; Mount Pilatus, 7546 feet; Mount Ætna, 10,871 feet, according to Captain Smyth; or 10,874 feet, according to the barometrical measurement made by Sir John Herschel, and communicated to me in writing in 1825, and 10,899 feet, according to angles of altitude taken by Cacciatore at Palermo (calculated by assuming the terrestrial refraction to be 0·076) ; the Schreck horn, 12,383 feet; the Jungfrau, 13,720 feet, according to Tralles; Mont Blanc, 15,775 feet, according to the different measurements considered by Roger (*Bibl. Univ.*, May, 1828, p. 24–53), 15,733 feet, according to the measurements taken from Mount Columbier by Carlini in 1821, and 15,748 feet, as measured by the Austrian engineers from Trelod and the Glacier d'Ambin.

The actual height of the Swiss mountains fluctuates, according to Eschman's observations, as much as 25 English feet, owing to the varying thickness of the stratum of snow that covers the summits. Chimborazo is, according to my trigonometrical measurements, 21,421 feet (see Humboldt, *Recueil d'Obs. Astr.*, tome i., p. 73), and Dhawalagiri, 28,074 feet. As there is a difference of 445 feet between the determinations of Blake and Webb, the elevation assigned to the Dhawalagiri (or white mountain, from the Sanscrit *dhawala*, white, and *giri*, mountain) can not be received with the same confidence as that of the Jawahir, 25,749 feet, since the latter rests on a complete trigonometrical measurement (see Herbert and Hodgson in the *Asiat. Res.*, vol. xiv., p. 189, and Suppl. to *Encycl. Brit.*, vol. iv., p. 643). I have shown elsewhere (*Ann. des Sciences Naturelles*, Mars, 1825) that the height of the Dhawalagiri (28,074 feet) depends on several elements that have not been ascertained with certainty, as azimuths and latitudes (Humboldt, *Asie Centrale*, t. iii., p. 282). It has been believed, but without foundation, that in the Tartaric chain, north of Thibet, opposite to the chain of Kuen-lun, there are several snowy summits, whose elevation is about 30,000 English feet (almost twice that of Mont Blanc), or, at any rate, 29,000 feet (see Captain Alexander Gerard's and John Gerard's *Journey to the Boorendo Pass*, 1840, vol. i., p. 143 and 311). Chimborazo is spoken of in the text only as *one* of the highest summits of the chain of the Andes; for in the year 1827, the learned and highly-gifted traveler, Pentland, in his memorable expedition to Upper Peru (Bolivia), measured the elevation of two mountains situated to the east of Lake Titicaca, viz., the Sorata, 25,200 feet, and the Illimani, 24,000 feet, both greatly exceeding the height of Chimborazo, which is only 21,421 feet, and being nearly equal in elevation to the Jawahir, which is the highest mountain in the Himalaya that has as yet been accurately measured.

not have attained to the height of that great Colossus of the Andes, the Chimborazo, whose height is twice that of Mount Ætna; and we must pile the Righi, or Mount Athos, on the summit of the Chimborazo, in order to form a just estimate of the elevation of the Dhawalagiri, the highest point of the Himalaya. But although the mountains of India greatly surpass the Cordilleras of South America by their astonishing elevation (which, after being long contested, has at last been confirmed by accurate measurements), they can not, from their geographical position, present the same inexhaustible variety of phenomena by which the latter are characterized. The impression produced by the grander aspects of nature does not depend exclusively on height. The chain of the Himalaya is placed far beyond the limits of the torrid zone, and scarcely is a solitary palm-tree to be found in the beautiful valleys of Kumaoun and Garhwal.* On the southern slope of the ancient Paropamisus, in the latitudes of 28° and 34°, nature no longer displays the same abundance of tree-ferns and arborescent grasses, heliconias and orchideous plants, which in tropic

Thus Mont Blanc is 5646 feet below Chimborazo; Chimborazo, 3779 feet below the Sorata; the Sorata, 549 feet below the Jawahir, and probably about 2880 feet below the Dhawalagiri. According to a new measurement of the Illimani, by Pentland, in 1838, the elevation of this mountain is given at 23,868 feet, varying only 133 feet from the measurement taken in 1827. The elevations have been given in this note with minute exactness, as erroneous numbers have been introduced into many maps and tables recently published, owing to incorrect reductions of the measurements.

[In the preceding note, taken from those appended to the Introduction in the French translation, rewritten by Humboldt himself, the measurements are given in meters, but these have been converted into English feet, for the greater convenience of the general reader.]—*Tr.*

* The absence of palms and tree-ferns on the temperate slopes of the Himalaya is shown in Don's *Flora Nepalensis*, 1825, and in the remarkable series of lithographs of Wallich's *Flora Indica*, whose catalogue contains the enormous number of 7683 Himalaya species, almost all phanerogamic plants, which have as yet been but imperfectly classified. In Nepaul (lat. 26½° to 27¼°) there has hitherto been observed only one species of palm, Chamærops martiana, Wall. (*Plantæ Asiat.*, lib. iii., p. 5, 211), which is found at the height of 5250 English feet above the level of the sea, in the shady valley of Bunipa. The magnificent tree-fern, Alsophila brunoniana, Wall. (of which a stem 48 feet long has been in the possession of the British Museum since 1831), does not grow in Nepaul, but is found on the mountains of Silhet, to the northwest of Calcutta, in lat. 24° 50′. The Nepaul fern, Paranema cyathöides, Don, formerly known as Sphæroptera barbata, Wall. (*Plantæ Asiat.*, lib. i., p. 42, 48), is, indeed, nearly related to Cyathea, a species of which I have seen in the South American Missions of Caripe, measuring 33 feet in height; this is not, however, properly speaking, a tree.

al regions are to be found even on the highest plateaux of the mountains. On the slope of the Himalaya, under the shade of the Deodora and the broad-leaved oak, peculiar to these Indian Alps, the rocks of granite and of mica schist are covered with vegetable forms almost similar to those which characterize Europe and Northern Asia. The species are not identical, but closely analogous in aspect and physiognomy, as, for instance, the juniper, the alpine birch, the gentian, the marsh parnassia, and the prickly species of Ribes.* The chain of the Himalaya is also wanting in the imposing phenomena of volcanoes, which in the Andes and in the Indian Archipelago often reveal to the inhabitants, under the most terrific forms, the existence of the forces pervading the interior of our planet.

Moreover, on the southern declivity of the Himalaya, where the ascending current deposits the exhalations rising from a vigorous Indian vegetation, the region of perpetual snow begins at an elevation of 11,000 or 12,000 feet above the level of the sea,† thus setting a limit to the development of organic

---

* Ribes nubicola, R. glaciale, R. grossularia. The species which compose the vegetation of the Himalaya are four pines, notwithstanding the assertion of the ancients regarding Eastern Asia (Strabo, lib. 11, p. 510, Cas.), twenty-five oaks, four birches, two chestnuts, seven maples, twelve willows, fourteen roses, three species of strawberry, seven species of Alpine roses (*rhododendra*), one of which attains a height of 20 feet, and many other northern genera. Large white apes, having black faces, inhabit the wild chestnut-tree of Kashmir, which grows to a height of 100 feet, in lat. 33° (see Carl von Hügel's *Kaschmir*, 1840, 2d pt. 249). Among the Coniferæ, we find the Pinus deodwara, or deodara (in Sanscrit, *déwa-daru*, the timber of the gods), which is nearly allied to Pinus cedrus. Near the limit of perpetual snow flourish the large and showy flowers of the Gentiana venusta, G. Moorcroftiana, Swertia purpurescens, S. speciosa, Parnassia armata, P. nubicola, Pœonia Emodi, Tulipa stellata; and, besides varieties of European genera peculiar to these Indian mountains, true European species, as Leontodon taraxacum, Prunella vulgaris, Galium aparine, and Thlaspi arvense. The heath mentioned by Saunders, in Turner's *Travels*, and which had been confounded with Calluna vulgaris, is an Andromeda, a fact of the greatest importance in the geography of Asiatic plants. If I have made use, in this work, of the unphilosophical expressions of *European* genera, *European* species, *growing wild in Asia*, &c., it has been in consequence of the old botanical language, which, instead of the idea of a large dissemination, or, rather, of the coexistence of organic productions, has dogmatically substituted the false hypothesis of a migration, which, from predilection for Europe, is further assumed to have been from west to east.

† On the southern declivity of the Himalaya, the limit of perpetual snow is 12,978 feet above the level of the sea; on the northern declivity, or, rather, on the peaks which rise above the Thibet, or Tartarian

life in a zone that is nearly 3000 feet lower than that to which it attains in the equinoctial region of the Cordilleras.

plateau, this limit is at 16,625 feet from 30½° to 32° of latitude, while at the equator, in the Andes of Quito, it is 15,790 feet. Such is the result I have deduced from the combination of numerous data furnished by Webb, Gerard, Herbert, and Moorcroft. (See my two memoirs on the mountains of India, in 1816 and 1820, in the *Ann. de Chimie et de Physique*, t. iii., p. 303; t. xiv., p. 6, 22, 50.) The greater elevation to which the limit of perpetual snow recedes on the Tartarian declivity is owing to the radiation of heat from the neighboring elevated plains, to the purity of the atmosphere, and to the infrequent formation of snow in an air which is both very cold and very dry. (Humboldt, *Asie Centrale*, t. iii., p. 281–326.) My opinion on the difference of height of the snow-line on the two sides of the Himalaya has the high authority of Colebrooke in its favor. He wrote to me in June, 1824, as follows: " I also find, from the data in my possession, that the elevation of the line of perpetual snow is 13,000 feet. On the southern declivity, and at latitude 31°, Webb's measurements give me 13,500 feet, consequently 500 feet more than the height deduced from Captain Hodgson's observations. Gerard's measurements fully confirm your opinion that the line of snow is higher on the northern than on the southern side.' It was not until the present year (1840) that we obtained the complete and collected journal of the brothers Gerard, published under the supervision of Mr. Lloyd. (*Narrative of a Journey from Cawnpoor to the Boorendo Pass, in the Himalaya, by Captain Alexander Gerard and John Gerard, edited by George Lloyd*, vol. i., p. 291, 311, 320, 327, and 341.) Many interesting details regarding some localities may be found in the narrative of *A Visit to the Shatool, for the Purpose of determining the Line of Perpetual Snow on the southern face of the Himalaya, in August*, 1822. Unfortunately, however, these travelers always confound the elevation at which sporadic snow falls with the maximum of the height that the snow-line attains on the Thibetian plateau. Captain Gerard distinguishes between the summits that rise in the middle of the plateau, where he states the elevation of the snow-line to be between 18,000 and 19,000 feet, and the northern slopes of the chain of the Himalaya, which border on the defile of the Sutledge, and can radiate but little heat, owing to the deep ravines with which they are intersected. The elevation of the village of Tangno is given at only 9300 feet, while that of the plateau surrounding the sacred lake of Mansa is 17,000 feet. Captain Gerard finds the snow-line 500 feet lower on the northern slopes, where the chain of the Himalaya is broken through, than toward the southern declivities facing Hindostan, and he there estimates the line of perpetual snow at 15,000 feet. The most striking differences are presented between the vegetation on the Thibetian plateau and that characteristic of the southern slopes of the Himalaya. On the latter the cultivation of grain is arrested at 9974 feet, and even there the corn has often to be cut when the blades are still green. The extreme limit of forests of tall oaks and deodars is 11,960 feet; that of dwarf birches, 12,983 feet. On the plains, Captain Gerard found pastures up to the height of 17,000 feet; the cereals will grow at 14,100 feet, or even at 18,540 feet; birches with tall stems at 14,100 feet, and copse or brush wood applicable for fuel is found at an elevation of upward of 17,000 feet, that is to say, 1280 feet above the lower limits of the snow-line at the equator, in the province of Quito. It is

But the countries bordering on the equator possess another advantage, to which sufficient attention has not hitherto been

very desirable that the *mean* elevation of the Thibetian plateau, which I have estimated at only about 8200 feet between the Himalaya and the Kuen-lun, and the difference in the height of the line of perpetual snow on the southern and on the northern slopes of the Himalaya, should be again investigated by travelers who are accustomed to judge of the general conformation of the land. Hitherto simple calculations have too often been confounded with actual measurements, and the elevations of isolated summits with that of the surrounding plateau. (Compare Carl Zimmerman's excellent Hypsometrical Remarks in his *Geographischen Analyse der Karte von Inner Asien*, 1841, s. 98.) Lord draws attention to the difference presented by the two faces of the Himalaya and those of the Alpine chain of Hindoo-Coosh, with respect to the limits of the snow-line. " The latter chain," he says, " has the table-land to the south, in consequence of which the snow-line is higher on the southern side, contrary to what we find to be the case with respect to the Himalaya, which is bounded on the south by sheltered plains, as Hindoo-Coosh is on the north." It must, however, be admitted that the hypsometrical data on which these statements are based require a critical revision with regard to several of their details ; but still they suffice to establish the main fact, that the remarkable configuration of the land in Central Asia affords man all that is essential to the mainte-nance of life, as habitation, food, and fuel, at an elevation above the level of the sea which in almost all other parts of the globe is covered with perpetual ice. We must except the very dry districts of Bolivia, where snow is so rarely met with, and where Pentland (in 1838) fixed the snow-line at 15,667 feet, between 16° and 17¾° south latitude. The opinion that I had advanced regarding the difference in the snow-line on the two faces of the Himalaya has been most fully confirmed by the barometrical observations of Victor Jacquemont, who fell an early sac-rifice to his noble and unwearied ardor. (See his *Correspondance pendant son Voyage dans l'Inde*, 1828 à 1832, liv. 23, p. 290, 296, 299.) " Perpetual snow," says Jacquemont, " descends lower on the southern than on the northern slopes of the Himalaya, and the limit constantly rises as we advance to the north of the chain bordering on India. On the Kioubrong, about 18,317 feet in elevation, according to Captain Gerard, I was still considerably below the limit of perpetual snow which I believe to be 19,690 feet in this part of Hindostan." (This estimate I consider much too high.)

The same traveler says, " To whatever height we rise on the south-ern declivity of the Himalaya, the climate retains the same character, and the same division of the seasons as in the plains of India ; the sum-mer solstice being every year marked by the same prevalence of rain, which continues to fall without intermission until the autumnal equi-nox. But a new, a totally different climate begins at Kashmir, whose elevation I estimate to be 5350 feet, nearly equal to that of the cities of Mexico and Popayan" (*Correspond. de Jacquemont*, t. ii., p. 58 et 74). The warm and humid air of the sea, as Leopold von Buch well observes, is carried by the monsoons across the plains of India to the skirts of the Himalaya, which arrest its course, and hinder it from diverging to the Thibetian districts of Ladak and Lassa. Carl von Hügel estimates the elevation of the Valley of Kashmir above the level of the sea at 5818 feet, and bases his observation on the determination of the boiling

directed. This portion of the surface of the globe affords in the smallest space the greatest possible variety of impressions from the contemplation of nature. Among the colossal mountains of Cundinamarca, of Quito, and of Peru, furrowed by deep ravines, man is enabled to contemplate alike all the families of plants, and all the stars of the firmament. There, at a single glance, the eye surveys majestic palms, humid forests of bambusa, and the varied species of Musaceæ, while above these forms of tropical vegetation appear oaks, medlars, the sweet-brier, and umbelliferous plants, as in our European homes. There, as the traveler turns his eyes to the vault of heaven, a single glance embraces the constellation of the Southern Cross, the Magellanic clouds, and the guiding stars of the constellation of the Bear, as they circle round the arctic pole. There the depths of the earth and the vaults of heaven display all the richness of their forms and the variety of their phenomena. There the different climates are ranged the one above the other, stage by stage, like the vegetable zones, whose succession they limit ; and there the observer may readily trace the laws that regulate the diminution of heat, as they stand indelibly inscribed on the rocky walls and abrupt declivities of the Cordilleras.

Not to weary the reader with the details of the phenomena which I long since endeavored graphically to represent,[*] I will here limit myself to the consideration of a few of the general results whose combination constitutes the *physical delineation of the torrid zone.* That which, in the vagueness of our

point of water (see theil 11, s. 155, and *Journal of Geog. Soc.*, vol. vi., p. 215). In this valley, where the atmosphere is scarcely ever agitated by storms, and in 34° 7' lat., snow is found, several feet in thickness, from December to March.

[*] See, generally, my *Essai sur la Géographie des Plantes, et le Tableau physique des Régions Equinoxiales,* 1807, p. 80–88. On the diurnal and nocturnal variations of temperature, see Plate 9 of my *Atlas Géogr. et Phys. du Nouveau Continent;* and the Tables in my work, entitled *De distributione Geographica Plantarum, secundum cœli temperiem, et altitudinem Montium,* 1817, p. 90–116 ; the meteorological portion of my *Asie Centrale,* t. iii., p. 212, 224; and, finally, the more recent and far more exact exposition of the variations of temperature experienced in correspondence with the increase of altitude on the chain of the Andes, given in Boussingault's Memoir, *Sur la profondeur à laquelle on trouve, sous les Tropiques, la couche de Temperature Invariable.* (Ann. de Chimie et de Physique, 1833, t. liii., p. 225–247.) This treatise contains the elevations of 128 points, included between the level of the sea and the declivity of the Antisana (17,900 feet), as well as the mean temperature of the atmosphere, which varies with the height between 81° and 35° F.

impressions, loses all distinctness of form, like some distant mountain shrouded from view by a vail of mist, is clearly revealed by the light of mind, which, by its scrutiny into the causes of phenomena, learns to resolve and analyze their different elements, assigning to each its individual character. Thus, in the sphere of natural investigation, as in poetry and painting, the delineation of that which appeals most strongly to the imagination, derives its collective interest from the vivid truthfulness with which the individual features are portrayed.

The regions of the torrid zone not only give rise to the most powerful impressions by their organic richness and their abundant fertility, but they likewise afford the inestimable advantage of revealing to man, by the uniformity of the variations of the atmosphere and the development of vital forces, and by the contrasts of climate and vegetation exhibited at different elevations, the invariability of the laws that regulate the course of the heavenly bodies, reflected, as it were, in terrestrial phenomena.   Let us dwell, then, for a few moments, on the proofs of this regularity, which is such that it may be submitted to numerical calculation and computation.

In the burning plains that rise but little above the level of the sea, reign the families of the banana, the cycas, and the palm, of which the number of species comprised in the flora of tropical regions has been so wonderfully increased in the present day by the zeal of botanical travelers.   To these groups succeed, in the Alpine valleys, and the humid and shaded clefts on the slopes of the Cordilleras, the tree-ferns, whose thick cylindrical trunks and delicate lace-like foliage stand out in bold relief against the azure of the sky, and the cinchona, from which we derive the febrifuge bark.   The medicinal strength of this bark is said to increase in proportion to the degree of moisture imparted to the foliage of the tree by the light mists which form the upper surface of the clouds resting over the plains.   Every where around, the confines of the forest are encircled by broad bands of social plants, as the delicate aralia, the thibaudia, and the myrtle-leaved Andromeda, while the Alpine rose, the magnificent befaria, weaves a purple girdle round the spiry peaks.   In the cold regions of the Paramos, which is continually exposed to the fury of storms and winds, we find that flowering shrubs and herbaceous plants, bearing large and variegated blossoms, have given place to monocotyledons, whose slender spikes constitute the sole covering of the soil.   This is the zone of the

grasses, one vast savannah extending over the immense mountain plateaux, and reflecting a yellow, almost golden tinge, to
the slopes of the Cordilleras, on which graze the lama and the
cattle domesticated by the European colonist. Where the
naked trachyte rock pierces the grassy turf, and penetrates into
those higher strata of air which are supposed to be less charged
with carbonic acid, we meet only with plants of an inferior organization, as lichens, lecideas, and the brightly-colored, dust-
like lepraria, scattered around in circular patches. Islets of
fresh-fallen snow, varying in form and extent, arrest the last
feeble traces of vegetable development, and to these succeeds
the region of perpetual snow, whose elevation undergoes but
little change, and may be easily determined. It is but rarely
that the elastic forces at work within the interior of our globe
have succeeded in breaking through the spiral domes, which,
resplendent in the brightness of eternal snow, crown the summits of the Cordilleras ; and even where these subterranean
forces have opened a permanent communication with the atmosphere, through circular craters or long fissures, they rarely
send forth currents of lava, but merely eject ignited scoriæ,
steam, sulphureted hydrogen gas, and jets of carbonic acid.

In the earliest stages of civilization, the grand and imposing
spectacle presented to the minds of the inhabitants of the tropics could only awaken feelings of astonishment and awe. It
might, perhaps, be supposed, as we have already said, that the
periodical return of the same phenomena, and the uniform manner in which they arrange themselves in successive groups,
would have enabled man more readily to attain to a knowledge of the laws of nature ; but, as far as tradition and history
guide us, we do not find that any application was made of the
advantages presented by these favored regions. Recent researches have rendered it very doubtful whether the primitive
seat of Hindoo civilization—one of the most remarkable phases
in the progress of mankind—was actually within the tropics
Airyana Vaedjo, the ancient cradle of the Zend, was situated
to the northwest of the upper Indus, and after the great re
ligious schism, that is to say, after the separation of the Ira
nians from the Brahminical institution, the language that had
previously been common to them and to the Hindoos assumed
among the latter people (together with the literature, habits,
and condition of society) an individual form in the Magodha or
Madhya Desa,* a district that is bounded by the great chain

* See, on the Madhjadêça, properly so called, Lassen's excellent
work, entitled *Indische Alterthumskunde*, bd. i., s. 92. The Chinese

of Himalaya and the smaller range of the Vindhya. In less ancient times the Sanscrit language and civilization advanced toward the southeast, penetrating further within the torrid zone, as my brother Wilhelm von Humboldt has shown in his great work on the Kavi and other languages of analogous structure.*

Notwithstanding the obstacles opposed in northern latitudes to the discovery of the laws of nature, owing to the excessive complication of phenomena, and the perpetual local variations that, in these climates, affect the movements of the atmosphere and the distribution of organic forms, it is to the inhabitants of a small section of the temperate zone that the rest of mankind owe the earliest revelation of an intimate and rational acquaintance with the forces governing the physical world. Moreover, it is from the same zone (which is apparently more favorable to the progress of reason, the softening of manners, and the security of public liberty) that the germs of civilization have been carried to the regions of the tropics, as much by the migratory movement of races as by the establishment of colonies, differing widely in their institution from those of the Phœnicians or Greeks.

In speaking of the influence exercised by the succession of phenomena on the greater or lesser facility of recognizing the causes producing them, I have touched upon that important stage of our communion with the external world, when the enjoyment arising from a knowledge of the laws, and the mutual connection of phenomena, associates itself with the charm of a simple contemplation of nature. That which for a long time remains merely an object of vague intuition, by degrees acquires the certainty of positive truth ; and man, as an immortal poet has said, in our own tongue—Amid ceaseless change seeks the unchanging pole.†

In order to trace to its primitive source the enjoyment derived from the exercise of thought, it is sufficient to cast a rapid glance on the earliest dawnings of the philosophy of nature, or of the ancient doctrine of the *Cosmos*.   We find even

give the name of Mo-kie-thi to the southern Bahar, situated to the south of the Ganges (see *Foe-Koue-Ki*, by *Chy-Fa-Hian*, 1836, p. 256). Djambu-dwipa is the name given to the whole of India ; but the words also indicate one of the four Buddhist continents.

* *Ueber die Kawi Sprache auf der Insel Java, nebst einer Einleitung über die Verschiedenheit des menschlichen Sprachbaues und ihren Einfluss auf die geistige Entwickelung des Menschengeschlecht's*, von Wilhelm v. Humboldt, 1836, bd. i., s. 5–510.

† This verse occurs in a poem of Schiller, entitled *Der Spaziergang* which first appeared in 1795, in the *Horen*.

among the most savage nations (as my own travels enable me to attest) a certain vague, terror-stricken sense of the all-powerful unity of natural forces, and of the existence of an invisible, spiritual essence manifested in these forces, whether in unfolding the flower and maturing the fruit of the nutrient tree, in upheaving the soil of the forest, or in rending the clouds with the might of the storm. We may here trace the revelation of a bond of union, linking together the visible world and that higher spiritual world which escapes the grasp of the senses. The two become unconsciously blended together, developing in the mind of man, as a simple product of ideal conception, and independently of the aid of observation, the first germ of a *Philosophy of Nature.*

Among nations least advanced in civilization, the imagination revels in strange and fantastic creations, and, by its predilection for symbols, alike influences ideas and language. Instead of examining, men are led to conjecture, dogmatize, and interpret supposed facts that have never been observed. The inner world of thought and of feeling does not reflect the image of the external world in its primitive purity. That which in some regions of the earth manifested itself as the rudiments of natural philosophy, only to a small number of persons endowed with superior intelligence, appears in other regions, and among entire races of men, to be the result of mystic tendencies and instinctive intuitions. An intimate communion with nature, and the vivid and deep emotions thus awakened, are likewise the source from which have sprung the first impulses toward the worship and deification of the destroying and preserving forces of the universe. But by degrees, as man, after having passed through the different gradations of intellectual development, arrives at the free enjoyment of the regulating power of reflection, and learns by gradual progress, as it were, to separate the world of ideas from that of sensations, he no longer rests satisfied merely with a vague presentiment of the harmonious unity of natural forces ; thought begins to fulfill its noble mission ; and observation, aided by reason, endeavors to trace phenomena to the causes from which they spring.

The history of science teaches us the difficulties that have opposed the progress of this active spirit of inquiry. Inaccurate and imperfect observations have led, by false inductions, to the great number of physical views that have been perpetuated as popular prejudices among all classes of society. Thus by the side of a solid and scientific knowledge of natural phenomena there has been preserved a system of the pretended results of observation, which is so much the more difficult to shake, as it denies the validity of the facts by which it may be refuted. This empiricism, the melancholy heritage transmitted to us from former times, invariably contends for the truth of its axioms with the arrogance of a narrow-minded spirit. Physical philosophy, on the other hand, when based upon science, doubts because it seeks to investigate, distinguishes between that which is certain and that which is merely probable, and strives incessantly to perfect theory by extending the circle of observation.

This assemblage of imperfect dogmas, bequeathed by one age to another—this physical philosophy, which is composed of popular prejudices—is not only injurious because it perpetuates error with the obstinacy engendered by the evidence of ill-observed facts, but also because it hinders the mind from attaining to higher views of nature. Instead of seeking to discover the *mean* or *medium* point, around which oscillate, in apparent independence of forces, all the phenomena of the external world, this system delights in multiplying exceptions to the law, and seeks, amid phenomena and in organic forms, for something beyond the marvel of a regular succession, and an internal and progressive development. Ever inclined to believe that the order of nature is disturbed, it refuses to rec·ognize in the present any analogy with the past, and, guided by its own varying hypotheses, seeks at hazard, either in the interior of the globe or in the regions of space, for the cause of these pretended perturbations.

It is the special object of the present work to combat those errors which derive their source from a vicious empiricism and from imperfect inductions. The higher enjoyments yielded by the study of nature depend upon the correctness and the depth of our views, and upon the extent of the subjects that may be comprehended in a single glance. Increased mental cultivation has given rise, in all classes of society, to an increased desire of embellishing life by augmenting the mass of ideas, and by multiplying means for their generalization ; and this sentiment fully refutes the vague accusations advanced against the age in which we live, showing that other interests, besides the material wants of life, occupy the minds of men.

It is almost with reluctance that I am about to speak of a sentiment, which appears to arise from narrow-minded views, or from a certain weak and morbid sentimentality—I allude to the *fear* entertained by some persons, that nature may by degrees lose a portion of the charm and magic of her power, ᴀs we learn more and more how to unvail her secrets, comprehend the mechanism of the movements of the heavenly bodies, and estimate numerically the intensity of natural forces. It is true that, properly speaking, the forces of nature can only exercise a magical power over us as long as their action is shrouded in mystery and darkness, and does not admit of being classed among the conditions with which experience has made us acquainted. The effect of such a power is, therefore, to excite the imagination, but that, assuredly, is not the faculty of mind we would evoke to preside over the laborious and elaborate observations by which we strive to attain to a knowledge of the greatness and excellence of the laws of the universe.

The astronomer who, by the aid of the heliometer or a double-refracting prism,* determines the diameter of planetary bodies ; who measures patiently, year after year, the meridian altitude and the relative distances of stars, or who seeks a tel escopic comet in a group of nebulæ, does not feel his imagination more excited—and this is the very guarantee of the precision of his labors—than the botanist who counts the divisions of the calyx, or the number of stamens in a flower, or examines the connected or the separate teeth of the peristoma surrounding the capsule of a moss. Yet the multiplied angular measurements on the one hand, and the detail of organic relations on the other, alike aid in preparing the way for the attainment of higher views of the laws of the universe.

We must not confound the disposition of mind in the observer at the time he is pursuing his labors, with the ulterior greatness of the views resulting from investigation and the exercise of thought. The physical philosopher measures with admirable sagacity the waves of light of unequal length which by interference mutually strengthen or destroy each other, even with respect to their chemical actions ; the astronomer, armed with powerful telescopes, penetrates the regions of space, contemplates, on the extremest confines of our solar system, the satellites of Uranus, or decomposes faintly sparkling points into double stars differing in color. The botanist discovers the constancy of the gyratory motion of the chara in the greater number of vegetable cells, and recognizes in the genera and natural families of plants the intimate relations of organic forms. The vault of heaven, studded with nebu-

---

* Arago's ocular micrometer, a happy improvement upon Rochon's prismatic or double-refraction micrometer. See M. Mathieu's note in Délambre's *Histoire de l'Astronomie au dix-huitième Siècle*, 1827.

læ and stars, and the rich vegetable mantle that covers the soil in the climate of palms, can not surely fail to produce on the minds of these laborious observers of nature an impression more imposing and more worthy of the majesty of creation than on those who are unaccustomed to investigate the great mutual relations of phenomena. I can not, therefore, agree with Burke when he says, " it is our ignorance of natural things that causes all our admiration, and chiefly excites our passions."

While the illusion of the senses would make the stars sta tionary in the vault of heaven, Astronomy, by her aspiring labors, has assigned indefinite bounds to space ; and if she have set limits to the great nebula to which our solar system belongs, it has only been to show us in those remote regions of space, which appear to expand in proportion to the increase of our optic powers, islet on islet of scattered nebulæ. The feeling of the sublime, so far as it arises from a contemplation of the distance of the stars, of their greatness and physical extent, reflects itself in the feeling of the infinite, which belongs to another sphere of ideas included in the domain of mind. The solemn and imposing impressions excited by this sentiment are owing to the combination of which we have spoken, and to the analogous character of the enjoyment and emotions awakened in us, whether we float on the surface of the great deep, stand on some lonely mountain summit enveloped in the half-transparent vapory vail of the atmosphere, or by the aid of powerful optical instruments scan the regions of space, and see the remote nebulous mass resolve itself into worlds of stars.

The mere accumulation of unconnected observations of details, devoid of generalization of ideas, may doubtlessly have tended to create and foster the deeply-rooted prejudice, that the study of the exact sciences must necessarily chill the feelings, and diminish the nobler enjoyments attendant upon a contemplation of nature. Those who still cherish such erro neous views in the present age, and amid the progress of public opinion, and the advancement of all branches of knowledge, fail in duly appreciating the value of every enlargement of the sphere of intellect, and the importance of the detail of isolated facts in leading us on to general results. The fear of sacri ficing the free enjoyment of nature, under the influence of scientific reasoning, is often associated with an apprehension that every mind may not be capable of grasping the truths of the philosophy of nature. It is certainly true that in the midst of the universal fluctuation of phenomena and vital forces—in that inextricable net-work of organisms by turns developed and destroyed—each step that we make in the more intimate knowledge of nature leads us to the entrance of new labyrinths ; but the excitement produced by a presentiment of discovery, the vague intuition of the mysteries to be unfolded, and the multiplicity of the paths before us, all tend to stimulate the exercise of thought in every stage of knowledge. The discovery of each separate law of nature leads to the establishment of some other more general law, or at least indicates to the intelligent observer its existence. Nature, as a celebrated physiologist* has defined it, and as the word was interpreted by the Greeks and Romans, is " that which is ever growing and ever unfolding itself in new forms."

The series of organic types becomes extended or perfected in proportion as hitherto unknown regions are laid open to our view by the labors and researches of travelers and observers ; as living organisms are compared with those which have disappeared in the great revolutions of our planet ; and as microscopes are made more perfect, and are more extensively and efficiently employed. In the midst of this immense variety, and this periodic transformation of animal and vegetable productions, we see incessantly revealed the primordial mystery of all organic development, that same great problem of *metamorphosis* which Göthe has treated with more than common sagacity, and to the solution of which man is urged by his desire of reducing vital forms to the smallest number of fundamental types. As men contemplate the riches of nature, and see the mass of observations incessantly increasing before them, they become impressed with the intimate conviction that the surface and the interior of the earth, the depths of the ocean, and the regions of air will still, when thousands and thousands of years have passed away, open to the scientific observer untrodden paths of discovery. The regret of Alexander can not be applied to the progress of observation and intelligence.† General considerations, whether they treat of the agglomeration of matter in the heavenly bodies, or of the geographical distribution of terrestrial organisms, are not only in themselves more attractive than special studies, but they also afford superior advantages to those who are unable to devote much time to occupations of this nature. The different branches of the study of natural history are only accessible in certain positions of social life, and do not, at every sea-

* Carus, *Von den Urtheilen des Knochen und Schalen Gerüstes*, 1828
§ 6   † Plut., in *Vita Alex. Magni*, cap. 7

son and in every climate, present like enjoyments. Thus, in
the dreary regions of the north, man is deprived for a long
period of the year of the spectacle presented by the activity
of the productive forces of organic nature ; and if the mind
be directed to one sole class of objects, the most animated
narratives of voyages in distant lands will fail to interest and
attract us, if they do not touch upon the subjects to which
we are most partial.

As the history of nations—if it were always able to trace
events to their true causes—might solve the ever-recurring
enigma of the oscillations experienced by the alternately progressive and retrograde movement of human society, so might
also the physical description of the world, the science of the
Cosmos, if it were grasped by a powerful intellect, and based
upon a knowledge of all the results of discovery up to a giv-
en period, succeed in dispelling a portion of the contradictions
which, at first sight, appear to arise from the complication of
phenomena and the multitude of the perturbations simultaneously manifested.

The knowledge of the laws of nature, whether we can
trace them in the alternate ebb and flow of the ocean, in the
measured path of comets, or in the mutual attractions of multiple stars, alike increases our sense of the calm of nature,
while the chimera so long cherished by the human mind in
its early and intuitive contemplations, the belief in a "discord
of the elements," seems gradually to vanish in proportion as
science extends her empire.   General views lead us habitually to consider each organism as a part of the entire creation,
and to recognize in the plant or the animal not merely an
isolated species, but a form linked in the chain of being to
other forms either living or extinct.   They aid us in comprehending the relations that exist between the most recent dis
coveries and those which have prepared the way for them.
Although fixed to one point of space, we eagerly grasp at a
knowledge of that which has been observed in different and
far-distant regions.   We delight in tracking the course of the
bold mariner through seas of polar ice, or in following him to
the summit of that volcano of the antarctic pole, whose fires
may be seen from afar, even at mid-day.   It is by an acquaintance with the results of distant voyages that we may
learn to comprehend some of the marvels of terrestrial magnetism, and be thus led to appreciate the importance of the
establishments of the numerous observatories which in the
present day cover both hemispheres, and are designed to note

the simultaneous occurrence of perturbations, and the frequency and duration of *magnetic storms.*

Let me be permitted here to touch upon a few points connected with discoveries, whose importance can only be estimated by those who have devoted themselves to the study of the physical sciences generally. Examples chosen from among the phenomena to which special attention has been directed in recent times, will throw additional light upon the preceding considerations. Without a preliminary knowledge of the orbits of comets, we should be unable duly to appreciate the importance attached to the discovery of one of these bodies, whose elliptical orbit is included in the narrow limits of our solar system, and which has revealed the existence of an ethereal fluid, tending to diminish its centrifugal force and the period of its revolution.

The superficial half-knowledge, so characteristic of the present day, which leads to the introduction of vaguely comprehended scientific views into general conversation, also gives rise, under various forms, to the expression of alarm at the supposed danger of a collision between the celestial bodies, or of disturbance in the climatic relations of our globe. These phantoms of the imagination are so much the more injurious as they derive their source from dogmatic pretensions to true science. The history of the atmosphere, and of the annual variations of its temperature, extends already sufficiently far back to show the recurrence of slight disturbances in the mean temperature of any given place, and thus affords sufficient guarantee against the exaggerated apprehension of a general and progressive deterioration of the climates of Europe. Encke's comet, which is one of the three *interior comets*, completes its course in 1200 days, but from the form and position of its orbit it is as little dangerous to the earth as Halley's great comet, whose revolution is not completed in less than seventy-six years (and which appeared less brilliant in 1835 than it had done in 1759): the interior comet of Biela intersects the earth's orbit, it is true, but it can only approach our globe when its proximity to the sun coincides with our winter solstice.

The quantity of heat received by a planet, and whose unequal distribution determines the meteorological variations of its atmosphere, depends alike upon the light-engendering force of the sun ; that is to say, upon the condition of its gaseous coverings, and upon the relative position of the planet and the central body.

There are variations, it is true, which, in obedience to the laws of universal gravitation, affect the form of the earth's orbit and the inclination of the ecliptic, that is, the angle which the axis of the earth makes with the plane of its orbit ; but these periodical variations are so slow, and are restricted within such narrow limits, that their thermic effects would hardly be appreciable by our instruments in many thousands of years. The astronomical causes of a refrigeration of our globe, and of the diminution of moisture at its surface, and the nature and frequency of certain epidemics—phenomena which are often discussed in the present day according to the benighted views of the Middle Ages—ought to be considered as beyond the range of our experience in physics and chemistry.

Physical astronomy presents us with other phenomena, which can not be fully comprehended in all their vastness without a previous acquirement of general views regarding the forces that govern the universe. Such, for instance, are the innumerable double stars, or rather suns, which revolve round one common center of gravity, and thus reveal in distant worlds the existence of the Newtonian law ; the larger or smaller number of spots upon the sun, that is to say, the openings formed through the luminous and opaque atmosphere surrounding the solid nucleus ; and the regular appearance, about the 13th of November and the 11th of August, of shooting stars, which probably form part of a belt of asteroids, intersecting the earth's orbit, and moving with planetary velocity.

Descending from the celestial regions to the earth, we would fain inquire into the relations that exist between the oscillations of the pendulum in air (the theory of which has been perfected by Bessel) and the density of our planet ; and how the pendulum, acting the part of a plummet, can, to a certain extent, throw light upon the geological constitution of strata at great depths ? By means of this instrument we are enabled to trace the striking analogy which exists between the formation of the granular rocks composing the lava currents ejected from active volcanoes, and those endogenous masses of granite, porphyry, and serpentine, which, issuing from the interior of the earth, have broken, as eruptive rocks, through the secondary strata, and modified them by contact, either in rendering them harder by the introduction of silex, or reducing them into dolomite, or, finally, by inducing within them the formation of crystals of the most varied composition. The elevation of sporadic islands, of domes of trachyte, and cones of basalt, by the elastic forces emanating from the fluid interior of our globe, has led one of the first geologists of the age, Leopold von Buch, to the theory of the elevation of continents, and of mountain chains generally. This action of subterranean forces in breaking through and elevating strata of sedimentary rocks, of which the coast of Chili, in consequence of a great earthquake, furnished a recent example, leads to the assumption that the pelagic shells found by M. Bonpland and myself on the ridge of the Andes, at an elevation of more than 15,000 English feet, may have been conveyed to so extraordinary a position, not by a rising of the ocean, but by the agency of volcanic forces capable of elevating into ridges the softened crust of the earth.

I apply the term *volcanic*, in the widest sense of the word, to every action exercised by the interior of a planet on its external crust. The surface of our globe, and that of the moon, manifest traces of this action, which in the former, at least, has varied during the course of ages. Those who are ignorant of the fact that the internal heat of the earth increases so rapidly with the increase of depth that granite is in a state of fusion about twenty or thirty geographical miles below the surface,* can not have a clear conception of the causes, and the simultaneous occurrence of volcanic eruptions at places widely removed from one another, or of the extent and intersection of *circles of commotion* in earthquakes, or of the uniformity of temperature, and equality of chemical composition observed in thermal springs during a long course of years. The quantity of heat peculiar to a planet is, however, a matter of such importance—being the result of its primitive condensation, and varying according to the nature and duration of the radiation—that the study of this subject may throw some degree of light on the history of the atmosphere, and the distribution of the organic bodies imbedded in the solid crust of the earth. This study enables us to understand how a tropical temperature, independent of latitude (that is, of the distance from the poles), may have been produced by deep fissures remaining open, and exhaling heat from the in-

* The determinations usually given of the point of fusion are in general much too high for refracting substances. According to the very accurate researches of Mitscherlich, the melting point of granite can hardly exceed 2372° F.

[Dr. Mantell states in *The Wonders of Geology*, 1848, vol. i., p. 34, that this increase of temperature amounts to 1° of Fahrenheit for every fifty-four feet of vertical depth.]—*Tr.*

terior of the globe, at a period when the earth's crust was
still furrowed and rent, and only in a state of semi-solidifica-
tion ; and a primordial condition is thus revealed to us, in
which the temperature of the atmosphere, and climates generally, were owing rather to a liberation of caloric and of different gaseous emanations (that is to say, rather to the energetic reaction of the interior on the exterior) than to the position of the earth with respect to the central body, the sun.

The cold regions of the earth contain, deposited in sedimentary strata, the products of tropical climates ; thus, in
the coal formations, we find the trunks of palms standing upright amid coniferæ, tree ferns, goniatites, and fishes having
rhomboidal osseous scales ;* in the Jura limestone, colossal
skeletons of crocodiles, plesiosauri, planulites, and stems of the
cycadeæ ; in the chalk formations, small polythalamia and
bryozoa, whose species still exist in our seas ; in tripoli, or
polishing slate, in the semi-opal and the farina-like opal or
mountain meal, agglomerations of siliceous infusoria, which
have been brought to light by the powerful microscope of
Ehrenberg ;† and, lastly, in transported soils, and in certain
caves, the bones of elephants, hyenas, and lions. An intimate
acquaintance with the physical phenomena of the universe
leads us to regard the products of warm latitudes that are
thus found in a fossil condition in northern regions not merely
as incentives to barren curiosity, but as subjects awakening
deep reflection, and opening new sources of study.

The number and the variety of the objects I have alluded
to give rise to the question whether general considerations of
physical phenomena can be made sufficiently clear to persons
who have not acquired a detailed and special knowledge of

---

* See the classical work on the fishes of the Old World by Agassiz,
Rech. sur les Poissons Fossiles, 1834, vol. i., p. 38 ; vol. ii., p. 3, 28,
34, App., p. 6. The whole genus of Amblypterus, Ag., nearly allied
to Palæoniscus (called also Palæothrissum), lies buried beneath the
Jura formations in the old carboniferous strata. Scales which, in some
fishes, as in the family of Lepidoides (order of Ganoides), are formed
like teeth, and covered in certain parts with enamel, belong, after the
Placoides, to the oldest forms of fossil fishes ; their living representatives are still found in two genera, the *Bichir* of the Nile and Senegal,
and the *Lepidosteus* of the Ohio.

† [The *polishing slate* of Bilin is stated by M. Ehrenberg to form a
series of strata fourteen feet in thickness, entirely made up of the siliceous shells of *Gaillonellæ*, of such extreme minuteness that a cubic
inch of the stone contains forty-one thousand millions ! The *Bergmehl*
(*mountain meal* or *fossil farina*) of San Fiora, in Tuscany, is one mass
of animalculites. See the interesting work of G. A. Mantell, *On the
Medals of Creation*, vol. i., p. 223.]—*Tr.*

descriptive natural history, geology, or mathematical astronomy? I think we ought to distinguish here between him whose task it is to collect the individual details of various observations, and study the mutual relations existing among them, and him to whom these relations are to be revealed, under the form of general results. The former should be acquainted with the specialities of phenomena, that he may arrive at a generalization of ideas as the result, at least in part, of his own observations, experiments, and calculations. It can not be denied, that where there is an absence of positive knowledge of physical phenomena, the general results which impart so great a charm to the study of nature can not all be made equally clear and intelligible to the reader, but still I venture to hope, that in the work which I am now preparing on the physical laws of the universe, the greater part of the facts advanced can be made manifest without the necessity of appealing to fundamental views and principles. The picture of nature thus drawn, notwithstanding the want of distinctness of some of its outlines, will not be the less able to enrich the intellect, enlarge the sphere of ideas, and nourish and vivify the imagination.

There is, perhaps, some truth in the accusation advanced against many German scientific works, that they lessen the value of general views by an accumulation of detail, and do not sufficiently distinguish between those great results which form, as it were, the beacon lights of science, and the long series of means by which they have been attained. This method of treating scientific subjects led the most illustrious of our poets* to exclaim with impatience, "The Germans have the art of making science inaccessible." An edifice can not produce a striking effect until the scaffolding is removed, that had of necessity been used during its erection. Thus the uniformity of figure observed in the distribution of continental masses, which all terminate toward the south in a pyramidal form, and expand toward the north (a law that determines the nature of climates, the direction of currents in the ocean and the atmosphere, and the transition of certain types of tropical vegetation toward the southern temperate zone), may be clearly apprehended without any knowledge of the geodesical and astronomical operations by means of which these pyramidal forms of continents have been determined. In like manner, physical geography teaches us by how many leagues

* Göthe, in *Die Aphorismen über Naturwissenschaft*, bd. l., s. 155 (*Werke kleine Ausgabe, von* 1833.)

the equatorial axis exceeds the polar axis of the globe, and shows us the mean equality of the flattening of the two hemispheres, without entailing on us the necessity of giving the detail of the measurement of the degrees in the meridian, or the observations on the pendulum, which have led us to know that the true figure of our globe is not exactly that of a regular ellipsoid of revolution, and that this irregularity is reflected in the corresponding irregularity of the movements of the moon.

The views of comparative geography have been specially enlarged by that admirable work, *Erdkunde im Verhältniss zur Natur und zur Geschichte*, in which Carl Ritter so ably delineates the physiognomy of our globe, and shows the influence of its external configuration on the physical phenomena on its surface, on the migrations, laws, and manners of nations, and on all the principal historical events enacted upon the face of the earth.

France possesses an immortal work, *L'Exposition du Système du Monde*, in which the author has combined the results of the highest astronomical and mathematical labors, and presented them to his readers free from all processes of demonstration. The structure of the heavens is here reduced to the simple solution of a great problem in mechanics ; yet Laplace's work has never yet been accused of incompleteness and want of profundity.

The distinction between dissimilar subjects, and the separation of the general from the special, are not only conducive to the attainment of perspicuity in the composition of a physical history of the universe, but are also the means by which a character of greater elevation may be imparted to the study of nature. By the suppression of all unnecessary detail, the great masses are better seen, and the reasoning faculty is enabled to grasp all that might otherwise escape the limited range of the senses.

The exposition of general results has, it must be owned, been singularly facilitated by the happy revolution experienced since the close of the last century, in the condition of all the special sciences, more particularly of geology, chemistry, and descriptive natural history. In proportion as laws admit of more general application, and as sciences mutually enrich each other, and by their extension become connected together in more numerous and more intimate relations, the development of general truths may be given with conciseness devoid of superficiality. On being first examined, all phenomena appear to be

isolated, and it is only by the result of a multiplicity of observations, combined by reason, that we are able to trace the
mutual relations existing between them. If, however, in the
present age, which is so strongly characterized by a brilliant
course of scientific discoveries, we perceive a want of connection in the phenomena of certain sciences, we may anticipate
the revelation of new facts, whose importance will probably
be commensurate with the attention directed to these branches
of study. Expectations of this nature may be entertained with
regard to meteorology, several parts of optics, and to radiating
heat, and electro-magnetism, since the admirable discoveries
of Melloni and Faraday. A fertile field is here opened to discovery, although the voltaic pile has already taught us the
intimate connection existing between electric, magnetic, and
chemical phenomena. Who will venture to affirm that we
have any precise knowledge, in the present day, of that part
of the atmosphere which is not oxygen, or that thousands of
gaseous substances affecting our organs may not be mixed with
the nitrogen, or, finally, that we have even discovered the whole
number of the forces which pervade the universe ?

It is not the purpose of this essay on the physical history of
the world to reduce all sensible phenomena to a small number
of abstract principles, based on reason only. The physical
history of the universe, whose exposition I attempt to develop,
does not pretend to rise to the perilous abstractions of a purely
rational science of nature, and is simply a *physical geography,
combined with a description of the regions of space and the
bodies occupying them.* Devoid of the profoundness of a purely
speculative philosophy, my essay on the *Cosmos* treats of the
contemplation of the universe, and is based upon a rational
empiricism, that is to say, upon the results of the facts regis-
tered by science, and tested by the operations of the intellect.
It is within these limits alone that the work, which I now
venture to undertake, appertains to the sphere of labor to
which I have devoted myself throughout the course of my
long scientific career. The path of inquiry is not unknown
to me, although it may be pursued by others with greater
success. The unity which I seek to attain in the development
of the great phenomena of the universe is analogous to that
which historical composition is capable of acquiring. All
points relating to the accidental individualities, and the essential variations of the actual, whether in the form and arrangement of natural objects in the struggle of man against the
elements, or of nations against nations, do not admit of being

based only on a *rational foundation*—that is to say, of being deduced from ideas alone.

It seems to me that a like degree of empiricism attaches to the Description of the Universe and to Civil History; but in reflecting upon physical phenomena and events, and tracing their causes by the process of reason, we become more and more convinced of the truth of the ancient doctrine, that the forces inherent in matter, and those which govern the moral world, exercise their action under the control of primordial necessity, and in accordance with movements occurring periodically after longer or shorter intervals.

It is this necessity, this occult but permanent connection, this periodical recurrence in the progressive development of forms, phenomena, and events, which constitute *nature*, obedient to the first impulse imparted to it. Physics, as the term signifies, is limited to the explanation of the phenomena of the material world by the properties of matter. The ultimate object of the experimental sciences is, therefore, to discover laws, and to trace their progressive generalization. All that exceeds this goes beyond the province of the physical description of the universe, and appertains to a range of higher speculative views.

Emanuel Kant, one of the few philosophers who have escaped the imputation of impiety, has defined with rare sagacity the limits of physical explanations, in his celebrated essay *On the Theory and Structure of the Heavens*, published at Königsberg in 1755.

The study of a science that promises to lead us through the vast range of creation may be compared to a journey in a far-distant land. Before we set forth, we consider, and often with distrust, our own strength, and that of the guide we have chosen. But the apprehensions which have originated in the abundance and the difficulties attached to the subjects we would embrace, recede from view as we remember that with the increase of observations in the present day there has also arisen a more intimate knowledge of the connection existing among all phenomena. It has not unfrequently happened, that the researches made at remote distances have often and unexpectedly thrown light upon subjects which had long resisted the attempts made to explain them within the narrow limits of our own sphere of observation. Organic forms that had long remained isolated, both in the animal and vegetable kingdom, have been connected by the discovery of intermediate links or stages of transition. The geography of beings endowed with life attains completeness as we see the species, genera, and entire families belonging to one hemisphere, reflected, as it were, in analogous animal and vegetable forms in the opposite hemisphere. These are, so to speak, the *equivalents* which mutually personate and replace one another in the great series of organisms. These connecting links and stages of transition may be traced, alternately, in a deficiency or an excess of development of certain parts, in the mode of junction of distinct organs, in the differences in the balance of forces, or in a resemblance to intermediate forms which are not permanent, but merely characteristic of certain phases of normal development. Passing from the consideration of beings endowed with life to that of inorganic bodies, we find many striking illustrations of the high state of advancement to which modern geology has attained. We thus see, according to the grand views of Elie de Beaumont, how chains of mountains dividing different climates and floras and different races of men, reveal to us their *relative age*, both by the character of the sedimentary strata they have uplifted, and by the directions which they follow over the long fissures with which the earth's crust is furrowed. Relations of superposition of trachyte and of syenitic porphyry, of diorite and of serpentine, which remain doubtful when considered in the auriferous soil of Hungary, in the rich platinum districts of the Oural, and on the southwestern declivity of the Siberian Altaï, are elucidated by the observations that have been made on the plateaux of Mexico and Antioquia, and in the unhealthy ravines of Choco. The most important facts on which the physical history of the world has been based in modern times, have not been accumulated by chance. It has at length been fully acknowledged, and the conviction is characteristic of the age, that the narratives of distant travels, too long occupied in the mere recital of hazardous adventures, can only be made a source of instruction where the traveler is acquainted with the condition of the science he would enlarge, and is guided by reason in his researches.

It is by this tendency to generalization, which is only dangerous in its abuse, that a great portion of the physical knowledge already acquired may be made the common property of all classes of society ; but, in order to render the instruction imparted by these means commensurate with the importance of the subject, it is desirable to deviate as widely as possible from the imperfect compilations designated, till the close of the eighteenth century, by the inappropriate term of *popular*

knowledge. I take pleasure in persuading myself that scientific subjects may be treated of in language at once dignified,
grave, and animated, and that those who are restricted within the circumscribed limits of ordinary life, and have long remained strangers to an intimate communion with nature,
may thus have opened to them one of the richest sources of
enjoyment, by which the mind is invigorated by the acquisition of new ideas. Communion with nature awakens within
us perceptive faculties that had long lain dormant ; and we
thus comprehend at a single glance the influence exercised by
physical discoveries on the enlargement of the sphere of intellect, and perceive how a judicious application of mechanics,
chemistry, and other sciences may be made conducive to national prosperity.

A more accurate knowledge of the connection of physical
phenomena will also tend to remove the prevalent error that
all branches of natural science are not equally important in
relation to general cultivation and industrial progress. An
arbitrary distinction is frequently made between the various
degrees of importance appertaining to mathematical sciences,
to the study of organized beings, the knowledge of electro-
magnetism, and investigations of the general properties of matter in its different conditions of molecular aggregation ; and it
is not uncommon presumptuously to affix a supposed stigma
upon researches of this nature, by terming them " purely theoretical," forgetting, although the fact has been long attested,
that in the observation of a phenomenon, which at first sight
appears to be wholly isolated, may be concealed the germ of a
great discovery. When Aloysio Galvani first stimulated the
nervous fiber by the accidental contact of two heterogeneous
metals, his cotemporaries could never have anticipated that
the action of the voltaic pile would discover to us, in the alkalies, metals of a silvery luster, so light as to swim on water, and eminently inflammable ; or that it would become a
powerful instrument of chemical analysis, and at the same
time a thermoscope and a magnet. When Huygens first observed, in 1678, the phenomenon of the polarization of light,
exhibited in the difference between the two rays into which
a pencil of light divides itself in passing through a doubly
refracting crystal, it could not have been foreseen that, a
century and a half later, the great philosopher Arago would,
by his discovery of *chromatic polarization,* be led to discern,
by means of a small fragment of Iceland spar, whether solar
light emanates from a solid body or a gaseous covering, or

whether comets transmit light directly or merely by reflection.*

An equal appreciation of all branches of the mathematical, physical, and natural sciences is a special requirement of the present age, in which the material wealth and the growing prosperity of nations are principally based upon a more enlightened employment of the products and forces of nature. The most superficial glance at the present condition of Europe shows that a diminution, or even a total annihilation of national prosperity, must be the award of those states who shrink with slothful indifference from the great struggle of rival nations in the career of the industrial arts. It is with nations as with nature, which, according to a happy expression of Göthe,† " knows no pause in progress and development, and attaches her curse on all inaction." The propagation of an earnest and sound knowledge of science can therefore alone avert the dangers of which I have spoken. Man can not act upon nature, or appropriate her forces to his own use, without comprehending their full extent, and having an intimate acquaintance with the laws of the physical world. Bacon has said that, in human societies, knowledge is power. Both must rise and sink together. But the knowledge that results from the free action of thought is at once the delight and the indestructible prerogative of man ; and in forming part of the wealth of mankind, it not unfrequently serves as a substitute for the natural riches, which are but sparingly scattered over the earth. Those states which take no active part in the general industrial movement, in the choice and preparation of natural substances, or in the application of mechanics and chemistry, and among whom this activity is not appreciated by all classes of society, will infallibly see their prosperity diminish in proportion as neighboring countries become strengthened and invigorated under the genial influence of arts and sciences.

As in nobler spheres of thought and sentiment, in philosophy, poetry, and the fine arts, the object at which we aim ought to be an inward one—an ennoblement of the intellect—so ought we likewise, in our pursuit of science, to strive after a knowledge of the laws and the principles of unity that pervade the vital forces of the universe ; and it is by such a course that

* Arago's Discoveries in the year 1811.—Delambre's *Histoire de l'Ast.*, p. 652. (Passage already quoted.)
† Göthe, in *Die Aphorismen über Naturwissenschaft.— Werke*, bd. 1., s. 4

physical studies may be made subservient to the progress of industry, which is a conquest of mind over matter. By a happy connection of causes and effects, we often see the useful linked to the beautiful and the exalted. The improvement of agriculture in the hands of freemen, and on properties of a moderate extent—the flourishing state of the mechanical arts freed from the trammels of municipal restrictions—the increased impetus imparted to commerce by the multiplied means of contact of nations with each other, are all brilliant results of the intellectual progress of mankind, and of the amelioration of political institutions, in which this progress is reflected. The picture presented by modern history ought to convince those who are tardy in awakening to the truth of the lesson it teaches.

Nor let it be feared that the marked predilection for the study of nature, and for industrial progress, which is so characteristic of the present age, should necessarily have a tendency to retard the noble exertions of the intellect in the domains of philosophy, classical history, and antiquity, or to deprive the arts by which life is embellished of the vivifying breath of imagination. Where all the germs of civilization are developed beneath the ægis of free institutions and wise legislation, there is no cause for apprehending that any one branch of knowledge should be cultivated to the prejudice of others. All afford the state precious fruits, whether they yield nourishment to man and constitute his physical wealth, or whether, more permanent in their nature, they transmit in the works of mind the glory of nations to remotest posterity. The Spartans, notwithstanding their Doric austerity, prayed the gods to grant them " the beautiful with the good."*

I will no longer dwell upon the considerations of the influence exercised by the mathematical and physical sciences on all that appertains to the material wants of social life, for the vast extent of the course on which I am entering forbids me to insist further upon the utility of these applications. Accustomed to distant excursions, I may, perhaps, have erred in describing the path before us as more smooth and pleasant than it really is, for such is wont to be the practice of those who delight in guiding others to the summits of lofty mountains : they praise the view even when great part of the distant plains lie hidden by clouds, knowing that this half-transparent vapory vail imparts to the scene a certain charm from

* Pseudo-Plato.—*Alcib.*, xi., p. 184, ed. Steph. ; Plut., *Instituta Laconica*, p. 253, ed. Hutten.

the power exercised by the imagination over the domain of the senses. In like manner, from the height occupied by the physical history of the world, all parts of the horizon will not appear equally clear and well defined. This indistinctness will not, however, be wholly owing to the present imperfect state of some of the sciences, but in part, likewise, to the unskillfulness of the guide who has imprudently ventured to ascend these lofty summits.

The object of this introductory notice is not, however, solely to draw attention to the importance and greatness of the phys ical history of the universe, for in the present day these are toc well understood to be contested, but likewise to prove how, without detriment to the stability of special studies, we may be enabled to generalize our ideas by concentrating them in one common focus, and thus arrive at a point of view from which all the organisms and forces of nature may be seen as one living, active whole, animated by one sole impulse. "Nature," as Schelling remarks in his poetic discourse on art, "is not an inert mass; and to him who can comprehend her vast sublimity, she reveals herself as the creative force of the universe—before all time, eternal, ever active, she calls to life all things, whether perishable or imperishable."

By uniting, under one point of view, both the phenomena of our own globe and those presented in the regions of space, we embrace the limits of the science of the *Cosmos*, and convert the physical history of the globe into the physical history of the universe, the one term being modeled upon that of the other. This science of the Cosmos is not, however, to be regarded as a mere encyclopedic aggregation of the most important and general results that have been collected together from special branches of knowledge. These results are nothing more than the materials for a vast edifice, and their combination can not constitute the physical history of the world, whose exalted part it is to show the simultaneous action and the connecting links of the forces which pervade the universe. The distribution of organic types in different climates and at different elevations—that is to say, the geography of plants and animals—differs as widely from botany and descriptive zoology as geology does from mineralogy, properly so called. The physical history of the universe must not, therefore, be confounded with the *Encyclopedias of the Natural Sciences*, as they have hitherto been compiled, and whose title is as vague as their limits are ill defined. In the work before us, partial facts will be considered only in relation to the whole

The higher the point of view, the greater is the necessity for a systematic mode of treating the subject in language at once animated and picturesque.

But thought and language have ever been most intimately allied. If language, by its originality of structure and its native richness, can, in its delineations, interpret thought with grace and clearness, and if, by its happy flexibility, it can paint with vivid truthfulness the objects of the external world, it reacts at the same time upon thought, and animates it, as it were, with the breath of life. It is this mutual reaction which makes words more than mere signs and forms of thought ; and the beneficent influence of a language is most strikingly manifested on its native soil, where it has sprung spontaneously from the minds of the people, whose character it embodies. Proud of a country that seeks to concentrate her strength in intellectual unity, the writer recalls with delight the advantages he has enjoyed in being permitted to express his thoughts in his native language ; and truly happy is he who, in attempting to give a lucid exposition of the great phenomena of the universe, is able to draw from the depths of a language, which, through the free exercise of thought, and by the effusions of creative fancy, has for centuries past exercised so powerful an influence over the destinies of man.

LIMITS AND METHOD OF EXPOSITION OF THE PHYSICAL DESCRIPTION OF THE UNIVERSE.

I HAVE endeavored, in the preceding part of my work, to explain and illustrate, by various examples, how the enjoyments presented by the aspect of nature, varying as they do in the sources from whence they flow, may be multiplied and ennobled by an acquaintance with the connection of phenomena and the laws by which they are regulated. It remains, then, for me to examine the spirit of the method in which the exposition of the *physical description of the universe* should be conducted, and to indicate the limits of this science in accordance with the views I have acquired in the course of my studies and travels in various parts of the earth. I trust I may flatter myself with a hope that a treatise of this nature will justify the title I have ventured to adopt for my work, and exonerate me from the reproach of a presumption that would be doubly reprehensible in a scientific discussion.

Before entering upon the delineation of the partial phenomena which are found to be distributed in various groups, I would consider a few general questions intimately connected together, and bearing upon the nature of our knowledge of the external world and its different relations, in all epochs of history and in all phases of intellectual advancement. Under this head will be comprised the following considerations :

1. The precise limits of the physical description of the universe, considered as a distinct science.

2. A brief enumeration of the totality of natural phenomena, presented under the form of *a general delineation of nature.*

3. The influence of the external world on the imagination and feelings, which has acted in modern times as a powerful impulse toward the study of natural science, by giving animation to the description of distant regions and to the delineation of natural scenery, as far as it is characterized by vegetable physiognomy and by the cultivation of exotic plants, and their arrangement in well-contrasted groups.

4. The history of the contemplation of nature, or the progressive development of the idea of the Cosmos, considered with reference to the historical and geographical facts that have led to the discovery of the connection of phenomena.

The higher the point of view from which natural phenomena may be considered, the more necessary it is to circumscribe the science within its just limits, and to distinguish it from all other analogous or auxiliary studies.

Physical cosmography is founded on the contemplation of all created things—all that exists in space, whether as substances or forces—that is, all the material beings that constitute the universe. The science which I would attempt to define presents itself, therefore, to man, as the inhabitant of the earth, under a two-fold form—as the earth itself and the regions of space. It is with a view of showing the actual character and the independence of the study of physical cosmography, and at the same time indicating the nature of its relations to *general physics, descriptive natural history, geology, and comparative geography*, that I will pause for a few moments to consider that portion of the science of the Cosmos which concerns the earth. As the history of philosophy does not consist of a mere material enumeration of the philosophical views entertained in different ages, neither should the physical description of the universe be a simple encyclopedic compilation of the sciences we have enumerated. The difficulty of defining the limits of intimately-connected studies has been increased, because for centuries it has been customary to designate various branches

of empirical knowledge by terms which admit either of too
wide or too limited a definition of the ideas which they were
intended to convey, and are, besides, objectionable from having had a different signification in those classical languages of
antiquity from which they have been borrowed. The terms
physiology, physics, natural history, geology, and geography
arose, and were commonly used, long before clear ideas were
entertained of the diversity of objects embraced by these
sciences, and consequently of their reciprocal limitation. Such
is the influence of long habit upon language, that by one of
the nations of Europe most advanced in civilization the word
" physic" is applied to medicine, while in a society of justly
deserved universal reputation, technical chemistry, geology,
and astronomy (purely experimental sciences) are comprised
under the head of " Philosophical Transactions."

An attempt has often been made, and almost always in vain,
to substitute new and more appropriate terms for these ancient
designations, which, notwithstanding their undoubted vagueness, are now generally understood. These changes have been
proposed, for the most part, by those who have occupied themselves with the general classification of the various branches
of knowledge, from the first appearance of the great encyclopedia (*Margarita Philosophica*) of Gregory Reisch,* prior of
the Chartreuse at Freiburg, toward the close of the fifteenth
century, to Lord Bacon, and from Bacon to D'Alembert ; and
in recent times to an eminent physicist, André Marie Ampère.†

* The *Margarita Philosophica* of Gregory Reisch, prior of the Chartreuse at Freiburg, first appeared under the following title : *Æpitome
omnis Philosophiæ, alias Margarita Philosophica, tractans de omni generi
scibili.* The Heidelberg edition (1486), and that of Strasburg (1504),
both bear this title, but the first part was suppressed in the Freiburg
edition of the same year, as well as in the twelve subsequent editions,
which succeeded one another, at short intervals, till 1535. This work
exercised a great influence on the diffusion of mathematical and physical sciences toward the beginning of the sixteenth century, and Chasles,
the learned author of *L'Aperçu Historique des Méthodes en Géométrie*
(1837), has shown the great importance of Reisch's *Encyclopedia* in
the history of mathematics in the Middle Ages. I have had recourse
to a passage in the *Margarita Philosophica*, found only in the edition
of 1513, to elucidate the important question of the relations between
the statements of the geographer of Saint-Die, Hylacomilus (Martin
Waldseemüller), the first who gave the name of America to the New
Continent, and those of Amerigo Vespucci, René, King of Jerusalem
and Duke of Lorraine, as also those contained in the celebrated editions
of Ptolemy of 1513 and 1522. See my *Examen Critique de la Géographie du Nouveau Continent, et des Progrès de l'Astronomie Nautique
aux 15e et 16e Siècles,* t. iv., p. 99–125.

† Ampère, *Essai sur la Phil. des Sciences,* 1834, p. 25. Whewell,

The selection of an inappropriate Greek nomenclature has perhaps been even more prejudicial to the last of these attempts than the injudicious use of binary divisions and the excessive multiplication of groups.

The physical description of the world, considering the universe as an object of the external senses, does undoubtedly require the aid of general physics and of descriptive natural history, but the contemplation of all created things, which are linked together, and form one *whole*, animated by internal forces, gives to the science we are considering a peculiar character. Physical science considers only the general properties of bodies ; it is the product of abstraction—a generalization of perceptible phenomena ; and even in the work in which were laid the first foundations of general physics, in the eight books on physics of Aristotle,* all the phenomena of nature are considered as depending upon the primitive and vital action of one sole force, from which emanate all the movements of the universe. The terrestrial portion of physical cosmography, for which I would willingly retain the expressive designation of *physical geography*, treats of the distribution of magnetism in our planet with relation to its intensity and direction, but does not enter into a consideration of the laws of attraction or repulsion of the poles, or the means of eliciting either permanent or transitory electro-magnetic currents. Physical geography depicts in broad outlines the even or irregular configuration of continents, the relations of superficial area, and the distribution of continental masses in the two hemispheres, a distribution which exercises a powerful influence on the diversity of climate and the meteorological modifications of the atmosphere ; this science defines the character of mountain chains, which, having been elevated at different epochs, constitute distinct systems, whether they run in parallel lines or intersect one another ; determines the mean height of continents above the level of the sea, the position of the center of gravity of their volume, and the relation of the highest summits of mountain chains to the mean elevation of their crests, or to their proximity with the sea-shore. It depicts the eruptive rocks as principles of movement, acting upon the sedimentary rocks by traversing, uplifting, and inclining them at various angles ; it

Philosophy of the Inductive Sciences, vol. ii., p. 277. Park, *Pantology*, p. 87.

* All changes in the physical world may be reduced to motion. Aristot., *Phys. Ausc.*, iii., 1 and 4, p. 200, 201. Bekker, viii., 1, 8, and 9, p. 250, 262, 265. *De Genere et Corr.*, ii., 10, p. 336. Pseudo-Aristot., *De Mundo*. cap. vi., p. 398.

considers volcanoes either as isolated, or ranged in single or in double series, and extending their sphere of action to various distances, either by raising long and narrow lines of rocks, or by means of circles of commotion, which expand or diminish in diameter in the course of ages. This terrestrial portion of the science of the Cosmos describes the strife of the liquid element with the solid land; it indicates the features possessed in common by all great rivers in the upper and lower portion of their course, and in their mode of bifurcation when their basins are unclosed; and shows us rivers breaking through the highest mountain chains, or following for a long time a course parallel to them, either at their base, or at a considerable distance, where the elevation of the strata of the mountain system and the direction of their inclination correspond to the configuration of the table-land. It is only the general results of comparative orography and hydrography that belong to the science whose true limits I am desirous of determining, and not the special enumeration of the greatest elevations of our globe, of active volcanoes, of rivers, and the number of their tributaries, these details falling rather within the domain of geography, properly so called. We would here only consider phenomena in their mutual connection, and in their relations to different zones of our planet, and to its physical constitution generally. The specialities both of inorganic and organized matter, classed according to analogy of form and composition, undoubtedly constitute a most interesting branch of study, but they appertain to a sphere of ideas having no affinity with the subject of this work.

The description of different countries certainly furnishes us with the most important materials for the composition of a physical geography; but the combination of these different descriptions, ranged in series, would as little give us a true image of the general conformation of the irregular surface of our globe, as a succession of all the floras of different regions would constitute that which I designate as a *Geography of Plants*. It is by subjecting isolated observations to the process of thought, and by combining and comparing them, that we are enabled to discover the relations existing in common between the climatic distribution of beings and the individuality of organic forms (in the morphology or descriptive natural history of plants and animals); and it is by induction that we are led to comprehend numerical laws, the proportion of natural families to the whole number of species, and to designate the latitude or geographical position of the zones in whose plains each organic form attains the maximum of its development. Considerations of this nature, by their tendency to generalization, impress a nobler character on the physical description of the globe, and enable us to understand how the aspect of the scenery, that is to say, the impression produced upon the mind by the physiognomy of the vegetation, depends upon the local distribution, the number, and the luxuriance of growth of the vegetable forms predominating in the general mass. The catalogues of organized beings, to which was formerly given the pompous title of *Systems of Nature*, present us with an admirably connected arrangement by analogies of structure, either in the perfected development of these beings, or in the different phases which, in accordance with the views of a spiral evolution, affect in vegetables the leaves, bracts, calyx, corolla, and fructifying organs ; and in animals, with more or less symmetrical regularity, the cellular and fibrous tissues, and their perfect or but obscurely developed articulations. But these pretended systems of nature, however ingenious their mode of classification may be, do not show us organic beings as they are distributed in groups throughout our planet, according to their different relations of latitude and elevation above the level of the sea, and to climatic influences, which are owing to general and often very remote causes. The ultimate aim of physical geography is, however, as we have already said, to recognize unity in the vast diversity of phenomena, and by the exercise of thought and the combination of observations, to discern the constancy of phenomena in the midst of apparent changes. In the exposition of the terrestrial portion of the Cosmos, it will occasionally be necessary to descend to very special facts ; but this will only be in order to recall the connection existing between the actual distribution of organic beings over the globe, and the laws of the ideal classification by natural families, analogy of internal organization, and progressive evolution.

It follows from these discussions on the limits of the various sciences, and more particularly from the distinction which must necessarily be made between descriptive botany (morphology of vegetables) and the geography of plants, that in the physical history of the globe, the innumerable multitude of organized bodies which embellish creation are considered rather according to *zones of habitation* or *stations*, and to differently inflected *isothermal bands*, than with reference to the principles of gradation in the development of internal organism. Notwithstanding this, botany and zoology, which constitute the descriptive natural history of all organized beings, are the fruitful sources whence we draw the materials necessary to give a solid basis to the study of the mutual relations and connection of phenomena.

We will here subjoin one important observation by way of elucidating the connection of which we have spoken. The first general glance over the vegetation of a vast extent of a continent shows us forms the most dissimilar—Gramineæ and Orchideæ, Coniferæ and oaks, in local approximation to one another; while natural families and genera, instead of being locally associated, are dispersed as if by chance. This dispersion is, however, only apparent. The physical description of the globe teaches us that vegetation every where presents numerically constant relations in the development of its forms and types; that in the same climates, the species which are wanting in one country are replaced in a neighboring one by other species of the same family; and that this *law of substitution*, which seems to depend upon some inherent mysteries of the organism, considered with reference to its origin, maintains in contiguous regions a numerical relation between the species of various great families and the general mass of the phanerogamic plants constituting the two floras. We thus find a principle of unity and a primitive plan of distribution revealed in the multiplicity of the distinct organizations by which these regions are occupied; and we also discover in each zone, and diversified according to the families of plants, a slow but continuous action on the aërial ocean, depending upon the influence of light—the primary condition of all organic vitality—on the solid and liquid surface of our planet. It might be said, in accordance with a beautiful expression of Lavoisier, that the ancient marvel of the myth of Prometheus was incessantly renewed before our eyes.

If we extend the course which we have proposed, following in the exposition of the physical description of the earth to the sidereal part of the science of the Cosmos, the delineation of the regions of space and the bodies by which they are occupied, we shall find our task simplified in no common degree. If, according to ancient but unphilosophical forms of nomenclature, we would distinguish between *physics*, that is to say, general considerations on the essence of matter, and the forces by which it is actuated, and *chemistry*, which treats of the nature of substances, their elementary composition, and those attractions that are not determined solely by the relations of mass, we must admit that the description of the earth comprises at

once *physical* and *chemical* actions. In addition to gravitation, which must be considered as a primitive force in nature,
we observe that attractions of another kind are at work around
us, both in the interior of our planet and on its surface. These
forces, to which we apply the term *chemical affinity*, act upon
molecules in contact, or at infinitely minute distances from one
another,* and which, being differently modified by electricity,
heat, condensation in porous bodies, or by the contact of an
intermediate substance, animate equally the inorganic world
and animal and vegetable tissues. If we except the small
asteroids, which appear to us under the forms of aërolites and
shooting stars, the regions of space have hitherto presented to
our direct observation physical phenomena alone ; and in the
case of these, we know only with certainty the effects depending upon the quantitative relations of matter or the distribution of masses. The phenomena of the regions of space may
consequently be considered as influenced by simple dynamical
laws—the laws of motion.

The effects that may arise from the specific difference and
the heterogeneous nature of matter have not hitherto entered
into our calculations of the mechanism of the heavens. The
only means by which the inhabitants of our planet can enter
into relation with the matter contained within the regions of
space, whether existing in scattered forms or united into large
spheroids, is by the phenomena of light, the propagation of
luminous waves, and by the influence universally exercised by
the force of gravitation or the attraction of masses. The existence of a periodical action of the sun and moon on the variations of terrestrial magnetism is even at the present day
extremely problematical. We have no direct experimental
knowledge regarding the properties and specific qualities of
the masses circulating in space, or of the matter of which they
are probably composed, if we except what may be derived from
the fall of aërolites or meteoric stones, which, as we have already observed, enter within the limits of our terrestrial sphere.
It will be sufficient here to remark, that the direction and the
excessive velocity of projection (a velocity wholly planetary)
manifested by these masses, render it more than probable that

* On the question already discussed by Newton, regarding the difference existing between the attraction of masses and molecular attraction,
see Laplace, *Exposition du Système du Monde*, p. 384, and supplement
to book x. of the *Mecanique Céleste*, p, 3, 4 ; Kant, *Metaph. Anfangs-
gründe der Naturwissenschaft, Säm. Werke*, 1839, bd. v., s. 309 (Meta-
physical Principles of the Natural Sciences) ; Pectet, *Physique*, 1838,
vol. i., p. 59–63.

they are small celestial bodies, which, being attracted by our planet, are made to deviate from their original course, and thus reach the earth enveloped in vapors, and in a high state of actual incandescence. The familiar aspect of these asteroids, and the analogies which they present with the minerals composing the earth's crust, undoubtedly afford ample grounds for surprise ;* but, in my opinion, the only conclusion to be drawn from these facts is, that, in general, planets and other sidereal masses, which, by the influence of a central body, have been agglomerated into rings of vapor, and subsequently into spheroids, being integrant parts of the same system, and having one common origin, may likewise be composed of substances chemically identical. Again, experiments with the pendulum, particularly those prosecuted with such rare precision by Bessel, confirm the Newtonian axiom, that bodies the most heterogeneous in their nature (as water, gold, quartz, granular limestone, and different masses of aërolites) experience a perfectly similar degree of acceleration from the attraction of the earth. To the experiments of the pendulum may be added the proofs furnished by purely astronomical observations. The almost perfect identity of the mass of Jupiter, deduced from the influence exercised by this stupendous planet on its own satellites, on Encke's comet of short period, and on the small planets Vesta, Juno, Ceres, and Pallas, indicates with equal certainty that within the limits of actual observation attraction is determined solely by the quantity of matter.†

This absence of any perceptible difference in the nature of matter, alike proved by direct observation and theoretical deductions, imparts a high degree of simplicity to the mechanism of the heavens. The immeasurable extent of the regions of space being subjected to laws of motion alone, the sidereal portion of the science of the Cosmos is based on the pure and abundant source of mathematical astronomy, as is the terrestrial portion on physics, chemistry, and organic morphology ; but the domain of these three last-named sciences embraces

---

* [The analysis of an aërolite which fell a few years since in Mary land, United States, and was examined by Professor Silliman, of New Haven, Connecticut, gave the following results: Oxyd of iron, 24; oxyd of nickel, 1·25; silica, with earthy matter, 3·46; sulphur, a trace $=28·71$. Dr. Mantell's *Wonders of Geology*, 1848, vol. i., p. 51.]—*Tr.*

† Poisson, *Connaissances des Temps pour l'Année* 1836, p. 64–66. Bessel, Poggendorf's *Annalen*, bd. xxv., s. 417. Encke, *Abhandlungen der Berliner Academie* (Trans. of the Berlin Academy), 1826, s. 257. Mitscherlich, *Lehrbuch der Chemie* (Manual of Chemistry), 1837 bd. i. s. 352.

the consideration of phenomena which are so complicated, and have, up to the present time, been found so little suscep- tible of the application of rigorous method, that the physical science of the earth can not boast of the same certainty and simplicity in the exposition of facts and their mutual connec- tion which characterize the celestial portion of the Cosmos. It is not improbable that the difference to which we allude may furnish an explanation of the cause which, in the earliest ages of intellectual culture among the Greeks, directed the natural philosophy of the Pythagoreans with more ardor to the heavenly bodies and the regions of space than to the earth and its productions, and how through Philolaüs, and subse- quently through the analogous views of Aristarchus of Samos, and of Seleucus of Erythrea, this science has been made more conducive to the attainment of a knowledge of the true system of the world than the natural philosophy of the Ionian school could ever be to the physical history of the earth. Giving but little attention to the properties and specific differences of matter filling space, the great Italian school, in its Doric gravity, turned by preference toward all that relates to meas- ure, to the form of bodies, and to the number and distances of the planets,* while the Ionian physicists directed their atten tion to the qualities of matter, its true or supposed metamor phoses. and to relations of origin. It was reserved for the powerful genius of Aristotle, alike profoundly speculative and practical, to sound with equal success the depths of abstraction and the inexhaustible resources of vital activity pervading the material world.

Several highly distinguished treatises on physical geography are prefaced by an introduction, whose purely astronomical sections are directed to the consideration of the earth in its planetary dependence, and as constituting a part of that great system which is animated by one central body, the sun. This course is diametrically opposed to the one which I propose following. In order adequately to estimate the dignity of the Cosmos, it is requisite that the sidereal portion, termed by Kant the *natural history of the heavens*, should not be made subordinate to the terrestrial. In the science of the Cosmos, according to the expression of Aristarchus of Samos, the pio- neer of the Copernican system, the sun, with its satellites, was nothing more than one of the innumerable stars by which space is occupied. The physical history of the world must, therefore, begin with the description of the heavenly bodies,

* Compare Otfried Müller's *Dorien*, bd. i., s. 365.

and with a geographical sketch of the universe, or, I would rather say, a true *map of the world*, such as was traced by the bold hand of the elder Herschel. If, notwithstanding the smallness of our planet, the most considerable space and the most attentive consideration be here afforded to that which exclusively concerns it, this arises solely from the disproportion in the extent of our knowledge of that which is accessible and of that which is closed to our observation. This subordination of the celestial to the terrestrial portion is met with in the great work of Bernard Varenius,* which appeared in the mid-

* *Geographia Generalis in qua affectiones generales telluris explicantur.* The oldest Elzevir edition bears date 1650, the second 1672, and the third 1681; these were published at Cambridge, under Newton's supervision. This excellent work by Varenius is, in the true sense of the words, a physical description of the earth. Since the work *Historia Natural de las Indias*, 1590, in which the Jesuit Joseph de Acosta sketched in so masterly a manner the delineation of the New Continent, questions relating to the physical history of the earth have never been considered with such admirable generality. Acosta is richer in original observations, while Varenius embraces a wider circle of ideas, since his sojourn in Holland, which was at that period the center of vast commercial relations, had brought him in contact with a great number of well-informed travelers. *Generalis sive Universalis Geographia dicitur quæ tellurem in genere considerat atque affectiones explicat, non habita particularium regionum ratione.* The general description of the earth by Varenius (*Pars Absoluta*, cap. i.–xxii.) may be considered as a treatise of comparative geography, if we adopt the term used by the author himself (*Geographia Comparativa*, cap. xxxiii.–xl.), although this must be understood in a limited acceptation. We may cite the following among the most remarkable passages of this book: the enumeration of the systems of mountains; the examination of the relations existing between their directions and the general form of continents (p. 66, 76, ed. Cantab., 1681); a list of extinct volcanoes, and such as were still in a state of activity; the discussion of facts relative to the general distribution of islands and archipelagoes (p. 220); the depth of the ocean relatively to the height of neighboring coasts (p. 103); the uniformity of level observed in all open seas (p. 97); the dependence of currents on the prevailing winds; the unequal saltness of the sea; the configuration of shores (p. 139); the direction of the winds as the result of differences of temperature, &c. We may further instance the remarkable considerations of Varenius regarding the equinoctial current from east to west, to which he attributes the origin of the Gulf Stream, beginning at Cape St. Augustin, and issuing forth between Cuba and Florida (p. 140). Nothing can be more accurate than his description of the current which skirts the western coast of Africa, between Cape Verde and the island of Fernando Po in the Gulf of Guinea. Varenius explains the formation of sporadic islands by supposing them to be " the raised bottom of the sea:" *magna spirituum inclusorum vi, sicut aliquando montes e terra protusos esse quidam scribunt* (p. 225). The edition published by Newton in 1681 (*auctior et emendatior*) unfortunately contains no additions from this great authority; and there is not even mention made of the polar compression of the globe, al-

die of the seventeenth century. He was the first to distinguish between *general and special geography*, the former of which he subdivides into an *absolute*, or, properly speaking, *terrestrial* part, and a *relative or planetary* portion, according to the mode of considering our planet either with reference to its surface in its different zones, or to its relations to the sun and moon. It redounds to the glory of Varenius that his work on *General and Comparative Geography* should in so high a degree have arrested the attention of Newton. The imperfect state of many of the auxiliary sciences from which this writer was obliged to draw his materials prevented his work from corresponding to the greatness of the design, and it was reserved for the present age, and for my own country, to see the delineation of comparative geography, drawn in its full extent, and in all its relations with the history of man, by the skillful hand of Carl Ritter.*

The enumeration of the most important results of the astronomical and physical sciences which in the history of the Cosmos radiate toward one common focus, may perhaps, to a certain degree, justify the designation I have given to my work, and, considered within the circumscribed limits I have proposed to myself, the undertaking may be esteemed less adventurous than the title. The introduction of new terms, especially with reference to the general results of a science which

though the experiments on the pendulum by Richer had been made nine years prior to the appearance of the Cambridge edition. Newton's *Principia Mathematica Philosophiæ Naturalis* were not communicated in manuscript to the Royal Society until April, 1686. Much uncertainty seems to prevail regarding the birth-place of Varenius. Jæcher says it was England, while, according to *La Biographie Universelle* (b. xlvii., p. 495), he is stated to have been born at Amsterdam; but it would appear, from the dedicatory address to the burgomaster of that city (see his *Geographia Comparativa*), that both suppositions are false. Varenius expressly says that he had sought refuge in Amsterdam, "because his native city had been burned and completely destroyed during a long war," words which appear to apply to the north of Germany, and to the devastations of the Thirty Years' War. In his dedication of another work, *Descriptio regni Japoniæ* (Amst., 1649), to the Senate of Hamburgh, Varenius says that he prosecuted his elementary mathematical studies in the gymnasium of that city. There is, therefore, every reason to believe that this admirable geographer was a native of Germany, and was probably born at Luneburg (*Witten. Mem. Theol.*, 1685, p. 2142; Zedler, *Universal Lexicon*, vol. xlvi., 1745, p. 187).

* Carl Ritter's *Erdkunde im Verhältniss zur Natur und zur Geschichte des Menschen, oder allgemeine vergleichende Geographie* (Geography in relation to Nature and the History of Man, or general Comparative Geography).

ought to be accessible to all, has always been greatly in oppo. sition to my own practice; and whenever I have enlarged upon the established nomenclature, it has only been in the specialities of descriptive botany and zoology, where the introduction of hitherto unknown objects rendered new names necessary. The denominations of physical descriptions of the universe, or physical cosmography, which I use indiscriminately, have been modeled upon those of *physical descriptions of the earth*, that is to say, *physical geography*, terms that have long been in common use. Descartes, whose genius was one of the most powerful manifested in any age, has left us a few fragments of a great work, which he intended publishing under the title of *Monde*, and for which he had prepared himself by special studies, including even that of human anatomy. The uncommon, but definite expression of the *science of the Cosmos* recalls to the mind of the inhabitant of the earth that we are treating of a more widely-extended horizon—of the assemblage of all things with which space is filled, from the remotest nebulæ to the climatic distribution of those delicate tissues of vegetable matter which spread a variegated covering over the surface of our rocks.

The influence of narrow-minded views peculiar to the earlier ages of civilization led in all languages to a confusion of ideas in the synonymic use of the words *earth* and *world*, while the common expressions *voyages round the world*, *map of the world*, and *new world*, afford further illustrations of the same confusion. The more noble and precisely-defined expressions of *system of the world*, *the planetary world*, and *creation and age of the world*, relate either to the totality of the substances by which space is filled, or to the origin of the whole universe.

It was natural that, in the midst of the extreme variability of phenomena presented by the surface of our globe, and the aërial ocean by which it is surrounded, man should have been impressed by the aspect of the vault of heaven, and the uniform and regular movements of the sun and planets. Thus the word Cosmos, which primitively, in the Homeric ages, indicated an idea of order and harmony, was subsequently adopted in scientific language, where it was gradually applied to the order observed in the movements of the heavenly bodies, to the whole universe, and then finally to the world in which this harmony was reflected to us. According to the assertion of Philolaüs, whose fragmentary works have been so ably commented upon by Böckh, and conformably to the general testimony of antiquity, Pythagoras was the first who used the word Cosmos to designate the order that reigns in the universe, or entire world.*

* Κόσμος, in the most ancient, and at the same time most precise, definition of the word, signified *ornament* (as an adornment for a man, a woman, or a horse); taken figuratively for εὐταξία, it implied the order or adornment of a discourse. According to the testimony of all the ancients, it was Pythagoras who first used the word to designate the order in the universe, and the universe itself. Pythagoras left no writings; but ancient attestation to the truth of this assertion is to be found in several passages of the fragmentary works of Philolaüs (Stob., *Eclog.*, p. 360 and 460, Heeren), p. 62, 90, in Böckh's German edition. I do not, according to the example of Näke, cite Timæus of Locris, since his authenticity is doubtful. Plutarch (*De plac. Phil.*, ii., 1) says, in the most express manner, that Pythagoras gave the name of Cosmos to the universe on account of the order which reigned throughout it; so likewise does Galen (*Hist. Phil.*, p. 429). This word, together with its novel signification, passed from the schools of philosophy into the language of poets and prose writers. Plato designates the heavenly bodies by the name of *Uranos*, but the order pervading the regions of space he too terms the Cosmos, and in his *Timæus* (p. 30, B.) he says *that the world is an animal endowed with a soul* (κόσμον ζῶον ἐμψύχον). Compare Anaxag. Claz., ed. Schaubach, p. 111, and Plut. (*De plac. Phil.*, ii., 3), on spirit apart from matter, as the ordaining power of nature. In Aristotle (*De Cœlo*, 1, 9), *Cosmos* signifies "the universe and the order pervading it," but it is likewise considered as divided in space into two parts—the sublunary world, and the world above the moon. (*Meteor.*, I., 2, 1, and I., 3, 13, p. 339, *a*, and 340, *b*, Bekk.) The definition of Cosmos, which I have already cited, is taken from Pseudo-Aristoteles *de Mundo*, cap. ii. (p. 391); the passage referred to is as follows: Κόσμος ἐστὶ σύστημα ἐς οὐρανοῦ καὶ γῆς καὶ τῶν ἐν τούτοις περιεχομένων φύσεων. Λέγεται δὲ καὶ ἐπέρως κόσλος ἡ τῶν ὅλων τάξις τε καὶ διακόσμησις, ὑπὸ θεῶν τε καὶ διά θεῶν φυλαττομένη. Most of the passages occurring in Greek writers on the word *Cosmos* may be found collected together in the controversy between Richard Bentley and Charles Boyle (*Opuscula Philologica*, 1781, p. 347, 445; *Dissertation upon the Epistles of Phalaris*, 1817, p. 254); on the historical existence of Zaleucus, legislator of Leucris, in Näke's excellent work, *Sched. Crit.*, 1812, p. 9, 15; and, finally, in Theophilus Schmidt, *ad Cleom. Cycl. Theor.*, met. I., 1, p. ix., 1, and 99. Taken in a more limited sense, the word Cosmos is also used in the plural (Plut., 1, 5), either to designate the stars (Stob., 1, p. 514; Plut., 11, 13), or the innumerable systems scattered like islands through the immensity of space, and each composed of a sun and a moon. (Anax. Claz., *Fragm.*, p. 89, 93, 120; Brandis, *Gesch. der Griechisch-Römischen Philosophie*, b. i., s. 252 (History of the Greco-Roman Philosophy). Each of these groups forming thus a *Cosmos*, the universe, τὸ πᾶν, the word must be understood in a wider sense (Plut., ii., 1). It was not until long after the time of the Ptolemies that the word was applied to the earth. Böckh has made known inscriptions in praise of Trajan and Adrian (*Corpus Inscr. Grœc.*, 1, n. 334 and 1036), in which Κόσμος occurs for οἰκουμένη, in the same manner as we still use the term *world* to signify the earth alone. We have already mentioned the singular division of the regions of space

From the Italian school of philosophy, the expression pass‐
ed, in this signification, into the language of those early poets
into three parts, the *Olympus*, *Cosmos*, and *Ouranos* (Stob., i., p. 488;
Philolaüs, p. 94, 202); this division applies to the different regions sur
rounding that mysterious focus of the universe, the Ἑστία τοῦ παντός
of the Pythagoreans. In the fragmentary passage in which this divi‐
sion is found, the term *Ouranos* designates the innermost region, situ‐
ated between the moon and earth; this is the domain of changing
things. The middle region, where the planets circulate in an invaria‐
ble and harmonious order, is, in accordance with the special concep‐
tions entertained of the universe, exclusively termed *Cosmos*, while the
word *Olympus* is used to express the exterior or igneous region. Bopp,
the profound philologist, has remarked, that we may deduce, as Pott
has done, *Etymol. Forschungen*, th. i., s. 39 and 252 (*Etymol. Research‐
es*), the word Κόσμος from the Sanscrit root '*sud*', *purificari*, by assum‐
ing two conditions; first, that the Greek κ in κόσμος comes from the
palatial ς, which Bopp represents by 's and Pott by ç (in the same man‐
ner as δέκα, *decem*, *taihun* in Gothic, comes from the Indian word *dá‐
san*), and, next, that the Indian *d*' corresponds, as a general rule, with
the Greek θ (*Vergleichende Grammatik*, § 99—Comparative Grammar),
which shows the relation of κόσμος (for κόθμος) with the Sanscrit root
'*sud*', whence is also derived καθαμός. Another Indian term for the
world is *gagat* (pronounced *dschagat*), which is, properly speaking, the
present participle of the verb *gagámi* (I go), the root of which is *gá*.
In restricting ourselves to the circle of Hellenic etymologies, we find
(*Etymol. M.*, p. 532, 12) that κόσμος is intimately associated with κάζω,
or rather with καίνυμαι, whence we have κεκασμένος or κεκαδμένος.
Welcker (*Eine Kretische Col. in Theben*, s. 23—A Cretan Colony in
Thebes) combines with this the name Κάδμος, as in Hesychius κάδμος
signifies a Cretan suit of arms. When the scientific language of Greece
was introduced among the Romans, the word *mundus*, which at first had
only the primary meaning of κόσμος (female ornament), was applied to
designate the entire universe. Ennius seems to have been the first
who ventured upon this innovation. In one of the fragments of this
poet, preserved by Macrobius, on the occasion of his quarrel with Vir‐
gil, we find the word used in its novel mode of acceptation: "*Mundus
cœli vastus constitit silentio*" (Sat., vi., 2). Cicero also says, "*Quem nos
lucentem mundum vocamus*" (Timæus, *S. de Univer.*, cap. x.). The
Sanscrit root *mand*, from which Pott derives the Latin *mundus* (*Etym.
Forsch.*, th. i., s. 240), combines the double signification of shining and
adorning. *Lôka* designates in Sanscrit the world and people in general,
in the same manner as the French word *monde*, and is derived, accord‐
ing to Bopp, from *lôk* (to see and shine); it is the same with the Scla‐
vonic root *swjet*, which means both *light* and *world*. (Grimm, *Deutsche
Gramm.*, b. iii., s. 394—German Grammar.) The word *welt*, which
the Germans make use of at the present day, and which was *weralt* in
old German, *worold* in old Saxon, and *vëruld* in Anglo-Saxon, was, ac‐
cording to James Grimm's interpretation, a period of time, an age (*sœ‐
culum*), rather than a term used for the world in space. The Etruscans
figured to themselves *mundus* as an inverted dome, symmetrically op‐
posed to the celestial vault (Otfried Müller's *Etrusken*, th. ii., s. 96,
&c.). Taken in a still more limited sense, the word appears to have
signified among the Goths the terrestrial surface girded by seas (*marei*,
meri), the *merigard*, literally, *garden of seas*.

of nature, Parmenides and Empedocles, and from thence into the works of prose writers. We will not here enter into a discussion of the manner in which, according to the Pythagorean views, Philolaüs distinguishes between Olympus, Uranus, or the heavens, and Cosmos, or how the same word, used in a plural sense, could be applied to certain heavenly bodies (the planets) revolving round one central focus of the world, or to groups of stars. In this work I use the word Cosmos in conformity with the Hellenic usage of the term subsequently to the time of Pythagoras, and in accordance with the precise definition given of it in the treatise entitled *De Mundo*, which was long erroneously attributed to Aristotle. It is the assemblage of all things in heaven and earth, the universality of created things constituting the perceptible world. If scientific terms had not long been diverted from their true verbal signification, the present work ought rather to have borne the title of *Cosmography*, divided into *Uranography* and *Geography*. The Romans, in their feeble essays on philosophy, imitated the Greeks by applying to the universe the term *mundus*, which, in its primary meaning, indicated nothing more than ornament, and did not even imply order or regularity in the disposition of parts. It is probable that the introduction into the language of Latium of this technical term as an equivalent for Cosmos, in its double signification, is due to Ennius,* who was a follower of the Italian school, and the translator of the writings of Epicharmus and some of his pupils on the Pythagorean philosophy.

We would first distinguish between the physical *history* and the physical *description* of the world. The former, conceived in the most general sense of the word, ought, if materials for writing it existed, to trace the variations experienced by the universe in the course of ages from the new stars which have suddenly appeared and disappeared in the vault of heaven, from nebulæ dissolving or condensing—to the first stratum of cryptogamic vegetation on the still imperfectly cooled surface of the earth, or on a reef of coral uplifted from the depths of ocean. *The physical description of the world* presents a picture of all that exists in space—of the simultaneous action of

---

* See, on Ennius, the ingenious researches of Leopold Krahner, in his *Grundlinien zur Geschichte des Verfalls der Römischen Staats-Reii gion*, 1837, s. 41-45 (Outlines of the History of the Decay of the Estab lished Religion among the Romans). In all probability, Ennius did not quote from writings of Epicharmus himself, but from poems composed in the name of that philosopher, and in accordance with his views.

natural forces, together with the phenomena which they produce.

But if we would correctly comprehend nature, we must not
entirely or absolutely separate the consideration of the present
state of things from that of the successive phases through
which they have passed. We can not form a just conception
of their nature without looking back on the mode of their formation. It is not organic matter alone that is continually undergoing change, and being dissolved to form new combinations. The globe itself reveals at every phase of its existence
the mystery of its former conditions.

We can not survey the crust of our planet without recognizing the traces of the prior existence and destruction of an
organic world. The sedimentary rocks present a succession
of organic forms, associated in groups, which have successively displaced and succeeded each other. The different super
imposed strata thus display to us the faunas and floras of different epochs. In this sense the description of nature is inti
mately connected with its history ; and the geologist, who is
guided by the connection existing among the facts observed,
can not form a conception of the present without pursuing,
through countless ages, the history of the past. In tracing
the physical delineation of the globe, we behold the present
and the past reciprocally incorporated, as it were, with one
another ; for the domain of nature is like that of languages, in
which etymological research reveals a successive development,
by showing us the primary condition of an idiom reflected in
the forms of speech in use at the present day. The study of
the material world renders this reflection of the past peculiarly manifest, by displaying in the process of formation rocks of
eruption and sedimentary strata similar to those of former
ages. If I may be allowed to borrow a striking illustration
from the geological relations by which the physiognomy of a
country is determined, I would say that domes of trachyte,
cones of basalt, lava streams (*coulées*) of amygdaloid with
elongated and parallel pores, and white deposits of pumice,
intermixed with black scoriæ, animate the scenery by the associations of the past which they awaken, acting upon the
imagination of the enlightened observer like traditional records
of an earlier world. Their form is their history.

The sense in which the Greeks and Romans originally employed the word *history* proves that they too were intimately
convinced that, to form a complete idea of the present state
of the universe, it was necessary to consider it in its successive

phases. It is not, however, in the definition given by Vale-rius Flaccus,* but in the zoological writings of Aristotle, that the word *history* presents itself as an exposition of the results of experience and observation. The physical description of the word by Pliny the elder bears the title of *Natural History*, while in the letters of his nephew it is designated by the nobler term of *History of Nature*. The earlier Greek his-torians did not separate the descriptions of countries from the narrative of events of which they had been the theater. With these writers, physical geography and history were long inti-mately associated, and remained simply but elegantly blended until the period of the development of political interests, when the agitation in which the lives of men were passed caused the geographical portion to be banished from the history of nations, and raised into an independent science.

It remains to be considered whether, by the operation of thought, we may hope to reduce the immense diversity of phenomena comprised by the Cosmos to the unity of a princi-ple, and the evidence afforded by rational truths. In the present state of empirical knowledge, we can scarcely flatter ourselves with such a hope. Experimental sciences, based on the observation of the external world, can not aspire to completeness ; the nature of things, and the imperfection of our organs, are alike opposed to it. We shall never succeed in exhausting the immeasurable riches of nature ; and no gen-eration of men will ever have cause to boast of having com-prehended the total aggregation of phenomena. It is only by distributing them into groups that we have been able, in the case of a few, to discover the empire of certain natural laws, grand and simple as nature itself. The extent of this empire will no doubt increase in proportion as physical sciences are more perfectly developed. Striking proofs of this advance-ment have been made manifest in our own day, in the phe-nomena of electro-magnetism, the propagation of luminous waves and radiating heat. In the same manner, the fruitful doctrine of evolution shows us how, in organic development, all that is formed is sketched out beforehand, and how the tissues of vegetable and animal matter uniformly arise from the multiplication and transformation of cells.

The generalization of laws, which, being at first bounded by narrow limits, had been applied solely to isolated groups of phenomena, acquires in time more marked gradations, and gains in extent and certainty as long as the process of reason-

* Aul. Gell., *Noct. Att.*, v., 18.

ing is applied strictly to analogous phenomena; but as soon
as dynamical views prove insufficient where the specific properties and heterogeneous nature of matter come into play, it is
to be feared that, by persisting in the pursuit of laws, we may
find our course suddenly arrested by an impassable chasm.
The principle of unity is lost sight of, and the guiding clew
is rent asunder whenever any specific and peculiar kind of
action manifests itself amid the active forces of nature. The
law of equivalents and the numerical proportions of composition, so happily recognized by modern chemists, and proclaimed
under the ancient form of atomic symbols, still remains isolated and independent of mathematical laws of motion and gravitation.

Those productions of nature which are objects of direct observation may be logically distributed in classes, orders, and
families. This form of distribution undoubtedly sheds some
light on descriptive natural history, but the study of organized
bodies, considered in their linear connection, although it may
impart a greater degree of unity and simplicity to the distribution of groups, can not rise to the height of a classification
based on one sole principle of composition and internal organization. As different gradations are presented by the laws
of nature according to the extent of the horizon, or the limits
of the phenomena to be considered, so there are likewise differently graduated phases in the investigation of the external
world. Empiricism originates in isolated views, which are
subsequently grouped according to their analogy or dissimilarity. To direct observation succeeds, although long afterward,
the wish to prosecute experiments; that is to say, to evoke
phenomena under different determined conditions. The rational experimentalist does not proceed at hazard, but acts
under the guidance of hypotheses, founded on a half indistinct
and more or less just intuition of the connection existing among
natural objects or forces. That which has been conquered
by observation or by means of experiments, leads, by analysis
and induction, to the discovery of empirical laws. These are
the phases in human intellect that have marked the different
epochs in the life of nations, and by means of which that great
mass of facts has been accumulated which constitutes at the
present day the solid basis of the natural sciences.

Two forms of abstraction conjointly regulate our knowledge, namely, relations of *quantity*, comprising ideas of number and size, and relations of *quality*, embracing the consideration of the specific properties and the heterogeneous nature

of matter. The former, as being more accessible to the exer
cise of thought, appertains to mathematics ; the latter, from
its apparent mysteries and greater difficulties, falls under the
domain of the chemical sciences. In order to submit phenomena to calculation, recourse is had to a hypothetical construction of matter by a combination of molecules and atoms,
whose number, form, position, and polarity determine, modify,
or vary phenomena.

The mythical ideas long entertained of the imponderable
substances and vital forces peculiar to each mode of organization, have complicated our views generally, and shed an uncertain light on the path we ought to pursue.

The most various forms of intuition have thus, age aftei
age, aided in augmenting the prodigious mass of empirical
knowledge, which in our own day has been enlarged with
ever-increasing rapidity. The investigating spirit of man
strives from time to time, with varying success, to break
through those ancient forms and symbols invented, to subject
rebellious matter to rules of mechanical construction.

We are still very far from the time when it will be possible for us to reduce, by the operation of thought, all that we
perceive by the senses, to the unity of a rational principle.
It may even be doubted if such a victory could ever be
achieved in the field of natural philosophy. The complication of phenomena, and the vast extent of the Cosmos, would
seem to oppose such a result ; but even a partial solution of
the problem—the tendency toward a comprehension of the
phenomena of the universe—will not the less remain the eternal and sublime aim of every investigation of nature.

In conformity with the character of my former writings, as
well as with the labors in which I have been engaged during
my scientific career, in measurements, experiments, and the
investigation of facts, I limit myself to the domain of empirical
ideas.

The exposition of mutually connected facts does not exclude
the classification of phenomena according to their rational connection, the generalization of many specialities in the great
mass of observations, or the attempt to discover laws. Conceptions of the universe solely based upon reason, and the
principles of speculative philosophy, would no doubt assign a
still more exalted aim to the science of the Cosmos. I am far
from blaming the efforts of others solely because their success
has hitherto remained very doubtful. Contrary to the wishes
and counsels of those profound and powerful thinkers who

have given new life to speculations which were already familiar to the ancients, systems of natural philosophy have in
our own country for some time past turned aside the minds
of men from the graver study of mathematical and physical
sciences. The abuse of better powers, which has led many
of our noble but ill-judging youth into the saturnalia of a pure
ly ideal science of nature, has been signalized by the intoxica-
tion of pretended conquests, by a novel and fantastically sym-
bolical phraseology, and by a predilection for the formulæ of
a scholastic rationalism, more contracted in its views than
any known to the Middle Ages. I use the expression " abuse
of better powers," because superior intellects devoted to philosophical pursuits and experimental sciences have remained
strangers to these saturnalia. The results yielded by an earnest investigation in the path of experiment can not be at variance with a true philosophy of nature. If there be any
contradiction, the fault must lie either in the unsoundness of
speculation, or in the exaggerated pretensions of empiricism,
which thinks that more is proved by experiment than is actually derivable from it.

External nature may be opposed to the intellectual world,
as if the latter were not comprised within the limits of the
former, or nature may be opposed to art when the latter is
defined as a manifestation of the intellectual power of man ;
but these contrasts, which we find reflected in the most cultivated languages, must not lead us to separate the sphere of
nature from that of mind, since such a separation would reduce the physical science of the world to a mere aggregation
of empirical specialities. Science does not present itself to
man until mind conquers matter in striving to subject the
result of experimental investigation to rational combinations.
Science is the labor of mind applied to nature, but the external world has no real existence for us beyond the image
reflected within ourselves through the medium of the senses.
As intelligence and forms of speech, thought and its verbal
symbols, are united by secret and indissoluble links, so does
the external world blend almost unconsciously to ourselves
with our ideas and feelings. " External phenomena," says
Hegel, in his *Philosophy of History*, " are in some degree
translated in our inner representations." The objective world,
conceived and reflected within us by thought, is subjected to
the eternal and necessary conditions of our intellectual being.
The activity of the mind exercises itself on the elements furnished to it by the perceptions of the senses. Thus, in the

early ages of mankind, there manifests itself in the simple intuition of natural facts, and in the efforts made to comprehend them, the germ of the philosophy of nature. These
ideal tendencies vary, and are more or less powerful, according to the individual characteristics and moral dispositions of
nations, and to the degrees of their mental culture, whether
attained amid scenes of nature that excite or chill the imagination.

History has preserved the record of the numerous attempts
that have been made to form a rational conception of the
whole world of phenomena, and to recognize in the universe
the action of one sole active force by which matter is penetrated, transformed, and animated.   These attempts are traced
in classical antiquity in those treatises on the principles of
things which emanated from the Ionian school, and in which
all the phenomena of nature were subjected to hazardous
speculations, based upon a small number of observations.   By
degrees, as the influence of great historical events has favored
the development of every branch of science supported by observation, that ardor has cooled which formerly led men to
seek the essential nature and connection of things by ideal
construction and in purely rational principles. In recent
times, the mathematical portion of natural philosophy has
been most remarkably and admirably enlarged.   The method
and the instrument (analysis) have been simultaneously perfected.   That which has been acquired by means so different
—by the ingenious application of atomic suppositions, by the
more general and intimate study of phenomena, and by the
improved construction of new apparatus—is the common property of mankind, and should not, in our opinion, now, more
than in ancient times, be withdrawn from the free exercise of
speculative thought.

It can not be denied that in this process of thought the
results of experience have had to contend with many disad-
vantages ; we must not, therefore, be surprised if, in the perpetual vicissitude of theoretical views, as is ingeniously expressed by the author of *Giordano Bruno*,* " most men see
nothing in philosophy but a succession of passing meteors,
while even the grander forms in which she has revealed herself share the fate of comets, bodies that do not rank in popular opinion among the eternal and permanent works of na-

---

* Schelling's Bruno, *Ueber das Göttliche und Natüraliche Princip.
der Dinge*, § 181 (Bruno, on the *Divine and Natural Principle of
Things*)

ture, but are regarded as mere fugitive apparitions of igneous vapor." We would here remark that the abuse of thought, and the false track it too often pursues, ought not to sanction an opinion derogatory to intellect, which would imply that the domain of mind is essentially a world of vague fantastic illusions, and that the treasures accumulated by laborious observations in philosophy are powers hostile to its own empire. It does not become the spirit which characterizes the present age distrustfully to reject every generalization of views and every attempt to examine into the nature of things by the process of reason and induction. It would be a denial of the dignity of human nature and the relative importance of the faculties with which we are endowed, were we to condemn at one time austere reason engaged in investigating causes and their mutual connections, and at another that exercise of the imagination which prompts and excites discoveries by its creative powers.

COSMOS.

WHEN the human mind first attempts to subject to its control the world of physical phenomena, and strives by meditative contemplation to penetrate the rich luxuriance of living
nature, and the mingled web of free and restricted natural
forces, man feels himself raised to a height from whence, as
he embraces the vast horizon, individual things blend together
in varied groups, and appear as if shrouded in a vapory vail.
These figurative expressions are used in order to illustrate the
point of view from whence we would consider the universe
both in its celestial and terrestrial sphere. I am not insen-
sible of the boldness of such an undertaking. Among all the
forms of exposition to which these pages are devoted, there
is none more difficult than the general delineation of nature,
which we purpose sketching, since we must not allow ourselves to be overpowered by a sense of the stupendous richness and variety of the forms presented to us, but must dwell
only on the consideration of masses either possessing actual
magnitude, or borrowing its semblance from the associations
awakened within the subjective sphere of ideas. It is by a
separation and classification of phenomena. by an intuitive insight into the play of obscure forces, and b.· animated expressions, in which the perceptible spectacle is reflected with vivid
truthfulness, that we may hope to comprehend and describe
the *universal all* (τὸ πᾶν) in a manner worthy of the dignity
of the word *Cosmos* in its signification of *universe, order of
the world*, and *adornment* of this universal order. May the
immeasurable diversity of phenomena which crowd into the
picture of nature in no way detract from that harmonious impression of rest and unity which is the ultimate object of every
literary or purely artistical composition.

Beginning with the depths of space and the regions of remotest nebulæ, we will gradually descend through the starry
zone to which our solar system belongs, to our own terrestrial
spheroid, circled by air and ocean, there to direct our atten-

tion to its form, temperature, and magnetic tension, and to
consider the fullness of organic life unfolding itself upon its
surface beneath the vivifying influence of light.   In this manner a picture of the world may, with a few strokes, be made
to include the realms of infinity no less than the minute microscopic animal and vegetable organisms which exist in standing waters and on the weather-beaten surface of our rocks.
All that can be perceived by the senses, and all that has been
accumulated up to the present day by an attentive and variously directed study of nature, constitute the materials from
which this representation is to be drawn, whose character is
an evidence of its fidelity and truth.   But the descriptive picture of nature which we purpose drawing must not enter too
fully into detail, since a minute enumeration of all vital forms,
natural objects, and processes is not requisite to the completeness of the undertaking.   The delineator of nature must resist the tendency toward endless division, in order to avoid
the dangers presented by the very abundance of our empirical
knowledge.   A considerable portion of the qualitative properties of matter—or, to speak more in accordance with the language of natural philosophy, of the qualitative expression of
forces—is doubtlessly still unknown to us, and the attempt
perfectly to represent unity in diversity must therefore necessarily prove unsuccessful.   Thus, besides the pleasure derived
from acquired knowledge, there lurks in the mind of man,
and tinged with a shade of sadness, an unsatisfied longing for
something beyond the present—a striving toward regions yet
unknown and unopened.   Such a sense of longing binds still
faster the links which, in accordance with the supreme laws
of our being, connect the material with the ideal world, and
animates the mysterious relation existing between that which
the mind receives from without, and that which it reflects
from its own depths to the external world.   If, then, nature
(understanding by the term all natural objects and phenomena)
be illimitable in extent and contents, it likewise presents itself to the human intellect as a problem which can not be
grasped, and whose solution is impossible, since it requires a
knowledge of the combined action of all natural forces.   Such
an acknowledgment is due where the actual state and pro-
spective development of phenomena constitute the sole objects
of direct investigation, which does not venture to depart from
the strict rules of induction.   But, although the incessant effort to embrace nature in its universality may remain unsatisfied, the history of the contemplation of the universe (which

will be considered in another part of this work) will teach us
how, in the course of ages, mankind has gradually attained
to a partial insight into the relative dependence of phenomena.
My duty is to depict the results of our knowledge in all their
bearings with reference to the present.   In all that is subject
to motion and change in space, the ultimate aim, the very expression of physical laws, depend upon *mean numerical values.*
which show us the constant amid change, and the stable amid
apparent fluctuations of phenomena.   Thus the progress of
modern physical science is especially characterized by the attainment and the rectification of the mean values of certain
quantities by means of the processes of weighing and measuring ; and it may be said, that the only remaining and wide-
ly-diffused hieroglyphic characters still in our writing—*numbers*—appear to us again, as powers of the Cosmos, although
in a wider sense than that applied to them by the Italian
School.
    The earnest investigator delights in the simplicity of nu-
merical relations, indicating the dimensions of the celestial
regions, the magnitudes and periodical disturbances of the
heavenly bodies, the triple elements of terrestrial magnetism,
the mean pressure of the atmosphere, and the quantity of heat
which the sun imparts in each year, and in every season of the
year, to all points of the solid and liquid surface of our planet.
These sources of enjoyment do not, however, satisfy the poet
of Nature, or the mind of the inquiring many.   To both of
these the present state of science appears as a blank, now that
she answers doubtingly, or wholly rejects as unanswerable,
questions to which former ages deemed they could furnish
satisfactory replies.   In her severer aspect, and clothed with
less luxuriance, she shows herself deprived of that seductive
charm with which a dogmatizing and symbolizing physical
philosophy knew how to deceive the understanding and give
the rein to imagination.   Long before the discovery of the
New World, it was believed that new lands in the Far West
might be seen from the shores of the Canaries and the Azores.
These illusive images were owing, not to any extraordinary
refraction of the rays of light, but produced by an eager longing for the distant and the unattained.   The philosophy of
the Greeks, the physical views of the Middle Ages, and even
those of a more recent period, have been eminently imbued
with the charm springing from similar illusive phantoms of
the imagination.   At the limits of circumscribed knowledge,
as from some lofty island shore, the eye delights to penetrate

to distant regions. The belief in the uncommon and the wonderful lends a definite outline to every manifestation of ideal
creation ; and the realm of fancy—a fairy-land of cosmolog-
ical, geognostical, and magnetic visions—becomes thus involuntarily blended with the domain of reality.

Nature, in the manifold signification of the word—whether
considered as the universality of all that is and ever will be—
as the inner moving force of all phenomena, or as their mysterious prototype—reveals itself to the simple mind and feelings of man as something earthly, and closely allied to himself. It is only within the animated circles of organic structure that we feel ourselves peculiarly at home. Thus,
wherever the earth unfolds her fruits and flowers, and gives
food to countless tribes of animals, there the image of nature
impresses itself most vividly upon our senses. The impression
thus produced upon our minds limits itself almost exclusively
to the reflection of the earthly. The starry vault and the
wide expanse of the heavens belong to a picture of the universe, in which the magnitude of masses, the number of congregated suns and faintly glimmering nebulæ, although they
excite our wonder and astonishment, manifest themselves to
us in apparent isolation, and as utterly devoid of all evidence
of their being the scenes of organic life. Thus, even in the
earliest physical views of mankind, heaven and earth have
been separated and opposed to one another as an upper and
lower portion of space. If, then, a picture of nature were to
correspond to the requirements of contemplation by the senses,
it ought to begin with a delineation of our native earth. It
should depict, first, the terrestrial planet as to its size and
form ; its increasing density and heat at increasing depths in
its superimposed solid and liquid strata ; the separation of sea
and land, and the vital forms animating both, developed in
the cellular tissues of plants and animals ; the atmospheric
ocean, with its waves and currents, through which pierce the
forest-crowned summits of our mountain chains. After this
delineation of purely telluric relations, the eye would rise to
the celestial regions, and the Earth would then, as the well-
known seat of organic development, be considered as a planet,
occupying a place in the series of those heavenly bodies which
circle round one of the innumerable host of self-luminous stars.
This succession of ideas indicates the course pursued in the
earliest stages of perceptive contemplation, and reminds us of
the ancient conception of the "sea-girt disk of earth," supporting the vault of heaven. It begins to exercise its action

at the spot where it originated, and passes from the consideration of the known to the unknown, of the near to the distant
It corresponds with the method pursued in our elementary
works on astronomy (and which is so admirable in a mathematical point of view), of proceeding from the apparent to the
real movements of the heavenly bodies.

Another course of ideas must, however, be pursued in a
work which proposes merely to give an exposition of what is
known—of what may in the present state of our knowledge
be regarded as certain, or as merely probable in a greater or
lesser degree—and does not enter into a consideration of the
proofs on which such results have been based. Here, therefore, we do not proceed from the subjective point of view of
human interests. The terrestrial must be treated only as a
part, subject to the whole. The view of nature ought to be
grand and free, uninfluenced by motives of proximity, social
sympathy, or relative utility. A physical cosmography—a
picture of the universe—does not begin, therefore, with the
terrestrial, but with that which fills the regions of space. But
as the sphere of contemplation contracts in dimension our perception of the richness of individual parts, the fullness of physical phenomena, and of the heterogeneous properties of matter becomes enlarged. From the regions in which we recognize only the dominion of the laws of attraction, we descend to our own planet, and to the intricate play of terrestrial
forces. The method here described for the delineation of nature is opposed to that which must be pursued in establishing conclusive results. The one enumerates what the other
demonstrates.

Man learns to know the external world through the organs
of the senses. Phenomena of light proclaim the existence of
matter in remotest space, and the eye is thus made the medium through which we may contemplate the universe. The
discovery of telescopic vision more than two centuries ago, has
transmitted to latest generations a power whose limits are as
yet unattained.

The first and most general consideration in the Cosmos is
that of the *contents of space*—the distribution of matter, or
of creation, as we are wont to designate the assemblage of all
that is and ever will be developed. We see matter either
agglomerated into rotating, revolving spheres of different density and size, or scattered through space in the form of self-
luminous vapor. If we consider first the cosmical vapor dispersed in definite nebulous spots, its state of aggregation will

appear constantly to vary, sometimes appearing separated into round or elliptical disks, single or in pairs, occasionally connected by a thread of light ; while, at another time, these nebulæ occur in forms of larger dimensions, and are either elongated, or variously branched, or fan-shaped, or appear like well-defined rings, inclosing a dark interior. It is conjectured that these bodies are undergoing variously developed formative processes, as the cosmical vapor becomes condensed in conformity with the laws of attraction, either round one or more of the nuclei. Between two and three thousand of such unresolvable nebulæ, in which the most powerful telescopes have hitherto been unable to distinguish the presence of stars, have been counted, and their positions determined.

The genetic evolution—that perpetual state of development which seems to affect this portion of the regions of space—has led philosophical observers to the discovery of the analogy existing among organic phenomena. As in our forests we see the same kind of tree in all the various stages of its growth, and are thus enabled to form an idea of progressive, vital development, so do we also, in the great garden of the universe, recognize the most different phases of sidereal formation. The process of condensation, which formed a part of the doctrines of Anaximenes and of the Ionian School, appears to be going on before our eyes. This subject of investigation and conjecture is especially attractive to the imagination, for in the study of the animated circles of nature, and of the action of all the moving forces of the universe, the charm that exercises the most powerful influence on the mind is derived less from a knowledge of that which *is* than from a perception of that which *will be*, even though the latter be nothing more than a new condition of a known material existence ; for of actual creation, of origin, the beginning of existence from non-existence, we have no experience, and can therefore form no conception.

A comparison of the various causes influencing the development manifested by the greater or less degree of condensation in the interior of nebulæ, no less than a successive course of direct observations, have led to the belief that changes of form have been recognized first in Andromeda, next in the constellation Argo, and in the isolated filamentous portion of the nebula in Orion. But want of uniformity in the power of the instruments employed, different conditions of our atmosphere, and other optical relations, render a part of the results invalid as historical evidence.

Nebulous stars must not be confounded either with irregularly-shaped nebulous spots, properly so called, whose separate parts have an unequal degree of brightness (and which may, perhaps, become concentrated into stars as their circumference contracts), nor with the so-called planetary nebulæ, whose circular or slightly oval disks manifest in all their parts a perfectly uniform degree of faint light. *Nebulous stars* are not merely accidental bodies projected upon a nebulous ground, but are a part of the nebulous matter constituting one mass with the body which it surrounds. The not unfrequently considerable magnitude of their apparent diameter, and the remote distance from which they are revealed to us, show that both the planetary nebulæ and the nebulous stars must be of enormous dimensions. New and ingenious considerations of the different influence exercised by distance* on the intensity of light of a disk of appreciable diameter, and of a single self-luminous point, render it not improbable that the planetary nebulæ are very remote nebulous stars, in which the difference between the central body and the surrounding nebulous covering can no longer be detected by our telescopic instruments.

The magnificent zones of the southern heavens, between 50° and 80°, are especially rich in nebulous stars, and in compressed unresolvable nebulæ. The larger of the two Magellanic clouds, which circle round the starless, desert pole of the south, appears, according to the most recent researches,† as " a collection of clusters of stars, composed of globular clusters and nebulæ of different magnitude, and of large nebulous spots

---

* The optical considerations relative to the difference presented by a single luminous point, and by a disk subtending an appreciable angle, in which the intensity of light is constant at every distance, are explained in Arago's *Analyse des Travaux de Sir William Herschel* (*Annuaire du Bureau des Long.*, 1842, p. 410–412, and 441).

† The two Magellanic clouds, Nubecula major and Nubecula minor, are very remarkable objects. The larger of the two is an accumulated mass of stars, and consists of clusters of stars of irregular form, either conical masses or nebulæ of different magnitudes and degrees of condensation. This is interspersed with nebulous spots, not resolvable into stars, but which are probably *star dust*, appearing only as a general radiance upon the telescopic field of a twenty-feet reflector, and forming a luminous ground on which other objects of striking and indescribable form are scattered. In no other portion of the heavens are so many nebulous and stellar masses thronged together in an equally small space. Nubecula minor is much less beautiful, has more unresolvable nebulous light, while the stellar masses are fewer and fainter in intensity.—(From a letter of Sir John Herschel, Feldhuysen, Cape of Good Hope, 13th June, 1836.)

not resolvable, which, producing a general brightness in the field of view, form, as it were, the back-ground of the picture." The appearance of these clouds, of the brightly-beaming constellation Argo, of the Milky Way between Scorpio, the Centaur, and the Southern Cross, the picturesque beauty, if one may so speak, of the whole expanse of the southern celestial hemisphere, has left upon my mind an ineffaceable impression. The zodiacal light, which rises in a pyramidal form, and constantly contributes, by its mild radiance, to the external beauty of the tropical nights, is either a vast nebulous ring, rotating between the Earth and Mars, or, less probably, the exterior stratum of the solar atmosphere. Besides these luminous clouds and nebulæ of definite form, exact and corresponding observations indicate the existence and the general distribution of an apparently non-luminous, infinitely-divided matter, which pos- sesses a force of resistance, and manifests its presence in Encke's, and perhaps also in Biela's comet, by diminishing their eccentricity and shortening their period of revolution. Of this impeding, ethereal, and cosmical matter, it may be supposed that it is in motion ; that it gravitates, notwithstanding its original tenuity ; that it is condensed in the vicinity of the great mass of the Sun ; and, finally, that it may, for myriads of ages, have been augmented by the vapor emanating from the tails of comets.

If we now pass from the consideration of the vaporous matter of the immeasurable regions of space (οὐρανοῦ χόρτος)* —whether, scattered without definite form and limits, it exists as a cosmical ether, or is condensed into nebulous spots, and becomes comprised among the solid agglomerated bodies of the universe—we approach a class of phenomena exclusively designated by the term of stars, or as the sidereal world.

* I should have made use, in the place of garden of the universe, of the beautiful expression χόρτος οὐρανοῦ, borrowed by Hesychius from an unknown poet, if χόρτος had not rather signified in general an inclosed space. The connection with the German *garten* and the English *garden, gards* in Gothic (derived, according to Jacob Grimm, from *gairdan, to gird*), is, however, evident, as is likewise the affinity with the Sclavonic *grad, gorod,* and as Pott remarks, in his *Etymol. Forschungen,* th. i., s. 144 (Etymol. Researches), with the Latin *chors,* whence we have the Spanish *corte,* the French *cour,* and the English word *court,* together with the Ossetic *khart.* To these may be further added the Scandinavian *gard,*[a] *gård,* a place inclosed, as a court, or a country seat, and the Persian *gerd, gird,* a district, a circle, a princely country seat, a castle or city, as we find the term applied to the names of places in Firdusi's Schahnameh, as *Siyawakschgird, Darabgird,* &c.

a [This word is written *gaard* in the Danish.]—*Tr.*

Here, too, we find differences existing in the solidity or density of the spheroidally agglomerated matter. Our own solar system presents all stages of *mean* density (or of the relation of *volume* to *mass*.) On comparing the planets from Mercury to Mars with the Sun and with Jupiter, and these two last named with the yet inferior density of Saturn, we arrive, by a descending scale—to draw our illustration from terrestrial substances—at the respective densities of antimony, honey, water, and pine wood. In comets, which actually constitute the most considerable portion of our solar system with respect to the number of individual forms, the concentrated part, usually termed the *head*, or *nucleus*, transmits sidereal light unimpaired. The mass of a comet probably in no case equals the five thousandth part of that of the earth, so dissimilar are the formative processes manifested in the original and perhaps still progressive agglomerations of matter. In proceeding from general to special considerations, it was particularly desirable to draw attention to this diversity, not merely as a possible, but as an actually proved fact.

The purely speculative conclusions arrived at by Wright, Kant, and Lambert, concerning the general structural arrangement of the universe, and of the distribution of matter in space, have been confirmed by Sir William Herschel, on the more certain path of observation and measurement. That great and enthusiastic, although cautious observer, was the first to sound the depths of heaven in order to determine the limits and form of the starry stratum which we inhabit, and he, too, was the first who ventured to throw the light of investigation upon the relations existing between the position and distance of remote nebulæ and our own portion of the sidereal universe. William Herschel, as is well expressed in the elegant inscription on his monument at Upton, broke through the inclosures of heaven (*cœlorum perrupit claustra*), and, like another Columbus, penetrated into an unknown ocean, from which he beheld coasts and groups of islands, whose true position it remains for future ages to determine.

Considerations regarding the different intensity of light in stars, and their relative number, that is to say, their numerical frequency on telescopic fields of equal magnitude, have led to the assumption of unequal distances and distribution in space in the strata which they compose. Such assumptions, in as far as they may lead us to draw the limits of the individual portions of the universe, can not offer the same degree of mathematical certainty as that which may be attained in all that

relates to our solar system, whether we consider the rotation
of double stars with unequal velocity round one common center of gravity, or the apparent or true movements of all the
heavenly bodies. If we take up the physical description of
the universe from the remotest nebulæ, we may be inclined
to compare it with the mythical portions of history. The one
begins in the obscurity of antiquity, the other in that of inaccessible space ; and at the point where reality seems to flee
before us, imagination becomes doubly incited to draw from
its own fullness, and give definite outline and permanence to
the changing forms of objects.

If we compare the regions of the universe with one of the
island-studded seas of our own planet, we may imagine matter to be distributed in groups, either as unresolvable nebulæ
of different ages, condensed around one or more nuclei, or as
already agglomerated into clusters of stars, or isolated spheroidal bodies. The cluster of stars, to which our cosmical isl-
and belongs, forms a lens-shaped, flattened stratum, detached
on every side, whose major axis is estimated at seven or eight
hundred, and its minor one at a hundred and fifty times the
distance of Sirius.   It would appear, on the supposition that
the parallax of Sirius is not greater than that accurately determined for the brightest star in the Centaur (0″·9128), that
light traverses one distance of Sirius in three years, while it
also follows, from Bessel's earlier excellent Memoir* on the
parallax of the remarkable star 61 Cygni (0″·3483), (whose
considerable motion might lead to the inference of great proximity), that a period of nine years and a quarter is required
for the transmission of light from this star to our planet.   Our
starry stratum is a disk of inconsiderable thickness, divided a

* See Maclear's " Results from 1839 to 1840," in the Trans. of the
Astronomical Soc., vol. xii., p. 370, on a Centauri, the probable mean
error being 0″·0640.   For 61 Cygni, see Bessel, in Schumacher's Jahrbuch, 1839, s. 47, and Schumacher's Astron. Nachr., bd. xviii., s. 401,
402, probable mean error, 0″·0141.   With reference to the relative
distances of stars of different magnitudes, how those of the third magnitude may probably be three times more remote, and the manner in
which we represent to ourselves the material arrangement of the starry
strata, I have found the following remarkable passage in Kepler's
Epitome Astronomiæ Copernicanæ, 1618, t. i., lib. 1, p. 34–39: "Sol
hic noster nil aliud est quam una ex fixis, nobis major et clarior visa,
quia propior quam fixa.   Pone terram stare ad latus, una semi-diametro
viæ lacteæ, tunc hæc via lactea apparebit circulus parvus, vel ellipsis par-
va, tota declinans ad latus alterum; eritque simul uno intuitu conspicua,
quæ nunc non potest nisi dimidia conspici quovis momento.   Itaque fix-
arum sphæra non tantum orbe stellarum, sed etiam circulo lactis versus
nos deorsum est terminata."

third of its length into two branches ; it is supposed that we are near this division, and nearer to the region of Sirius than to the constellation Aquila, almost in the middle of the stratum in the line of its thickness or minor axis.

This position of our solar system, and the form of the whole discoidal stratum, have been inferred from sidereal scales, that is to say, from that method of counting the stars to which I have already alluded, and which is based upon the equidistant subdivision of the telescopic field of view. The relative depth of the stratum in all directions is measured by the greater or smaller number of stars appearing in each division. These divisions give the length of the ray of vision in the same manner as we measure the depth to which the plummet has been thrown, before it reaches the bottom, although in the case of a starry stratum there can not, correctly speaking, be any idea of depth, but merely of outer limits. In the direction of the longer axis, where the stars lie behind one another, the more remote ones appear closely crowded together, united, as it were, by a milky-white radiance or luminous vapor, and are perspectively grouped, encircling, as in a zone, the visible vault of heaven. This narrow and branched girdle, studded with radiant light, and here and there interrupted by dark spots, deviates only by a few degrees from forming a perfect large circle round the concave sphere of heaven, owing to our being near the center of the large starry cluster, and almost on the plane of the Milky Way. If our planetary system were far *outside* this cluster, the Milky Way would appear to telescopic vision as a ring, and at a still greater distance as a resolvable discoidal nebula.

Among the many self-luminous moving suns, erroneously called *fixed stars*, which constitute our cosmical island, our own sun is the only one known by direct observation to be a *central body* in its relations to spherical agglomerations of matter directly depending upon and revolving round it, either in the form of planets, comets, or aërolite asteroids. As far as we have hitherto been able to investigate *multiple* stars (double stars or suns), these bodies are not subject, with respect to relative motion and illumination, to the same planetary dependence that characterizes our own solar system. Two or more self-luminous bodies, whose planets and moon, if such exist, have hitherto escaped our telescopic powers of vision, certainly revolve around one common center of gravity ; but this is in a portion of space which is probably occupied merely by unagglomerated matter or cosmical vapor, while in our systein the center of gravity is often comprised within the inner-most limits of a *visible* central body. If, therefore, we regard the Sun and the Earth, or the Earth and the Moon, as double stars, and the whole of our planetary solar system as a multi-ple cluster of stars, the analogy thus suggested must be limit-ed to the universality of the laws of attraction in different sys-tems, being alike applicable to the independent processes of light and to the method of illumination.

For the generalization of cosmical views, corresponding with the plan we have proposed to follow in giving a delineation of nature or of the universe, the solar system to which the Earth belongs may be considered in a two-fold relation : first, with respect to the different classes of individually agglomerated matter, and the relative size, conformation, density, and dis-tance of the heavenly bodies of this system ; and, secondly, with reference to other portions of our starry cluster, and of the changes of position of its central body, the Sun.

The solar system, that is to say, the variously-formed matter circling round the Sun, consists, according to the present state of our knowledge, of *eleven primary planets*,* eighteen satel-

* [Since the publication of Baron Humboldt's work in 1845, several other planets have been discovered, making the number of those be-longing to our planetary system *sixteen* instead of *eleven*. Of these, Astrea, Hebe, Flora, and Iris are members of the remarkable group of asteroids between Mars and Jupiter. Astrea and Hebe were dis-covered by Hencke at Driesen, the one in 1846 and the other in 1847 ; Flora and Iris were both discovered in 1847 by Mr. Hind, at the South Villa Observatory, Regent's Park. It would appear from the latest de-terminations of their elements, that the small planets have the following order with respect to mean distance from the Sun : Flora, Iris, Vesta, Hebe, Astrea, Juno, Ceres, Pallas. Of these, Flora has the shortest period (about 3¼ years). The planet Neptune, which, after having been predicted by several astronomers, was actually observed on the 25th of September, 1846, is situated on the confines of our planetary system beyond Uranus. The discovery of this planet is not only highly interesting from the importance attached to it as a question of science, but also from the evidence it affords of the care and unremitting labor evinced by modern astronomers in the investigation and comparison of the older calculations, and the ingenious application of the results thus obtained to the observation of new facts. The merit of having paved the way for the discovery of the planet Neptune is due to M. Bouvard, who, in his persevering and assiduous efforts to deduce the entire orbit of Uranus from observations made during the forty years that succeed-ed the discovery of that planet in 1781, found the results yielded by theory to be at variance with fact, in a degree that had no parallel in the history of astronomy. This startling discrepancy, which seemed only to gain additional weight from every attempt made by M. Bouvard to correct his calculations, led Leverrier, after a careful modification of the tables of Bouvard, to establish the proposition that there was " a

lites or secondary planets, and myriads of comets, three of which, known as the "planetary comets," do not pass beyond the narrow limits of the orbits described by the principal planets. We may, with no inconsiderable degree of probability, include within the domain of our Sun, in the immediate sphere of its central force, a rotating ring of vaporous matter, lying probably between the orbits of Venus and Mars, but certainly beyond that of the Earth,* which appears to us in

formal incompatibility between the observed motions of Uranus and *the hypothesis that he was acted on only* by the Sun and known planets, according to the law of universal gravitation." Pursuing this idea, Leverrier arrived at the conclusion that the disturbing cause must be a *planet*, and, finally, after an amount of labor that seems perfectly overwhelming, he, on the 31st of August, 1846, laid before the French Institute a paper, in which he indicated the exact spot in the heavens where this new planetary body would be found, giving the following data for its various elements : mean distance from the Sun, 36·154 times that of the Earth ; period of revolution, 217·387 years ; mean long., Jan. 1st, 1847, 318° 47' ; mass, $\frac{1}{9300}$th ; heliocentric long., Jan. 1st, 1847, 326° 32'. Essential difficulties still intervened, however, and as the remoteness of the planet rendered it improbable that its disk would be discernible by any telescopic instrument, no other means remained for detecting the suspected body but its planetary motion, which could only be ascertained by mapping, after every observation, the quarter of the heavens scanned, and by a comparison of the various maps. Fortunately for the verification of Leverrier's predictions, Dr. Bremiker had just completed a map of the precise region in which it was expected the new planet would appear, this being one of a series of maps made for the Academy of Berlin, of the small stars along the entire zodiac. By means of this valuable assistance, Dr. Galle, of the Berlin Observatory, was led, on the 25th of September, 1846, by the discovery of a star of the eighth magnitude, not recorded in Dr. Bremiker's map, to make the first observation of the planet predicted by Leverrier. By a singular coincidence, Mr. Adams, of Cambridge, had predicted the appearance of the planet simultaneously with M. Leverrier ; but by the concurrence of several circumstances much to be regretted, the world at large were not made acquainted with Mr. Adams's valuable discovery until subsequently to the period at which Leverrier published his observations. As the data of Leverrier and Adams stand at present, there is a discrepancy between the predicted and the true distance, and in some other elements of the planet ; it remains, therefore, for these or future astronomers to reconcile theory with fact, or perhaps, as in the case of Uranus, to make the new planet the means of leading to yet greater discoveries. It would appear from the most recent observations, that the mass of Neptune, instead of being, as at first stated, $\frac{1}{9300}$th, is only about $\frac{1}{23000}$th that of the Sun, while its periodic time is now given with a greater probability at 166 years, and its mean distance from the Sun nearly 30. The planet appears to have a ring, but as yet no accurate observations have been made regarding its system of satellites. See *Trans. Astron. Soc.*, and *The Planet Neptune*, 1848, by J. P. Nicholl.] —*Tr.*

* " If there should be molecules in the zones diffused by the atmos

a pyramidal form, and is known as the *Zodiacal Light;* and a host of very small asteroids, whose orbits either intersect, or very nearly approach, that of our earth, and which present us with the phenomena of aërolites and falling or shooting stars. When we consider the complication of variously-formed bodies which revolve round the Sun in orbits of such dissimilar eccentricity—although we may not be disposed, with the immortal author of the *Mécanique Céleste*, to regard the larger number of comets as nebulous stars, passing from one central system to another,* we yet can not fail to acknowledge that the planetary system, especially so called (that is, the group of heavenly bodies which, together with their satellites, revolve with but slightly eccentric orbits round the Sun), constitutes but a small portion of the whole system with respect to individual numbers, if not to mass.

It has been proposed to consider the telescopic planets, Vesta, Juno, Ceres, and Pallas, with their more closely intersecting, inclined, and eccentric orbits, as a zone of separation, or as a middle group in space ; and if this view be adopted, we shall discover that the interior planetary group (consisting of Mercury, Venus, the Earth, and Mars) presents several very striking contrasts† when compared with the exterior group, comprising Jupiter, Saturn, and Uranus.   The planets nearest the Sun, and consequently included in the inner group, are of more moderate size, denser, rotate more slowly and with nearly equal velocity (their periods of revolution being almost all about 24 hours), are less compressed at the poles, and, with the exception of one, are without satellites.   The exterior planets, which are further removed from the Sun, are very considerably larger, have a density five times less, more than twice as great a velocity in the period of their rotation round their axes, are more compressed at the poles, and if six satellites may be ascribed to Uranus, have a quantitative preponderance in the number of their attendant moons, which is as seventeen to one.

phere of the Sun of too volatile a nature either to combine with one another or with the planets, we must suppose that they would, in circling round that luminary, present all the appearances of zodiacal light, without opposing any appreciable resistance to the different bodies composing the planetary system, either owing to their extreme rarity, or to the similarity existing between their motion and that of the planets with which they come in contact."—Laplace, *Expos. du Syst. du Monde* (ed. 5), p. 415.

* Laplace, *Exp. du Syst. du Monde*, p. 396, 414.

† Littrow, *Astronomie*, 1825, bd. xi., § 107.   Mädler, *Astron.*, 1841, § 212.   Laplace, *Exp. du Syst. du Monde*, p. 210.

Such general considerations regarding certain characteristic properties appertaining to whole groups, can not, however, be applied with equal justice to the individual planets of every group, nor to the relations between the distances of the revolving planets from the central body, and their absolute size, density, period of rotation, eccentricity, and the inclination of their orbits and the axes. We know as yet of no inherent necessity, no mechanical natural law, similar to the one which teaches us that the squares of the periodic times are proportional to the cubes of the major axes, by which the above-named six elements of the planetary bodies and the form of their orbit are made dependent either on one another, or on their mean distance from the Sun. Mars is smaller than the Earth and Venus, although further removed from the Sun than these last-named planets, approaching most nearly in size to Mercury, the nearest planet to the Sun. Saturn is smaller than Jupiter, and yet much larger than Uranus. The zone of the telescopic planets, which have so inconsiderable a vol ume, immediately precede Jupiter (the greatest in size of any of the planetary bodies), if we consider them with regard to distance from the Sun ; and yet the disks of these small asteroids, which scarcely admit of measurement, have an areal surface not much more than half that of France, Madagascar, or Borneo. However striking may be the extremely small density of all the colossal planets, which are furthest removed from the Sun, we are yet unable in this respect to recognize any regular succession.* Uranus appears to be denser than Saturn, even if we adopt the smaller mass, $\frac{1}{24063}$, assumed by Lamont ; and, notwithstanding the inconsiderable difference of density observed in the innermost planetary group,† we find both Venus and Mars less dense than the Earth, which lies between them. The time of rotation certainly diminishes with increasing solar distance, but yet it is greater in Mars than in the Earth, and in Saturn than in Jupiter. The el-

---

* See Kepler, on the increasing density and volume of the planets in proportion with their increase of distance from the Sun, which is described as the densest of all the heavenly bodies; in the *Epitome Astron. Copern. in* vii. *libros digesta,* 1618–1622, p. 420. Leibnitz also inclined to the opinions of Kepler and Otto von Guericke, that the planets increase in volume in proportion to their increase of distance from the Sun. See his letter to the Magdeburg Burgomaster (Mayence, 1671), in Leibnitz, *Deutschen Schriften, herausg. von Guhrauer,* th. i., § 264.

† On the arrangement of masses, see Encke, in Schum., *Astr. Nachr,* 1843 Nr. 488, § 114.

liptic orbits of Juno, Pallas, and Mercury have the greatest
degree of eccentricity, and Mars and Venus, which immediately follow each other, have the least.   Mercury and Venus
exhibit the same contrasts that may be observed in the four
smaller planets, or asteroids, whose paths are so closely interwoven.

The eccentricities of Juno and Pallas are very nearly identical, and are each three times as great as those of Ceres and
Vesta.   The same may be said of the inclination of the orbits
of the planets toward the plane of projection of the ecliptic, or
in the position of their axes of rotation with relation to their
orbits, a position on which the relations of climate, seasons of
the year, and length of the days depend more than on eccentricity.   Those planets that have the most elongated elliptic
orbits, as Juno, Pallas, and Mercury, have also, although not
to the same degree, their orbits most strongly inclined toward
the ecliptic.   Pallas has a comet-like inclination nearly twenty-six times greater than that of Jupiter, while in the little
planet Vesta, which is so near Pallas, the angle of inclination
scarcely by six times exceeds that of Jupiter.   An equally irregular succession is observed in the position of the axes of
the few planets (four or five) whose planes of rotation we
know with any degree of certainty.   It would appear from
the position of the satellites of Uranus, two of which, the second and fourth, have been recently observed with certainty,
that the axis of this, the outermost of all the planets, is scarcely inclined as much as 11° toward the plane of its orbit, while
Saturn is placed between this planet, whose axis almost coincides with the plane of its orbit, and Jupiter, whose axis of
rotation is nearly perpendicular to it.

In this enumeration of the forms which compose the world
in space, we have delineated them as possessing an actual existence, and not as objects of intellectual contemplation, or as
mere links of a mental and causal chain of connection.   The
planetary system, in its relations of absolute size and relative
position of the axes, density, time of rotation, and different de-
gress of eccentricity of the orbits, does not appear to offer to
our apprehension any stronger evidence of a natural necessity
than the proportion observed in the distribution of land and
water on the Earth, the configuration of continents, or the
height of mountain chains.   In these respects we can discover
no common law in the regions of space or in the inequalities
of the earth's crust.   They are *facts* in nature that have
arisen from the conflict of manifold forces acting under un-

known conditions, although man considers as *accidental* whatever he is unable to explain in the planetary formation on purely genetic principles. If the planets have been formed out of
separate rings of vaporous matter revolving round the Sun,
we may conjecture that the different thickness, unequal density, temperature, and electro-magnetic tension of these rings
may have given occasion to the most various agglomerations
of matter, in the same manner as the amount of tangential
velocity and small variations in its direction have produced so
great a difference in the forms and inclinations of the elliptic
orbits. Attractions of mass and laws of gravitation have no
doubt exercised an influence here, no less than in the geognostic relations of the elevations of continents ; but we are unable from present forms to draw any conclusions regarding the
series of conditions through which they have passed. Even
the so-called law of the distances of the planets from the Sun,
the law of progression (which led Kepler to conjecture the existence of a planet supplying the link that was wanting in the
chain of connection between Mars and Jupiter), has been found
numerically inexact for the distances between Mercury, Venus,
and the Earth, and at variance with the conception of a series,
owing to the necessity for a supposition in the case of the first
member.

The hitherto discovered principal planets that revolve round
our Sun are attended certainly by fourteen, and probably by
eighteen secondary planets (moons or satellites). The principal planets are, therefore, themselves the central bodies of subordinate systems. We seem to recognize in the fabric of the
universe the same process of arrangement so frequently exhibited in the development of organic life, where we find in
the manifold combinations of groups of plants or animals the
same typical form repeated in the *subordinate classes.* The
secondary planets or satellites are more frequent in the external region of the planetary system, lying beyond the intersecting orbits of the smaller planets or asteroids ; in the inner region none of the planets are attended by satellites, with the
exception of the Earth, whose moon is relatively of great magnitude, since its diameter is equal to a fourth of that of the
Earth, while the diameter of the largest of all known secondary planets—the sixth satellite of Saturn—is probably about
one seventeenth, and the largest of Jupiter's moons, the third,
only about one twenty-sixth part that of the primary planet
or central body. The planets which are attended by the
largest number of satellites are most remote from the Sun,

and are at the same time the largest, most compressed at the poles, and the least dense. According to the most recent measurements of Mädler, Uranus has a greater planetary compression than any other of the planets, viz., $\frac{1}{9 \cdot 0 \cdot 2}$d. In our Earth and her moon, whose mean distance from one another amounts to 207,200 miles, we find that the differences of mass* and diameter between the two are much less considerable than are usually observed to exist between the principal planets and their attendant satellites, or between bodies of different orders in the solar system. While the density of the Moon is five ninths less than that of the Earth, it would appear, if we may sufficiently depend upon the determinations of their magnitudes and masses, that the second of Jupiter's moons is actually denser than that great planet itself. Among the fourteen satellites that have been investigated with any degree of certainty, the system of the seven satellites of Saturn presents an instance of the greatest possible contrast, both in absolute magnitude and in distance from the central body. The sixth of these satellites is probably not much smaller than Mars, while our moon has a diameter which does not amount to more than half that of the latter planet. With respect to volume, the two outer, the sixth and seventh of Saturn's satellites, approach the nearest to the third and brightest of Jupiter's moons. The two innermost of these satellites belong perhaps, together with the remote moons of Uranus, to the smallest cosmical bodies of our solar system, being only made visible under favorable circumstances by the most powerful instruments. They were first discovered by the forty-foot telescope of William Herschel in 1789, and were seen again by John Herschel at the Cape of Good Hope, by Vico at Rome, and by Lamont at Munich. Determinations of the *true* diameter of satellites, made by the measurement of the apparent size of their small disks, are subjected to many optical difficulties; but numerical astronomy, whose task it is to predetermine by calculation the motions of the heavenly bodies as they will appear when viewed from the Earth, is directed al-

---

* If, according to Burckhardt's determination, the Moon's radius be 0.2725 and its volume $\frac{1}{49 \cdot 0 \cdot 0}$th, its density will be 0·5596, or nearly five ninths. Compare, also, Wilh. Beer und H. Mädler, *der Mond*, § 2, 10, and Mädler, *Ast.*, § 157. The material contents of the Moon are, according to Hausen, nearly $\frac{1}{34}$th (and according to Mädler $\frac{1}{40 \cdot 6}$th) that of the Earth, and its mass equal to $\frac{1}{87 \cdot 73}$d that of the Earth. In the largest of Jupiter's moons, the third, the relations of volume to the central body are $\frac{1}{15370}$th, and of mass $\frac{1}{11300}$th. On the polar flattening of Uranus, see Schum., *Astron. Nachr.*, 1844, No. 493.

most exclusively to motion and mass, and but little to volume. The absolute distance of a satellite from its central body is greatest in the case of the outermost or seventh satellite of Saturn, its distance from the body round which it revolves amounting to more than two millions of miles, or ten times as great a distance as that of our moon from the Earth. In the case of Jupiter we find that the outermost or fourth attendant moon is only 1,040,000 miles from that planet, while the distance between Uranus and its sixth satellite (if the latter really exist) amounts to as much as 1,360,000 miles. If we compare, in each of these subordinate systems, the volume of the main planet with the distance of the orbit of its most remote satellite, we discover the existence of entirely new numerical relations. The distances of the outermost satellites of Uranus, Saturn, and Jupiter are, when expressed in semi-diameters of the main planets, as 91, 64, and 27. The outermost satellite of Saturn appears, therefore, to be removed only about one fifteenth further from the center of that planet than our moon is from the Earth. The first or innermost of Saturn's satellites is nearer to its central body than any other of the secondary planets, and presents, moreover, the only instance of a period of revolution of less than twenty-four hours. Its distance from the center of Saturn may, according to Mädler and Wilhelm Beer, be expressed as 2·47 semi-diameters of that planet, or as 80,088 miles. Its distance from the surface of the main planet is therefore 47,480 miles, and from the outermost edge of the ring only 4916 miles. The traveler may form to himself an estimate of the smallness of this amount by remembering the statement of an enterprising navigator, Captain Beechey, that he had in three years passed over 72,800 miles. If, instead of absolute distances, we take the semi-diameters of the principal planets, we shall find that even the first or nearest of the moons of Jupiter (which is 26,000 miles further removed from the center of that planet than our moon is from that of the Earth) is only six semi-diameters of Jupiter from its center, while our moon is removed from us fully 60⅓d semi-diameters of the Earth.

In the subordinate systems of satellites, we find that the same laws of gravitation which regulate the revolutions of the principal planets round the Sun likewise govern the mutual relations existing between these planets among one another and with reference to their attendant satellites. The twelve moons of Saturn, Jupiter, and the Earth all move like the primary planets from west to east, and in elliptic orbits, de-

viating but little from circles.   It is only in the case of our
moon, and perhaps in that of the first and innermost of the
satellites of Saturn (0·068), that we discover an eccentricity
greater than that of Jupiter ; according to the very exact observations of Bessel, the eccentricity of the sixth of Saturn's
satellites (0·029) exceeds that of the Earth.   On the extremest
limits of the planetary system, where, at a distance nineteen
times greater than that of our Earth, the centripetal force of
the Sun is greatly diminished, the satellites of Uranus (which
have certainly been but imperfectly investigated) exhibit the
most striking contrasts from the facts observed with regard to
other secondary planets.   Instead, as in all other satellites, of
having their orbits but slightly inclined toward the ecliptic
and (not excepting even Saturn's ring, which may be regarded as a fusion of agglomerated satellites) moving from west to
east, the satellites of Uranus are almost perpendicular to the
ecliptic, and move retrogressively from east to west, as Sir
John Herschel has proved by observations continued during
many years   If the primary and secondary planets have been
formed by the condensation of rotating rings of solar and planetary atmospheric vapor, there must have existed singular
causes of retardation or impediment in the vaporous rings revolving round Uranus, by which, under relations with which
we are unacquainted, the revolution of the second and fourth
of its satellites was made to assume a direction opposite to that
of the rotation of the central planet.

It seems highly probable that the period of rotation of *all*
secondary planets is equal to that of their revolution round
the main planet, and therefore that they always present to
the latter the same side.   Inequalities, occasioned by slight
variations in the revolution, give rise to fluctuations of from
6° to 8°, or to an apparent libration in longitude as well as
in latitude.   Thus, in the case of our moon, we sometimes
observe more than the half of its surface, the eastern and
northern edges being more visible at one time, and the western or southern at another.   By means of this libration* we
are enabled to see the annular mountain Malapert (which occasionally conceals the Moon's south pole), the arctic landscape round the crater of Gioja, and the large gray plane near
Endymion, which exceeds in superficial extent the *Mare Va-
porum*.   Three sevenths of the Moon's surface are entirely

* Beer and Mädler, op. cit., § 185, s. 208, and § 347, s 333; and in
their *Phys. Kenntniss der himml. Körper*, s. 4 und 69, Tab. 1 (Physic
al History of the Heavenly Bodies).

concealed from our observation, and must always remain so, unless new and unexpected disturbing causes come into play. These cosmical relations involuntarily remind us of nearly similar conditions in the intellectual world, where, in the domain of deep research into the mysteries and the primeval creative forces of nature, there are regions similarly turned away from us, and apparently unattainable, of which only a narrow margin has revealed itself, for thousands of years, to the human mind, appearing, from time to time, either glimmering in true or delusive light. We have hitherto considered the primary planets, their satellites, and the concentric rings which belong to one, at least, of the outermost planets, as products of tangential force, and as closely connected together by mutual attraction; it therefore now only remains for us to speak of the unnumbered host of *comets* which constitute a portion of the cosmical bodies revolving in independent orbits round the Sun. If we assume an equable distribution of their orbits, and the limits of their perihelia, or greatest proximities to the Sun, and the possibility of their remaining invisible to the inhabitants of the Earth, and base our estimates on the rules of the calculus of probabilities, we shall obtain as the result an amount of myriads perfectly astonishing. Kepler, with his usual animation of expression, said that there were more comets in the regions of space than fishes in the depths of the ocean. As yet, however, there are scarcely one hundred and fifty whose paths have been calculated, if we may assume at six or seven hundred the number of comets whose appearance and passage through known constellations have been ascertained by more or less precise observations. While the so-called classical nations of the West, the Greeks and Romans, although they may occasionally have indicated the position in which a comet first appeared, never afford any information regarding its apparent path, the copious literature of the Chinese (who observed nature carefully, and recorded with accuracy what they saw) contains circumstantial notices of the constellations through which each comet was observed to pass. These notices go back to more than five hundred years before the Christian era, and many of them are still found to be of value in astronomical observations.*

* The first comets of whose orbits we have any knowledge, and which were calculated from Chinese observations, are those of 240 (under Gordian III.), 539 (under Justinian), 565, 568, 574, 837, 1337, and 1385. See John Russell Hind, in Schum., *Astron. Nachr.*, 1843, No. 498. While the comet of 837 (which, according to Du Séjour, continued dur

Although comets have a smaller mass than any other cosmical bodies—being, according to our present knowledge, probably not equal to $\frac{1}{3000}$th part of the Earth's mass—yet they occupy the largest space, as their tails in several instances extend over many millions of miles. The cone of luminous vapor which radiates from them has been found, in some cases (as in 1680 and 1811), to equal the length of the Earth's distance from the Sun, forming a line that intersects both the orbits of Venus and Mercury. It is even probable that the vapor of the tails of comets mingled with our atmosphere in the years 1819 and 1823.

Comets exhibit such diversities of form, which appear rather to appertain to the individual than the class, that a description of one of these " wandering light-clouds," as they were already called by Xenophanes and Theon of Alexandria, cotemporaries of Pappus, can only be applied with caution to another. The faintest telescopic comets are generally devoid of visible tails, and resemble Herschel's nebulous stars. They appear like circular nebulæ of faintly-glimmering vapor, with the light concentrated toward the middle. This is the most simple type; but it can not, however, be regarded as rudimentary, since it might equally be the type of an older cosmical body, exhausted by exhalation. In the larger comets we may distinguish both the so-called " head" or " nucleus," and the single or multiple tail, which is characteristically denominated by the Chinese astronomers " the brush" (*sui*). The nucleus generally presents no definite outline, although, in a few rare cases, it appears like a star of the first or second magnitude, and has even been seen in bright sunshine ;* as,

ing twenty-four hours within a distance of 2,000,000 miles from the Earth) terrified Louis I. of France to that degree that he busied himself in building churches and founding monastic establishments, in the hope of appeasing the evils threatened by its appearance, the Chinese astronomers made observations on the path of this cosmical body, whose tail extended over a space of 60°, appearing sometimes single and sometimes multiple. The first comet that has been calculated solely from European observations was that of 1456, known as Halley's comet, from the belief long, but erroneously, entertained that the period when it was first observed by that astronomer was its first and only well-attested appearance. See Arago, in the *Annuaire*, 1836, p. 204, and Laugier, *Comptes Rendus des Séances de l'Acad.*, 1843, t. xvi., 1006.

* Arago, *Annuaire*, 1832, p. 209, 211. The phenomenon of the tail of a comet being visible in bright sunshine, which is recorded of the comet of 1402, occurred again in the case of the large comet of 1843, whose nucleus and tail were seen in North America on the 28th of February (according to the testimony of J. G. Clarke, of Portland, state of

for instance, in the large comets of 1402, 1532, 1577, 1744, and 1843. This latter circumstance indicates, in particular individuals, a denser mass, capable of reflecting light with greater intensity. Even in Herschel's large telescope, only two comets, that discovered in Sicily in 1807, and the splendid one of 1811, exhibited well-defined disks ;* the one at an angle of 1″, and the other at 0″·77, whence the true diameters are assumed to be 536 and 428 miles. The diameters of the less well-defined nuclei of the comets of 1798 and 1805 did not appear to exceed 24 or 28 miles.

In several comets that have been investigated with great care, especially in the above-named one of 1811, which continued visible for so long a period, the nucleus and its nebulous envelope were entirely separated from the tail by a darker space. The intensity of light in the nucleus of comets does not augment toward the center in any uniform degree, brightly shining zones being in many cases separated by concentric nebulous envelopes. The tails sometimes appear single, sometimes, although more rarely, double ; and in the comets of 1807 and 1843 the branches were of different lengths ; in one instance (1744) the tail had six branches, the whole forming an angle of 60°. The tails have been sometimes straight, sometimes curved, either toward both sides, or toward the side appearing to us as the exterior (as in 1811), or convex toward the direction in which the comet is moving (as in that of 1618) ; and sometimes the tail has even appeared like a flame in motion. The tails are always turned away from the sun, so that their line of prolongation passes through its center ; a fact which, according to Edward Biot, was noticed by the Chinese astronomers as early as 837, but was first generally made known in Europe by Fracastoro and Peter Apian in the sixteenth century. These emanations may be regarded as conoidal envelopes of greater or less thick-

Maine), between 1 and 3 o'clock in the afternoon.[a] The distance of the very dense nucleus from the sun's light admitted of being measured with much exactness. The nucleus and tail appeared like a very pure white cloud, a darker space intervening between the tail and the nucleus. (*Amer. Journ. of Science*, vol. xlv., No. 1, p. 229.)

* *Phil. Trans.* for 1808, Part ii., p. 155, and for 1812, Part i., p. 118. The diameters found by Herschel for the nuclei were 538 and 428 English miles. For the magnitudes of the comets of 1798 and 1805, see Arago, *Annuaire*, 1832, p. 203.

[a] [The translator was at New Bedford, Massachusetts, U. S., on the 28th February, 1843, and distinctly saw the comet, between 1 and 2 in the afternoon. The sky at the time was intensely blue, and the sun shining with a dazzling brightness unknown in European climates.]—*Tr*

ness, and, considered in this manner, they furnish a simple explanation of many of the remarkable optical phenomena already spoken of.

Comets are not only characteristically different in form, some being entirely without a visible tail, while others have a tail of immense length (as in the instance of the comet of 1618, whose tail measured 104°), but we also see the same comets undergoing successive and rapidly-changing processes of configuration. These variations of form have been most accurately and admirably described in the comet of 1744, by Hensius, at St. Petersburg, and in Halley's comet, on its last reappearance in 1835, by Bessel, at Königsberg. A more or less well-defined tuft of rays emanated from that part of the nucleus which was turned toward the Sun; and the rays being bent backward, formed a part of the tail. The nucleus of Halley's comet, with its emanations, presented the appearance of a burning rocket, the end of which was turned sideways by the force of the wind. The rays issuing from the head were seen by Arago and myself, at the Observatory at Paris, to assume very different forms on successive nights.* The great Königsberg astronomer concluded from many measurements, and from theoretical considerations, " that the cone of light issuing from the comet deviated considerably both to the right and the left of the true direction of the Sun, but that it always returned to that direction, and passed over to the opposite side, so that both the cone of light and the body of the comet from whence it emanated experienced a rotatory, or, rather, a vibratory motion in the plane of the orbit." He finds that " the attractive force exercised by the Sun on heavy bodies is inadequate to explain such vibrations, and is of opinion that they indicate a polar force, which turns one semi-diameter of the comet toward the Sun, and strives to turn the opposite side away from that luminary. The magnetic polarity possessed by the Earth may present some analogy to this, and, should the Sun have an opposite polarity, an influence might be manifested, resulting in the precession of the equinoxes." This is not the place to enter more fully upon the grounds on which explanations of this subject have been based; but observations so remarkable,† and views of so exalted

* Arago, *Des Changements physiques de la Comète de Halley du* 15–23 *Oct.*, 1835. *Annuaire*, 1836, p. 218, 221. The ordinary direction of the emanations was noticed even in Nero's time. " *Comæ radios solis effugiunt.*"—Seneca, *Nat. Quæst.*, vii., 20.

† Bessel, in Schumacher, *Astr. Nachr.*, 1836, No. 300–302, s. 188, 192,

ι character, regarding the most wonderful class of the cosmic-al bodies belonging to our solar system, ought not to be entirely passed over in this sketch of a general picture of nature.

Although, as a rule, the tails of comets increase in magnitude and brilliancy in the vicinity of the sun, and are directed away from that central body, yet the comet of 1823 offered the remarkable example of two tails, one of which was turned toward the sun, and the other away from it, forming with each other an angle of 160°. Modifications of polarity and the unequal manner of its distribution, and of the direction in which it is conducted, may in this rare instance have occasioned a double, unchecked, continuous emanation of nebulous matter.*

Aristotle, in his *Natural Philosophy*, makes these emanations the means of bringing the phenomena of comets into a singular connection with the existence of the Milky Way. According to his views, the innumerable quantity of stars which compose this starry zone give out a self-luminous, incandescent matter. The nebulous belt which separates the different portions of the vault of heaven was therefore regarded by the Stagirite as a large comet, the substance of which was incessantly being renewed.†

197, 200, 202, und 230. Also in Schumacher, *Jahrb.*, 1837, s. 149, 168. William Herschel, in his observations on the beautiful comet of 1811, believed that he had discovered evidences of the rotation of the nucleus and tail (*Phil. Trans.* for 1812, Part i., p. 140). Dunlop, at Paramatta, thought the same with reference to the third comet of 1825.

* Bessel, in *Astr. Nachr.*, 1836, No. 302, s. 231. Schum., *Jahrb.*, 1837, s. 175. See, also, Lehmann, *Ueber Cometenschweife* (On the Tails of Comets), in Bode, *Astron. Jahrb. für* 1826, s. 168.

† Aristot., *Meteor.*, i., 8, 11–14, und 19–21 (ed. Ideler, t. i., p. 32–34). Biese, *Phil. des Aristoteles*, bd. ii., s. 86. Since Aristotle exercised so great an influence throughout the whole of the Middle Ages, it is very much to be regretted that he was so averse to those grander views of the elder Pythagoreans, which inculcated ideas so nearly approximating to truth respecting the structure of the universe. He asserts that comets are transitory meteors belonging to our atmosphere in the very book in which he cites the opinion of the Pythagorean school, according to which these cosmical bodies are supposed to be planets having long periods of revolution. (Aristot., i., 6, 2.) This Pythagorean doctrine, which, according to the testimony of Apollonius Myndius, was still more ancient, having originated with the Chaldeans, passed over to the Romans, who in this instance, as was their usual practice, were merely the copiers of others. The Myndian philosopher describes the path of comets as directed toward the upper and remote regions of heaven. Hence Seneca says, in his *Nat. Quæst.*, vii., 17 : "*Cometes non est species falsa, sed proprium sidus sicut solis et lunæ : altiora mundi secat et tunc demum apparet quum in imum cursum sui venit;*" and again (at vii., 27), "*Cometes æternos esse et sortis ejusdem, cujus cætera*

The occultation of the fixed stars by the nucleus of a com
et, or by its innermost vaporous envelopes, might throw some
light on the physical character of these wonderful bodies ; but
we are unfortunately deficient in observations by which we
may be assured* that the occultation was perfectly central ;
for, as it has already been observed, the parts of the envelope
contiguous to the nucleus are alternately composed of layers
of dense or very attenuated vapor. On the other hand, the
carefully conducted measurements of Bessel prove, beyond all
doubt, that on the 29th of September, 1835, the light of a
star of the tenth magnitude, which was then at a distance of
$7''\cdot78$ from the central point of the head of Halley's comet,
passed through very dense nebulous matter, without experiencing any deflection during its passage.† If such an absence
of refracting power must be ascribed to the nucleus of a comet, we can scarcely regard the matter composing comets as a
gaseous fluid. The question here arises whether this absence
of refracting power may not be owing to the extreme tenuity
of the fluid ; or does the comet consist of separated particles,
constituting a cosmical stratum of clouds, which, like the
clouds of our atmosphere, that exercise no influence on the

(*sidera*), *etiamsi faciem illis non habent similem.*" Pliny (ii., 25) also refers to Apollonius Myndius, when he says, "*Sunt qui et hæc sidera perpetua esse credant suoque ambitu ire, sed non nisi relicta a sole cerni.*"

* Olbers, in *Astr. Nachr.*, 1828, s. 157, 184. Arago, *De la Constitution physique des Comètes; Annuaire de* 1832, p. 203, 208. The ancients were struck by the phenomenon that it was possible to see
through comets as through a flame. The earliest evidence to be met
with of stars having been seen through comets is that of Democritus
(Aristot., *Meteor.*, i., 6, 11), and the statement leads Aristotle to make
the not unimportant remark, that he himself had observed the occultation of one of the stars of Gemini by Jupiter. Seneca only speaks decidedly of the transparence of the tail of comets. " We may see," says
he, "stars through a comet as through a cloud (*Nat. Quæst.*, vii., 18);
but we can only see through the rays of the tail, and not through the
body of the comet itself: *non in ea parte qua sidus ipsum est spissi et
solidi ignis, sed qua rarus splendor occurrit et in crines dispergitur. Per
intervalla ignium, non per ipsos, vides*" (vii., 26). The last remark is
unnecessary, since, as Galileo observed in the *Saggiatore* (*Lettera a
Monsignor Cesarini*, 1619), we can certainly see through a flame when
it is not of too great a thickness.

† Bessel, in the *Astron. Nachr.*, 1836, No. 301, s. 204, 206. Struve,
in *Recueil des Mém. de l'Acad. de St. Petersb.*, 1836, p. 140, 143, and
Astr. Nachr., 1836, No. 303, s. 238, writes as follows: "At Dorpat the
star was in conjunction only $2''\cdot2$ from the brightest point of the comet.
The star remained continually visible, and its light was not perceptibly
diminished, while the nucleus of the comet seemed to be almost extinguished before the radiance of the small star of the ninth or tenth mag
nitude."

zenith distance of the stars, does not affect the ray of light passing through it ? In the passage of a comet over a star, a more or less considerable diminution of light has often been observed ; but this has been justly ascribed to the brightness of the ground from which the star seems to stand forth during the passage of the comet.

The most important and decisive observations that we possess on the nature and the light of comets are due to Arago's polarization experiments. His polariscope instructs us regarding the physical constitution of the Sun and comets, indicating whether a ray that reaches us from a distance of many millions of miles transmits light directly or by reflection ; and if the former, whether the source of light is a solid, a liquid, or a gaseous body. His apparatus was used at the Paris Observatory in examining the light of Capella and that of the great comet of 1819. The latter showed polarized, and therefore reflected light, while the fixed star, as was to be expected, appeared to be a self-luminous sun.[*] The existence of polarized cometary light announced itself not only by the inequality of the images, but was proved with greater certainty on the reappearance of Halley's comet, in the year 1835, by the more striking contrast of the complementary colors, deduced from the laws of chromatic polarization discovered by Arago in 1811. These beautiful experiments still leave it undecided whether, in addition to this reflected solar light, comets may not have light of their own. Even in the case of the planets, as, for instance, in Venus, an evolution of independent light seems very probable.

The variable intensity of light in comets can not always be

[*] On the 3d of July, 1819, Arago made the first attempt to analyze the light of comets by polarization, on the evening of the sudden appearance of the great comet. I was present at the Paris Observatory, and was fully convinced, as were also Matthieu and the late Bouvard, of the dissimilarity in the intensity of the light seen in the polariscope, when the instrument received cometary light. When it received light from Capella, which was near the comet, and at an equal altitude, the images were of equal intensity. On the reappearance of Halley's comet in 1835, the instrument was altered so as to give, according to Arago's chromatic polarization, two images of complementary colors (green and red). (*Annales de Chimie*, t. xiii., p. 108; *Annuaire*, 1832, p. 216.) "We must conclude from these observations," says Arago, "that the cometary light was not entirely composed of rays having the properties of direct light, there being light which was reflected specularly or polarized, that is, coming from the sun. It can not be stated with absolute certainty that comets shine only with borrowed light, for bodies, in becoming self-luminous, do not, on that account, lose the power of reflecting foreign light."

explained by the position of their orbits and their distance from the Sun. It would seem to indicate, in some individuals, the existence of an inherent process of condensation, and an increased or diminished capacity of reflecting borrowed light. In the comet of 1618, and in that which has a period of three years, it was observed first by Hevelius that the nucleus of the comet diminished at its perihelion and enlarged at its aphelion, a fact which, after remaining long unheeded, was again noticed by the talented astronomer Valz at Nismes. The regularity of the change of volume, according to the different degrees of distance from the Sun, appears very striking. The physical explanation of the phenomenon can not, however, be sought in the condensed layers of cosmical vapor occurring in the vicinity of the Sun, since it is difficult to imagine the nebulous envelope of the nucleus of the comet to be vesicular and impervious to the ether.[*]

The dissimilar eccentricity of the orbits of comets has, in recent times (1819), in the most brilliant manner enriched our knowledge of the solar system. Encke has discovered the existence of a comet of so short a period of revolution that it remains entirely within the limits of our planetary system, attaining its aphelion between the orbits of the smaller planets and that of Jupiter. Its eccentricity must be assumed at $0·845$, that of Juno (which has the greatest eccentricity of any of the planets) being $0·255$. Encke's comet has several times, although with difficulty, been observed by the naked eye, as in Europe in 1819, and, according to Rümker, in New Holland in 1822. Its period of revolution is about $3\frac{1}{3}$d years; but, from a careful comparison of the epochs of its return to its perihelion, the remarkable fact has been discovered that these periods have diminished in the most regular manner between the years 1786 and 1838, the diminution amounting, in the course of 52 years, to about $1\frac{8}{10}$th days. The attempt to bring into unison the results of observation and calculation in the investigation of all the planetary disturbances, with the view of explaining this phenomenon, has led to the adoption of the very probable hypothesis that there exists dispersed in space a vaporous substance capable of acting as a resisting medium. This matter diminishes the tangential force, and with it the major axis of the comet's orbit. The value of the constant of the resistance appears to be somewhat different before and after the perihelion; and this may, perhaps, be as

[*] Arago, in the *Annuaire*, 1832, p. 217–220. Sir John Herschel, *Astron.*, § 488.

ɛribed to the altered form of the small nebulous star in the vicinity of the Sun, and to the action of the unequal density of the strata of cosmical ether.* These facts, and the investigations to which they have led, belong to the most interesting results of modern astronomy. Encke's comet has been the means of leading astronomers to a more exact investigation of Jupiter's mass (a most important point with reference to the calculation of perturbations); and, more recently, the course of this comet has obtained for us the first determination, although only an approximative one, of a smaller mass for Mercury.

The discovery of Encke's comet, which had a period of only $3\frac{1}{3}$d years, was speedily followed, in 1826, by that of another, Biela's comet, whose period of revolution is $6\frac{3}{4}$th years, and which is likewise planetary, having its aphelion beyond the orbit of Jupiter, but within that of Saturn. It has a fainter light than Encke's comet, and, like the latter, its motion is direct, while Halley's comet moves in a course opposite to that pursued by the planets. Biela's comet presents the first certain example of the orbit of a comet intersecting that of the Earth. This position, with reference to our planet, may therefore be productive of danger, if we can associate an idea of danger with so extraordinary a natural phenomenon, whose history presents no parallel, and the results of which we are consequently unable correctly to estimate. Small masses endowed with enormous velocity may certainly exercise a considerable power; but Laplace has shown that the mass of the comet of 1770 is probably not equal to $\frac{1}{5000}$th of that of the Earth, estimating farther with apparent correctness the *mean* mass of comets as much below $\frac{1}{100000}$th that of the Earth, or about $\frac{1}{1200}$th that of the Moon.† We must not confound the passage of Biela's comet through the Earth's orbit with its proximity to, or collision with, our globe. When this passage took place, on the 29th of October, 1832, it required a full month before the Earth would reach the point of intersection of the two orbits. These two comets of short periods of revolution also intersect each other, and it has been justly observed,‡ that amid the many perturbations experienced by

* Encke, in the *Astronomische Nachrichten*, 1843, No. 489, s. 130–132.

† Laplace, *Expos. du Syst. du Monde*, p. 216, 237.

‡ Littrow, *Beschreibende Astron.*, 1835, s. 274. On the inner comet recently discovered by M. Faye, at the Observatory of Paris, and whose eccentricity is 0·551, its distance at its perihelion 1·690, and its distance at its aphelion 5·832, see Schumacher, *Astron. Nachr.*, 1844, No. 495. Regarding the supposed identity of the comet of 1766 with the third

such small bodies from the larger planets, there is a *possibility*
—supposing a meeting of these comets to occur in October—
that the inhabitants of the Earth may witness the extraordinary spectacle of an encounter between two cosmical bodies,
and possibly of their reciprocal penetration and amalgamation,
or of their destruction by means of exhausting emanations.
Events of this nature, resulting either from deflection occasioned by disturbing masses or primevally intersecting orbits,
must have been of frequent occurrence in the course of millions of years in the immeasurable regions of ethereal space ;
but they must be regarded as isolated occurrences, exercising
no more general or alterative effects on cosmical relations than
the breaking forth or extinction of a volcano within the limited sphere of our Earth.

A third interior comet, having likewise a short period of
revolution, was discovered by Faye on the 22d of November,
1843, at the Observatory at Paris. Its elliptic path, which
approaches much more nearly to a circle than that of any
other known comet, is included within the orbits of Mars and
Saturn. This comet, therefore, which, according to Gold-
schmidt, passes beyond the orbit of Jupiter, is one of the few
whose perihelia are beyond Mars. Its period of revolution is
$7\frac{29}{100}$ years, and it is not improbable that the form of its present orbit may be owing to its great approximation to Jupiter
at the close of the year 1839.

If we consider the comets in their inclosed elliptic orbits as
members of our solar system, and with respect to the length
of their major axes, the amount of their eccentricity, and their
periods of revolution, we shall probably find that the three
planetary comets of Encke, Biela, and Faye are most nearly
approached in these respects, first, by the comet discovered in
1766 by Messier, and which is regarded by Clausen as identical with the third comet of 1819 ; and, next, by the fourth
comet of the last-mentioned year, discovered by Blaupain, but
considered by Clausen as identical with that of the year 1743,
and whose orbit appears, like that of Lexell's comet, to have
suffered great variations from the proximity and attraction of
Jupiter. The two last-named comets would likewise seem to
have a period of revolution not exceeding five or six years, and
their aphelia are in the vicinity of Jupiter's orbit. Among
the comets that have a period of revolution of from seventy to

comet of 1819, see *Astr. Nachr.*, 1833, No. 239 ; and on the identity of
the comet of 1743 and the fourth comet of 1819, see No. 237 of the last
mentioned work.

seventy-six years, the first in point of importance with respect to theoretical and physical astronomy is Halley's comet, whose last appearance, in 1835, was much less brilliant than was to be expected from preceding ones; next we would notice Olbers's comet, discovered on the 6th of March, 1815; and, lastly, the comet discovered by Pons in the year 1812, and whose elliptic orbit has been determined by Encke. The two latter comets were invisible to the naked eye. We now know with certainty of nine returns of Halley's large comet, it having recently been proved by Laugier's calculations,* that in the Chinese table of comets, first made known to us by Edward Biot, the comet of 1378 is identical with Halley's; its periods of revolution have varied in the interval between 1378 and 1835 from 74·91 to 77·58 years, the mean being 76·1.

A host of other comets may be contrasted with the cosmical bodies of which we have spoken, requiring several thousand years to perform their orbits, which it is difficult to determine with any degree of certainty. The beautiful comet of 1811 requires, according to Argelander, a period of 3065 years for its revolution, and the colossal one of 1680 as much as 8800 years, according to Encke's calculation. These bodies respectively recede, therefore, 21 and 44 times further than Uranus from the Sun, that is to say, 33,600 and 70,400 millions of miles. At this enormous distance the attractive force of the Sun is still manifested; but while the velocity of the comet of 1680 at its perihelion is 212 miles in a second, that is, thirteen times greater than that of the Earth, it scarcely moves ten feet in the second when at its aphelion. This velocity is only three times greater than that of water in our most sluggish European rivers, and equal only to half that which I have observed in the Cassiquiare, a branch of the Orinoco. It is highly probable that, among the innumerable host of uncalculated or undiscovered comets, there are many whose major axes greatly exceed that of the comet of 1680. In order to form some idea by numbers, I do not say of the sphere of attraction, but of the distance in space of a fixed star, or other sun, from the aphelion of the comet of 1680 (the furthest receding cosmical body with which we are acquainted in our solar system), it must be remembered that, according to the most recent determinations of parallaxes, the nearest fixed star is full 250 times further removed from our sun than the comet in its aphelion. The comet's distance is only 44

* Laugier, in the *Comptes Rendus des Séances de l'Academie*, 1843, t. xvi., p. 1006,

times that of Uranus, while α Centauri is 11,000, and 61 Cygni 31,000 times that of Uranus, according to Bessel's determinations.

Having considered the greatest distances of comets from the central body, it now remains for us to notice instances of the greatest proximity hitherto measured. Lexell and Burckhardt's comet of 1770, so celebrated on account of the disturbances it experienced from Jupiter, has approached the Earth within a smaller distance than any other comet. On the 28th of June, 1770, its distance from the Earth was only six times that of the Moon. The same comet passed twice, viz., in 1769 and 1779, through the system of Jupiter's four satellites without producing the slightest notable change in the well-known orbits of these bodies. The great comet of 1680 approached at its perihelion eight or nine times nearer to the surface of the Sun than Lexell's comet did to that of our Earth, being on the 17th of December a sixth part of the Sun's diameter, or seven tenths of the distance of the Moon from that luminary. Perihelia occurring beyond the orbit of Mars can seldom be observed by the inhabitants of the Earth, owing to the faintness of the light of distant comets; and among those already calculated, the comet of 1729 is the only one which has its perihelion between the orbits of Pallas and Jupiter; it was even observed beyond the latter.

Since scientific knowledge, although frequently blended with vague and superficial views, has been more extensively diffused through wider circles of social life, apprehensions of the possible evils threatened by comets have acquired more weight as their direction has become more definite. The certainty that there are within the known planetary orbits comets which revisit our regions of space at short intervals—that great disturbances have been produced by Jupiter and Saturn in their orbits, by which such as were apparently harmless have been converted into dangerous bodies—the intersection of the Earth's orbit by Biela's comet—the cosmical vapor, which, acting as a resisting and impeding medium, tends to contract all orbits —the individual difference of comets, which would seem to indicate considerable decreasing gradations in the quantity of the mass of the nucleus, are all considerations more than equivalent, both as to number and variety, to the vague fears entertained in early ages of the general conflagration of the world by *flaming swords*, and stars with *fiery streaming hair*. As the consolatory considerations which may be derived from the calculus of probabilities address themselves to reason and to meditative understanding only, and not to the imagination or to a desponding condition of mind, modern science has been accused, and not entirely without reason, of not attempting to allay apprehensions which it has been the very means of exciting. It is an inherent attribute of the human mind to experience fear, and not hope or joy, at the aspect of that which is unexpected and extraordinary.* The strange form of a large comet, its faint nebulous light, and its sudden appearance in the vault of heaven, have in all regions been almost invariably regarded by the people at large as some new and formidable agent inimical to the existing state of things. The sudden occurrence and short duration of the phenomenon lead to the belief of some equally rapid reflection of its agency in terrestrial matters, whose varied nature renders it easy to find events that may be regarded as the fulfillment of the evil foretold by the appearance of these mysterious cosmical bodies. In our own day, however, the public mind has taken another and more cheerful, although singular, turn with regard to comets; and in the German vineyards in the beautiful valleys of the Rhine and Moselle, a belief has arisen, ascribing to these once ill-omened bodies a beneficial influence on the ripening of the vine. The evidence yielded by experience, of which there is no lack in these days, when comets may so frequently be observed, has not been able to shake the common belief in the meteorological myth of the existence of wandering stars capable of radiating heat.

From comets I would pass to the consideration of a far more enigmatical class of agglomerated matter—the smallest of all asteroids, to which we apply the name *aërolites*, or *meteoric stones*,† when they reach our atmosphere in a fragmentary condition. If I should seem to dwell on the specific enumeration of these bodies, and of comets, longer than the general nature of this work might warrant, I have not done so undesignedly. The diversity existing in the individual characteristics of comets has already been noticed. The imperfect knowledge we possess of their physical character renders it

---

* Fries, *Vorlesungen über die Sternkunde*, 1833, s. 262–267 (Lectures on the Science of Astronomy). An infelicitously chosen instance of the good omen of a comet may be found in Seneca, *Nat. Quæst.*, vii., 17 and 21. The philosopher thus writes of the comet: " *Quem nos Neronis principatu lætissimo vidimus et qui cometis detraxit infamiam.*"

† [Much valuable information may be obtained regarding the origin and composition of aërolites or meteoric stones in Memoirs on the subject, by Baumbeer and other writers, in the numbers of Poggendorf's *Annalen*, from 1845 to the present time.]—*Tr.*

difficult, in a work like the present, to give the proper degree
of circumstantiality to the phenomena, which, although of
frequent recurrence, have been observed with such various degrees of accuracy, or to separate the necessary from the accidental.   It is only with respect to measurements and compu-
tations that the astronomy of comets has made any marked
advancement, and, consequently, a scientific consideration of
these bodies must be limited to a specification of the differences
of physiognomy and conformation in the nucleus and tail, the
instances of great approximation to other cosmical bodies, and
of the extremes in the length of their orbits and in their periods
of revolution.   A faithful delineation of these phenomena, as
well as of those which we proceed to consider, can only be
given by sketching individual features with the animated circumstantiality of reality.

Shooting stars, fire-balls, and meteoric stones are, with great
probability, regarded as small bodies moving with planetary
velocity, and revolving in obedience to the laws of general
gravity in conic sections round the Sun.   When these masses
meet the Earth in their course, and are attracted by it, they
enter within the limits of our atmosphere in a luminous condition, and frequently let fall more or less strongly heated stony
fragments, covered with a shining black crust.   When we
enter into a careful investigation of the facts observed at those
epochs when showers of shooting stars fell periodically in Cumana in 1799, and in North America during the years 1833
and 1834, we shall find that *fire-balls* can not be considered
separately from shooting stars.   Both these phenomena are
frequently not only simultaneous and blended together, but
they likewise are often found to merge into one another, the
one phenomenon gradually assuming the character of the other
alike with respect to the size of their disks, the emanation of
sparks, and the velocities of their motion.   Although explod-
ing smoking luminous fire-balls are sometimes seen, even in
the brightness of tropical daylight,* equaling in size the ap-

---

* A friend of mine, much accustomed to exact trigonometrical measurements, was in the year 1788 at Popayan, a city which is 2° 26'
north latitude, lying at an elevation of 5583 feet above the level of the
sea, and at noon, when the sun was shining brightly in a cloudless sky,
saw his room lighted up by a fire-ball.   He had his back to the window
at the time, and on turning round, perceived that great part of the path
traversed by the fire-ball was still illuminated by the brightest radiance.
Different nations have had the most various terms to express these phenomena: the Germans use the word *Sternschnuppe*, literally *star snuff*
—an expression well suited to the physical views of the vulgar in former

parent diameter of the Moon, innumerable quantities of shooting stars have, on the other hand, been observed to fall in
forms of such extremely small dimensions that they appear
only as moving points or *phosphorescent lines.**

It still remains undetermined whether the many luminous
bodies that shoot across the sky may not vary in their nature.
On my return from the equinoctial zones, I was impressed
with an idea that in the torrid regions of the tropics I had
more frequently than in our colder latitudes seen shooting
stars fall as if from a height of twelve or fifteen thousand feet ;
that they were of brighter colors, and left a more brilliant line
of light in their track ; but this impression was no doubt owing
to the greater transparency of the tropical atmosphere,† which

times, according to which, the lights in the firmament were said to under
go a process of *snuffing* or cleaning ; and other nations generally adopt a
term expressive of a *shot* or *fall* of stars, as the Swedish *stjernffall*, the
Italian *stella cadente*, and the English *star shoot.* In the woody district
of the Orinoco, on the dreary banks of the Cassiquiare, I heard the natives in the Mission of Vasiva use terms still more inelegant than the
German *star snuff.* (*Relation Historique du Voy. aux Régions Equinox.*,
t. ii., p. 513.) These same tribes term the pearly drops of dew which
cover the beautiful leaves of the heliconia *star spit.* In the Lithuanian
mythology, the imagination of the people has embodied its ideas of the
nature and signification of falling stars under nobler and more graceful
symbols. The Parcæ, *Werpeja*, weave in heaven for the new-born
child its thread of fate, attaching each separate thread to a star. When
death approaches the person, the thread is rent, and the star wanes and
sinks to the earth. Jacob Grimm, *Deutsche Mythologie*, 1843, s. 685.

* According to the testimony of Professor Denison Olmsted, of Yale
College, New Haven, Connecticut. (See Poggend., *Annalen der Physik*,
bd. xxx., s. 194.) Kepler, who excluded fire-balls and shooting stars
from the domain of astronomy, because they were, according to his
views, "meteors arising from the exhalations of the earth, and blending with the higher ether," expresses himself, however, generally with
much caution. He says : " *Stellæ cadentes sunt materia viscida inflam-
mata. Earum aliquæ inter cadendum absumuntur, aliquæ verè in terram
cadunt, pondere suo tractæ. Nec est dissimile vero, quasdam conglobatas
esse ex materia fæculentâ, in ipsam auram æthereum immixta : exque
ætheris regione, tractu rectilineo, per aërem trajicere, ceu minutos com-
etas, occultâ causa motus utrorumque.*"—Kepler, *Epit. Astron. Copernicanæ*, t. i., p. 80.

† *Relation Historique*, t. i., p. 80, 213, 527. If in falling stars, as in
comets, we distinguish between the head or nucleus and the tail, we
shall find that the greater transparency of the atmosphere in tropical
climates is evinced in the greater length and brilliancy of the tail which
may be observed in those latitudes. The phenomenon is therefore not
necessarily more frequent there, because it is oftener seen and continues longer visible. The influence exercised on shooting stars by the
character of the atmosphere is shown occasionally even in our temperate zone, and at very small distances apart. Wartmann relates that on
the occasion of a November phenomenon at two places lying very near

enables the eye to penetrate further into distance.   Sir Alex·
ander Burnes likewise extols as a consequence of the purity of
the atmosphere in Bokhara the enchanting and constantly-re-
curring spectacle of variously-colored shooting stars.
    The connection of meteoric stones with the grander phe-
nomenon of fire-balls—the former being known to be projected from the latter with such force as to penetrate from ten
to fifteen feet into the earth—has been proved, among many
other instances, in the falls of aërolites at Barbotan, in the
Department des Landes (24th July, 1790), at Siena (16th
June, 1794), at Weston, in Connecticut, U. S. (14th December, 1807), and at Juvenas, in the Department of Ardèche
(15th June, 1821).   Meteoric stones are in some instances
thrown from dark clouds suddenly formed in a clear sky, and
fall with a noise resembling thunder.   Whole districts have
thus occasionally been covered with thousands of fragmentary
masses, of uniform character but unequal magnitudes, that

each other, Geneva and Aux Planchettes, the number of the meteors
counted were as 1 to 7.   (Wartmann, *Mém. sur les Etoiles filantes*, p.
17.)   The tail of a shooting star (or its *train*), on the subject of which
Brandes has made so many exact and delicate observations, is in no
way to be ascribed to the continuance of the impression produced by
light on the retina.   It sometimes continues visible a whole minute,
and in some rare instances longer than the light of the nucleus of the
shooting star; in which case the luminous track remains motionless.
(Gilb., *Ann.*, bd. xiv., s. 251.)   This circumstance further indicates the
analogy between large shooting stars and fire-balls.   Admiral Krusenstern saw, in his voyage round the world, the train of a fire-ball shine
for an hour after the luminous body itself had disappeared, and scarcely move throughout the whole time.   (*Reise*, th. i., s. 58.)   Sir Alexander Burnes gives a charming description of the transparency of the
clear atmosphere of Bokhara, which was once so favorable to the pursuit of astronomical observations.   Bokhara is situated in 39° 43' north
latitude, and at an elevation of 1280 feet above the level of the sea.
"There is a constant serenity in its atmosphere, and an admirable clearness in the sky.   At night, the stars have uncommon luster, and the
Milky Way shines gloriously in the firmament.   There is also a never-
ceasing display of the most brilliant meteors, which dart like rockets
in the sky; ten or twelve of them are sometimes seen in an hour, assuming every color—fiery red, blue, pale, and faint.   It is a noble
country for astronomical science, and great must have been the advantage enjoyed by the famed observatory of Samarkand."   (Burnes,
Travels into Bokhara, vol. ii. (1834), p. 158.)   A mere traveler must
not be reproached for calling ten or twelve shooting stars in an hour
"many," since it is only recently that we have learned, from careful
observations on this subject in Europe, that eight is the mean number
which may be seen in an hour in the field of vision of one individual
(Quetelet, *Corresp. Mathém.*, Novem., 1837, p. 447); this number is,
however, limited to five or six by that diligent observer, Olbers.
(Schum., *Jahrb.*, 1838, s. 325.)

nave been hurled from one of these moving clouds. In less
frequent cases, as in that which occurred on the 16th of Sep·
tember, 1843, at Kleinwenden, near Mühlhausen, a large
aërolite fell with a thundering crash while the sky was clear
and cloudless. The intimate affinity between fire-balls and
shooting stars is further proved by the fact that fire-balls, from
which meteoric stones have been thrown, have occasionally
been found, as at Angers, on the 9th of June, 1822, having a
liameter scarcely equal to that of the small fire-works called
Roman candles.

The formative power, and the nature of the physical and
·hemical processes involved in these phenomena, are questions
ıll equally shrouded in mystery, and we are as yet ignorant
whether the particles composing the dense mass of meteoric
₁tones are originally, as in comets, separated from one another
m the form of vapor, and only condensed within the fiery ball
when they become luminous to our sight, or whether, in the
ₑase of smaller shooting stars, any compact substance actually
falls, or, finally, whether a meteor is composed only of a smoke-
like dust, containing iron and nickel; while we are wholly
ignorant of what takes place within the dark cloud from which
a noise like thunder is often heard for many minutes before
the stones fall.*

* On *meteoric dust*, see Arago, in the *Annuaire* for 1832, p. 254. I
have very recently endeavored to show, in another work (*Asie Centrale*,
t. i., p. 408), how the Scythian saga of the sacred gold, which fell burning from heaven, and remained in the possession of the Golden Horde
of the Paralatæ (Herod., iv., 5–7), probably originated in the vague recollection of the fall of an aërolite. The ancients had also some strange
fictions (Dio Cassius, lxxv., 1259) of silver which had fallen from heav-
en, and with which it had been attempted, under the Emperor Seve-
rus, to cover bronze coins; metallic iron was, however, known to exist
in meteoric stones. (Plin., ii., 56.) The frequently-recurring expression *lapidibus pluit* must not always be understood to refer to falls of
aërolites. In Liv., xxv., 7, it probably refers to pumice (*rapilli*) ejected from the volcano, Mount Albanus (Monte Cavo), which was not
wholly extinguished at the time. (See Heyne, *Opuscula Acad.*, t. iii.,
p. 261; and my *Relation Hist.*, t. i., p. 394.) The contest of Hercules
with the Ligyans, on the road from the Caucasus to the Hesperides,
belongs to a different sphere of ideas, being an attempt to explain myth-
ically the origin of the round quartz blocks in the Ligyan field of stones
at the mouth of the Rhone, which Aristotle supposes to have been ejected from a fissure during an earthquake, and Posidonius to have been
caused by the force of the waves of an inland piece of water. In the
fragments that we still possess of the play of Æschylus, the *Prometheus
Delivered*, every thing proceeds, however, in part of the narration, as
in a fall of aërolites, for Jupiter draws together a cloud, and causes the
"district around to be covered by a shower of round stones" Posido-

We can ascertain by measurement the enormous, wonderful, and wholly planetary velocity of shooting stars, fire-balls,
and meteoric stones, and we can gain a knowledge of what is
the general and uniform character of the phenomenon, but
not of the genetically cosmical process and the results of the
metamorphoses. If meteoric stones while revolving in space
are already consolidated into dense masses,* less dense, how-

nius even ventured to deride the geognostic myth of the blocks and
stones. The Lygian field of stones was, however, very naturally and
well described by the ancients. The district is now known as *La Crau.*
(See Guerin, *Mesures Barométriques dans les Alpes, et Météorologie
d'Avignon,* 1829, chap. xii., p. 115.)

* The specific weight of aërolites varies from 1·9 (Alais) to 4·3
(Tabor). Their general density may be set down as 3, water being 1.
As to what has been said in the text of the actual diameters of fire-balls,
we must remark, that the numbers have been taken from the few
measurements that can be relied upon as correct. These give for the
fire-ball of Weston, Connecticut (14th December, 1807), only 500; for
that observed by Le Roi (10th July, 1771) about 1000, and for that
estimated by Sir Charles Blagden (18th January, 1783) 2600 feet in
diameter. Brandes (*Unterhaltungen,* bd. i., s. 42) ascribes a diameter
varying from 80 to 120 feet to shooting stars, and a luminous train extending from 12 to 16 miles. There are, however, ample optical causes for supposing that the apparent diameter of fire-balls and shooting
stars has been very much overrated. The volume of the largest fire-
ball yet observed can not be compared with that cf Ceres, estimating
this planet to have a diameter of only 7J English miles. (See the
generally so exact and admirable treatise, *On the Connection of the
Physical Sciences,* 1835, p. 411.) With the view of elucidating what
has been stated in the text regarding the large aërolite that fell into
the bed of the River Narni, but has not again been found, I will give
the passage made known by Pertz, from the *Chronicon Benedicti, Mon-
achi Sancti Andreæ in Monte Soracte,* a MS. belonging to the tenth
century, and preserved in the Chigi Library at Rome. The barbarous
Latin of that age has been left unchanged. "*Anno* 921, *temporibus
domini Johannis Decimi pape, in anno pontificatus illius* 7 *visa sunt sig-
na. Nam juxta urbem Romam lapides plurimi de cælo cadere visi sunt.
In civitate quæ vocatur Narnia tam diri ac tetri, ut nihil aliud credatur,
quam de infernalibus locis deducti essent. Nam ita ex illis lapidibus
unus omnium maximus est, ut dicidens in flumen Narnus, ad mensuram
unius cubiti super aquas flumini usque hodie videretur. Nam et ignitæ
faculæ de cælo plurimæ omnibus in hac civitate Romani populi visæ sunt,
ita ut pene terra contingeret. Aliæ cadentes,*" &c. (Pertz, *Monum.
Germ. Hist. Scriptores,* t. iii., p. 715.) On the aërolites of Ægos Potamos, which fell, according to the Parian Chronicle, in the 78 1 Olympiad, see Böckh, *Corp. Inscr. Graec.,* t. ii., p. 302, 320, 340; also Aristot., *Meteor.,* i., 7 (Ideler's *Comm.,* t. i., p. 404–407); Stob., *Ecl. Phys.,*
i., 25, p. 508 (Heeren); Plut., *Lys.,* c. 12; Diog. Laert., ii., 10; and
see, also, subsequent notes in this work. According to a Mongolian
tradition, a black fragment of a rock, forty feet in height, fell from
heaven on a plain near the source of the Great Yellow River in Western China. (Abel Rémusat, in Lamétherie, *Jour. de Phys.,* 1819, Mai
p. 264.)

ever, than the mean density of the earth, they must be very small nuclei, which, surrounded by inflammable vapor or gas, form the innermost part of fire-balls, from the height and apparent diameter of which we may, in the case of the largest, estimate that the actual diameter varies from 500 to about 2800 feet. The largest meteoric masses as yet known are those of Otumpa, in Chaco, and of Bahia, in Brazil, described by Rubi de Celis as being from 7 to 7½ feet in length. The meteoric stone of Ægos Potamos, celebrated in antiquity, and even mentioned in the Chronicle of the Parian Marbles, which fell about the year in which Socrates was born, has been described as of the size of two mill-stones, and equal in weight to a full wagon load. Notwithstanding the failure that has attended the efforts of the African traveler, Brown, I do not wholly relinquish the hope that, even after the lapse of 2312 years, this Thracian meteoric mass, which it would be so difficult to destroy, may be found, since the region in which it fell is now become so easy of access to European travelers. The huge aërolite which in the beginning of the tenth century fell into the river at Narni, projected between three and four feet above the surface of the water, as we learn from a document lately discovered by Pertz. It must be remarked that these meteoric bodies, whether in ancient or modern times, can only be regarded as the principal fragments of masses that have been broken up by the explosion either of a fire-ball or a dark cloud.

On considering the enormous velocity with which, as has been mathematically proved, meteoric stones reach the earth from the extremest confines of the atmosphere, and the lengthened course traversed by fire-balls through the denser strata of the air, it seems more than improbable that these metalliferous stony masses, containing perfectly-formed crystals of olivine, labradorite, and pyroxene, should in so short a period of time have been converted from a vaporous condition to a solid nucleus. Moreover, that which falls from meteoric masses, even where the internal composition is chemically different, exhibits almost always the peculiar character of a fragment, being of a prismatic or truncated pyramidal form, with broad, somewhat curved faces, and rounded angles. But whence comes this form, which was first recognized by Schreiber as characteristic of the *severed* part of a rotating planetary body? Here, as in the sphere of organic life, all that appertains to the history of development remains hidden in obscurity. Meteoric masses become luminous and kindle at heights which must be regarded as almost devoid of air, or occupied by an atmosphere that does not even contain $\frac{1}{100000}$th part of oxy gen. The recent investigations of Biot on the important phe nomenon of twilight* have considerably lowered the lines which had, perhaps with some degree of temerity, been usual ly termed the boundaries of the atmosphere; but processes of light may be evolved independently of the presence of oxygen, and Poisson conjectured that aërolites were ignited far beyond the range of our atmosphere. Numerical calculation and geo metrical measurement are the only means by which, as in the case of the larger bodies of our solar system, we are enabled to impart a firm and safe basis to our investigations of meteoric stones. Although Halley pronounced the great fire-ball of 1686, whose motion was opposite to that of the earth in its orbit,† to be a cosmical body, Chladni, in 1794, first recognized, with ready acuteness of mind, the connection between fire-balls and the stones projected from the atmosphere, and the motions of the former bodies in space.‡ A brilliant confirmation of the cos mical origin of these phenomena has been afforded by Denison Olmsted, at New Haven, Connecticut, who has shown, on the concurrent authority of all eye-witnesses, that during the cele brated fall of shooting stars on the night between the 12th

---

* Biot, *Traité d'Astronomie Physique* (3ème éd.), 1841, t. i., p. 149, 177, 238, 312. My lamented friend Poisson endeavored, in a singular manner, to solve the difficulty attending an assumption of the sponta neous ignition of meteoric stones at an elevation where the density of the atmosphere is almost null. These are his words : " It is difficult to attribute, as is usually done, the incandescence of aërolites to friction against the molecules of the atmosphere at an elevation above the earth where the density of the air is almost null. May we not suppose that the electric fluid, in a neutral condition, forms a kind of atmosphere, ex tending far beyond the mass of our atmosphere, yet subject to terres trial attraction, although physically imponderable, and consequently following our globe in its motion? According to this hypothesis, the bodies of which we have been speaking would, on entering this im ponderable atmosphere, decompose the neutral fluid by their unequal action on the two electricities, and they would thus be heated, and in a state of incandescence, by becoming electrified." (Poisson, *Rech. sur la Probabilité des Jugements*, 1837, p. 6.)

† *Philos. Transact.*, vol. xxix., p. 161–163.

‡ The first edition of Chladni's important treatise, *Ueber den Ur sprung der von Pallas gefundenen und anderen Eisenmassen* (On the Origin of the masses of Iron found by Pallas, and other similar masses), appeared two months prior to the shower of stones at Siena, and two years before Lichtenberg stated, in the *Göttingen Taschenbuch*, tha " stones reach our atmosphere from the remoter regions of space.' Comp., also, Olbers's letter to Benzenberg, 18th Nov., 1837, in Ben zenberg's *Treatise on Shooting Stars*, p. 186.

and 13th of November, 1833, the fire-balls and shooting stars all emerged from one and the same quarter of the heavens, namely, in the vicinity of the star γ in the constellation Leo, and did not deviate from this point, although the star changed its apparent height and azimuth during the time of the observation. Such an independence of the Earth's rotation shows that the luminous body must have reached our atmosphere from *without*. According to Encke's computation* of the whole

* Encke, in Poggend., *Annalen*, bd. xxxiii. (1834), s. 213. Arago, in the *Annuaire* for 1836, p. 291. Two letters which I wrote to Benzenberg, May 19 and October 22, 1837, on the conjectural precession of the nodes in the orbit of periodical falls of shooting stars. (Benzenberg's *Sternsch.*, s. 207 and 209.) Olbers subsequently adopted this opinion of the gradual retardation of the November phenomenon. (*Astron. Nachr.*, 1838, No. 372, s. 180.) If I may venture to combine two of the falls of shooting stars mentioned by the Arabian writers with the epochs found by Boguslawski for the fourteenth century, I obtain the following more or less accordant elements of the movements of the nodes :

In Oct., 902, on the night in which King Ibrahim ben Ahmed died, there fell a heavy shower of shooting stars, "like a fiery rain;" and this year was, therefore, called the year of stars. (Conde, *Hist. de la Domin. de los Arabes*, p. 346.)

On the 19th of Oct., 1202, the stars were in motion all night. "They fell like locusts." (*Comptes Rendus*, 1837, t. i., p. 294; and Fræhn, in the *Bull. de l'Académie de St. Pétersbourg*, t. iii., p. 308.)

On the 21st Oct., O.S., 1366, "*die sequente post festum XI. millia Virginum ab hora matutina usque ad horam primam visæ sunt quasi stellæ de cœlo cadere continuo, et in tanta multitudine, quod nemo narrare suf ficit.*" This remarkable notice, of which we shall speak more fully in the subsequent part of this work, was found by the younger Von Boguslawski, in Benesse (de Horowic) de Weitmil or Weithmül, *Chron. icon Ecclesiæ Pragensis*, p. 389. This chronicle may also be found in the second part of *Scriptores rerum Bohemicarum*, by Pelzel and Dobrowsky, 1784. (Schum., *Astr. Nachr.*, Dec., 1839.)

On the night between the 9th and 10th of November, 1787, many falling stars were observed at Manheim, Southern Germany, by Hemmer (Kämtz, *Meteor.*, th. iii., s. 237.)

After midnight, on the 12th of November, 1799, occurred the extraordinary fall of stars at Cumana, which Bonpland and myself have described, and which was observed over a great part of the earth. (*Relat Hist.*, t. i., p. 519-527.)

Between the 12th and 13th of November, 1822, shooting stars, intermingled with fire-balls, were seen in large numbers by Kloden, at Potsdam. (Gilbert's *Ann.*, bd. lxxii., s. 291.)

On the 13th of November, 1831, at 4 o'clock in the morning, a great shower of falling stars was seen by Captain Bérard, on the Spanish coast, near Carthagena del Levante. (*Annuaire*, 1836, p. 297.)

In the night between the 12th and 13th of November, 1833, occurred the phenomenon so admirably described by Professor Olmsted, in North America.

In the night of the 13-14th of November, 1834, a similar fall of shoot-

number of observations made in the United States of North America, between the thirty-fifth and the forty-second degrees of latitude, it would appear that all these meteors came from the same point of space in the direction in which the Earth was moving at the time. On the recurrence of falls of shoot-ing stars in North America, in the month of November of the years 1834 and 1837, and in the analogous falls observed at Bremen in 1838, a like general parallelism of the orbits, and the same direction of the meteors from the constellation Leo, were again noticed. It has been supposed that a greater parallelism was observable in the direction of periodic falls of shooting stars than in those of sporadic occurrence; and it has further been remarked, that in the periodically-recurring falls in the month of August, as, for instance, in the year 1839, the meteors came principally from one point between Perseus and Taurus, toward the latter of which constellations the Earth was then moving. This peculiarity of the phenomenon, mani-fested in the retrograde direction of the orbits in November and August, should be thoroughly investigated by accurate observations, in order that it may either be fully confirmed or refuted.

The heights of shooting stars, that is to say, the heights of the points at which they begin and cease to be visible, vary exceedingly, fluctuating between 16 and 140 miles. This important result, and the enormous velocity of these problem-atical asteroids, were first ascertained by Benzenberg and Brandes, by simultaneous observations and determinations of parallax at the extremities of a base line of 49,020 feet in length.* The relative velocity of motion is from 18 to 36 miles in a second, and consequently equal to planetary velocity. This planetary velocity,† as well as the direction of the orbits

ing stars was seen in North America, although the numbers were not quite so considerable. (Poggend., *Annalen*, bd. xxxiv., s. 129.)

On the 13th of November, 1835, a barn was set on fire by the fall of a sporadic fire-ball, at Belley, in the Department de l'Ain. (*Annuaire*, 1836, p. 296.)

In the year 1838, the stream showed itself most decidedly on the night of the 13–14th of November. (*Astron. Nachr.*, 1838, No. 372.)

* I am well aware that, among the 62 shooting stars simultaneously observed in Silesia, in 1823, at the suggestion of Professor Brandes some appeared to have an elevation of 183 to 240, or even 400 miles. (Brandes, *Unterhaltungen für Freunde der Astronomie und Physik*, heft i., s. 48. Instructive Narratives for the Lovers of Astronomy and Phys-ics.) But Olbers considered that all determinations for elevations be-yond 120 miles must be doubtful, owing to the smallness of the parallax.

† The planetary velocity of translation, the movement in the orbit, is in Mercury 26·4, in Venus 19·2, and in the Earth 16·4 miles in a second.

of fire-balls and shooting stars, which has frequently been observed to be opposite to that of the Earth, may be considered
as conclusive arguments against the hypothesis that aërolites
derive their origin from the so-called active *lunar volcanoes.*
Numerical views regarding a greater or lesser volcanic force
on a small cosmical body, not surrounded by any atmosphere,
must, from their nature, be wholly arbitrary. We may imagine the reaction of the interior of a planet on its crust ten or
even a hundred times greater than that of our present terrestrial volcanoes ; the direction of masses projected from a satellite revolving from west to east might appear retrogressive,
owing to the Earth in its orbit subsequently reaching that
point of space at which these bodies fall. If we examine the
whole sphere of relations which I have touched upon in this
work, in order to escape the charge of having made unproved
assertions, we shall find that the hypothesis of the selenic origin of meteoric stones* depends upon a number of conditions

* Chladni states that an Italian physicist, Paolo Maria Terzago, on
the occasion of the fall of an aërolite at Milan in 1660, by which a Fran-
ciscan monk was killed, was the first who surmised that aërolites were
of selenic origin. He says, in a memoir entitled *Musæum Septalianum,
Manfredi Septalæ, Patricii Mediolanensis, industrioso labore constructum*
(Tortona, 1664, p. 44), *"Labant philosophorum mentes sub horum lapidum
ponderibus; ni dicire velimus, lunam terram alteram, sine mundum esse,
ex cujus montibus divisa frustra in inferiorem nostrum hunc orbem dela
bantur."* Without any previous knowledge of this conjecture, Olbers
was led, in the year 1795 (after the celebrated fall at Siena on the 16th
of June, 1794), into an investigation of the amount of the initial tangential force that would be requisite to bring to the Earth masses projected from the Moon. This ballistic problem occupied, during ten or
twelve years, the attention of the geometricians Laplace, Biot, Brandes,
and Poisson. The opinion which was then so prevalent, but which has
since been abandoned, of the existence of active volcanoes in the Moon,
where air and water are absent, led to a confusion in the minds of the
generality of persons between mathematical possibilities and physical
probabilities. Olbers, Brandes, and Chladni thought "that the velocity
of 16 to 32 miles, with which fire-balls and shooting stars entered our
atmosphere," furnished a refutation to the view of their selenic origin.
According to Olbers, it would require to reach the Earth, setting aside
the resistance of the air, an initial velocity of 8292 feet in the second ;
according to Laplace, 7862 ; to Biot, 8282; and to Poisson, 7595.   Laplace states that this velocity is only five or six times as great as that of
a cannon ball; but Olbers has shown "that, with such an initial velocity as 7500 or 8000 feet in a second, meteoric stones would arrive at the
surface of our earth with a velocity of only 35,000 feet (or 1·53 German
geographical mile). But the measured velocity of meteoric stones av-
erages five such miles, or upward of 114,000 feet to a second ; and,
consequently, the original velocity of projection from the Moon must
be almost 110,000 feet, and therefore fourteen times greater than Laplace asserted." (Olbers, in Schum., *Jahrb.*, 1837, p. 52–58; and in

whose accidental coincidence could alone convert a possible into an actual fact. The view of the original existence of

Gehler, *Neues Physik. Wörterbuche*, bd. vi., abth. 3, s. 2129–2136.) If we could assume volcanic forces to be still active on the Moon's surface, the absence of atmospheric resistance would certainly give to their projectile force an advantage over that of our terrestrial volcanoes; but even in respect to the measure of the latter force (the projectile force of our own volcanoes), we have no observations on which any reliance can be placed, and it has probably been exceedingly overrated. Dr. Peters, who accurately observed and measured the phenomena presented by Ætna, found that the greatest velocity of any of the stones projected from the crater was only 1250 feet to a second. Observations on the Peak of Teneriffe, in 1798, gave 3000 feet. Although Laplace, at the end of his work (*Expos. du Syst. du Monde*, ed. de 1824, p. 399), cautiously observes, regarding aërolites, "that in all probability they come from the depths of space," yet we see from another passage (chap. vi., p. 233) that, being probably unacquainted with the extraordinary planetary velocity of meteoric stones, he inclines to the hypothesis of their lunar origin, always, however, assuming that the stones projected from the Moon "become satellites of our Earth, describing around it more or less eccentric orbits, and thus not reaching its atmosphere until several or even many revolutions have been accomplished." As an Italian at Tortona had the fancy that aërolites came from the Moon, so some of the Greek philosophers thought they came from the Sun. This was the opinion of Diogenes Laertius (ii., 9) regarding the origin of the mass that fell at Ægos Potamos (see note, p. 116). Pliny, whose labors in recording the opinions and statements of preceding writers are astonishing, repeats the theory, and derides it the more freely, because he, with earlier writers (Diog. Laert., 3 and 5, p. 99, Hübner), accuses Anaxagoras of having predicted the fall of aërolites from the Sun: "Celebrant Græci Anaxagoram Clazomenium Olympiadis septuagesimæ octavæ secundo anno prædixisse cælestium litterarum scientia, quibus diebus saxum casurum esse e sole, idque factum interdiu in Thraciæ parte ad Ægos flumen. Quod si quis prædictum credat, simul fateatur necesse est, majoris miraculi divinitatem Anaxagoræ fuisse, solvique rerum naturæ intellectum, et confundi omnia, si aut ipse Sol lapis esse aut unquam lapidem in eo fuisse credatur; decidere tamen crebro non erit dubium." The fall of a moderate-sized stone, which is preserved in the Gymnasium at Abydos, is also reported to have been foretold by Anaxagoras. The fall of aërolites in bright sunshine, and when the Moon's disk was invisible, probably led to the idea of sun-stones. Moreover, according to one of the physical dogmas of Anaxagoras, which brought on him the persecution of the theologians (even as they have attacked the geologists of our own times), the Sun was regarded as "a molten fiery mass" ($μύδρος \ διάπυρος$). In accordance with these views of Anaxagoras, we find Euripides, in *Phaëton*, terming the Sun "a golden mass;" that is to say, a fire-colored, brightly-shining matter, but not leading to the inference that aërolites are golden sun-stones. (See note to page 115.) Compare Valckenaer, *Diatribe in Eurip. perd. Dram. Reliquias*, 1767, p. 30. Diog. Laert., ii., 40. Hence, among the Greek philosophers, we find four hypotheses regarding the origin of falling stars: a telluric origin from ascending exhalations; masses of stone raised by hurricane (see Aristot., *Meteor.*, lib. i., cap. iv., 2–13, and cap. vii., 9); a solar origin; and, lastly, an

small planetary masses in space is simpler, and, at the same
time, more analogous with those entertained concerning the
formation of other portions of the solar system.

It is very probable that a large number of these cosmical
bodies traverse space undestroyed by the vicinity of our atmosphere, and revolve round the Sun without experiencing
any alteration but a slight increase in the eccentricity of their
orbits, occasioned by the attraction of the Earth's mass.   We
may, consequently, suppose the possibility of these bodies remaining invisible to us during many years and frequent revolutions.   The supposed phenomenon of ascending shooting
stars and fire-balls, which Chladni has unsuccessfully endeavored to explain on the hypothesis of the *reflection* of strongly
compressed air, appears at first sight as the consequence of
some unknown tangential force propelling bodies from the
earth; but Bessel has shown by theoretical deductions, confirmed by Feldt's carefully-conducted calculations, that, owing
to the absence of any proofs of the simultaneous occurrence
of the observed disappearances, the assumption of an ascent
of shooting stars was rendered wholly improbable, and inad-
missible as a result of observation.*   The opinion advanced
by Olbers that the explosion of shooting stars and ignited fire-
balls not moving in straight lines may impel meteors upward
in the manner of rockets, and influence the direction of their
orbits, must be made the subject of future researches.

Shooting stars fall either separately and in inconsiderable
numbers, that is, sporadically, or in swarms of many thou-

origin in the regions of space, as heavenly bodies which had long remained invisible. Respecting this last opinion, which is that of Diogenes of Apollonia, and entirely accords with that of the present day,
see pages 124 and 125. It is worthy of remark, that in Syria, as I have
been assured by a learned Orientalist, now resident at Smyrna, Andrea
de Nericat, who instructed me in Persian, there is a popular belief that
aërolites chiefly fall on clear moonlight nights. The ancients, on the
contrary, especially looked for their fall during lunar eclipses. (See
Pliny, xxxvii., 10, p. 164. Solinus, c. 37. Salm., *Exerc.*, p. 531; and
the passages collected by Ukert, in his *Geogr. der Griechen und Römer*,
th. ii., 1, s. 131, note 14.) On the improbability that meteoric masses
are formed from metal-dissolving gases, which, according to Fusinieri,
may exist in the highest strata of our atmosphere, and, previously diffused through an almost boundless space, may suddenly assume a solid
condition, and on the penetration and misceability of gases, see my
Relat. Hist., t. i., p. 525.

* Bessel, in Schum., *Astr. Nachr.*, 1839, No 380 und 381, s. 222 und
346. At the conclusion of the Memoir there is a comparison of the
Sun's longitudes with the epochs of the November phenomenon, from
the period of the first observations in Cumana in 1799.

sands. The latter, which are compared by Arabian authors to swarms of locusts, are periodic in their occurrence, and move in streams, generally in a parallel direction. Among periodic falls, the most celebrated are that known as the November phenomenon, occurring from about the 12th to the 14th of November, and that of the festival of St. Lawrence (the 10th of August), whose "fiery tears" were noticed in former times in a church calendar of England, no less than in old traditionary legends, as a meteorological event of constant recurrence.* Notwithstanding the great quantity of shooting stars and fire-balls of the most various dimensions, which, according to Klöden, were seen to fall at Potsdam on the night between the 12th and 13th of November, 1822, and on the same night of the year in 1832 throughout the whole of Europe, from Portsmouth to Orenburg on the Ural River, and even in the southern hemisphere, as in the Isle of France, no attention was directed to the *periodicity* of the phenomenon, and no idea seems to have been entertained of the connection existing between the fall of shooting stars and the recurrence of certain days, until the prodigious swarm of shooting stars which occurred in North America between the 12th and 13th of November, 1833, and was observed by Olmsted and Palmer. The stars fell, on this occasion, like flakes of snow, and it was calculated that at least 240,000 had fallen during a period of nine hours. Palmer, of New Haven, Connecticut, was led, in consequence of this splendid phenomenon, to the recollection of the fall of meteoric stones in 1799, first described by Ellicot and myself,† and which, by

* Dr. Thomas Forster (*The Pocket Encyclopedia of Natural Phenomena*, 1827, p. 17) states that a manuscript is preserved in the library of Christ's College, Cambridge,ᵃ written in the tenth century by a monk, and entitled *Ephemerides Rerum Naturalium*, in which the natural phenomena for each day of the year are inscribed, as, for instance, the first flowering of plants, the arrival of birds, &c.; the 10th of August is distinguished by the word "meteorodes." It was this indication, and the tradition of the fiery tears of St. Lawrence, that chiefly induced Dr. Forster to undertake his extremely zealous investigation of the August phenomena. (Quetelet, *Correspond. Mathém.*, Série III., t. i., 1837, p. 433.)

† Humb., *Rel. Hist.*, t. i., p. 519–527. Ellicot, in the *Transactions of the American Society*, 1804, vol. vi., p. 29. Arago makes the following observations in reference to the November phenomena: "We thus become more and more confirmed in the belief that there exists a zone composed of millions of small bodies, whose orbits cut the plane of the

ᵃ [No such manuscript is at present known to exist in the library of that college. For this information I am indebted to the inquiries of Mr. Cory, of Pembroke College, the learned editor of *Hieroglyphics of Horapollo Nilous*, Greek and English, 1840.]—*Tr.*

a comparison of the facts I had adduced, showed that the phenomenon had been simultaneously seen in the New Continent, from the equator to New Herrnhut in Greenland (64° 14' north latitude), and between 46° and 82° longitude. The identity of the epochs was recognized with astonishment. The stream, which had been seen from Jamaica to Boston (40° 21' north latitude) to traverse the whole vault of heaven on the 12th and 13th of November, 1833, was again observed in the United States in 1834, on the night between the 13th and 14th of November, although on this latter occasion it showed itself with somewhat less intensity. In Europe the periodicity of the phenomenon has since been manifested with great regularity.

Another and a like regularly recurring phenomenon is that noticed in the month of August, the meteoric stream of St. Lawrence, appearing between the 9th and 14th of August. Muschenbroek,* as early as in the middle of the last century, drew attention to the frequency of meteors in the month of August; but their certain periodic return about the time of St. Lawrence's day was first shown by Quetelet, Olbers, and Benzenberg. We shall, no doubt, in time, discover other periodically appearing streams,† probably about the 22d to the

ecliptic at about the point which our Earth annually occupies between the 11th and 13th of November. It is a new planetary world beginning to be revealed to us." (*Annuaire*, 1836, p. 296.)

* Compare Muschenbroek, *Introd. ad Phil. Nat.*, 1762, t. ii., p. 1061; Howard, *On the Climate of London*, vol. ii., p. 23, observations of the year 1806; seven years, therefore, after the earliest observations of Brandes (Benzenberg, *über Sternschnuppen*, s. 240–244); the August observations of Thomas Forster, in Quetelet, op. cit., p. 438–453; those of Adolph Erman, Boguslawski, and Kreil, in Schum., *Jahrb.*, 1838, s. 317–330. Regarding the point of origin in Perseus, on the 10th of August, 1839, see the accurate measurements of Bessel and Erman (Schum., *Astr. Nachr.*, No. 385 und 428); but on the 10th of August, 1837, the path does not appear to have been retrograde; see Arago, in *Comptes Rendus*, 1837, t. ii., p. 183.

† On the 25th of April, 1095, "innumerable eyes in France saw stars falling from heaven as thickly as hail" (*ut grando, nisi lucerent, pro densitate putaretur;* Baldr., p. 88), and this occurrence was regarded by the Council of Clermont as indicative of the great movement in Christendom. (Wilken, *Gesch. der Kreuzzüge*, bd. i., s. 75.) On the 25th of April, 1800, a great fall of stars was observed in Virginia and Massachusetts; it was "a fire of rockets that lasted two hours." Arago was the first to call attention to this "traînée d'asteroïdes," as a recurring phenomenon. (*Annuaire*, 1836, p. 297.) The falls of aërolites in the beginning of the month of December are also deserving of notice. In reference to their periodic recurrence as a meteoric stream, we may mention the early observation of Brandes on the night of the 6th and 7th of December, 1798 (when he counted 2000 falling stars), and very

25th of April, between the 6th and 12th of December, and, to judge by the number of true falls of aërolites enumerated by Capocci, also between the 27th and 29th of November, or about the 17th of July.

Although the phenomena hitherto observed appear to have been independent of the distance from the pole, the temperature of the air, and other climatic relations, there is, however, one perhaps accidentally coincident phenomenon which must not be wholly disregarded. The Northern Light, the Aurora Borealis, was unusually brilliant on the occurrence of the splendid fall of meteors of the 12th and 13th November, 1833, described by Olmsted. It was also observed at Bremen in 1838, where the periodic meteoric fall was, however, less remarkable than at Richmond, near London. I have mentioned in another work the singular fact observed by Admiral Wrangel, and frequently confirmed to me by himself,* that when he

probably the enormous fall of aërolites that occurred at the Rio Assu, near the village of Macao, in the Brazils, on the 11th of December, 1836. (Brandes, *Unterhalt. für Freunde der Physik*, 1825, heft i., s. 65, and *Comptes Rendus*, t. v., p. 211.) Capocci, in the interval between 1809 and 1839, a space of thirty years, has discovered twelve authenticated cases of aërolites occurring between the 27th and 29th of November, besides others on the 13th of November, the 10th of August, and the 17th of July. (*Comptes Rendus*, t. xi., p. 357.) It is singular that in the portion of the Earth's path corresponding with the months of January and February, and probably also with March, no *periodic* streams of falling stars or aërolites have as yet been noticed; although, when in the South Sea in the year 1803, I observed on the 15th of March a remarkably large number of falling stars, and they were seen to fall as in a swarm in the city of Quito, shortly before the terrible earthquake of Riobamba on the 4th of February, 1797. From the phenomena hitherto observed, the following epochs seem especially worthy of remark:

22d to the 25th of April.
17th of July (17th to the 26th of July ?). (Quet., *Corr.*, 1837, p. 435.)
10th of August.
12th to the 14th of November.
27th to the 29th of November.
6th to the 12th of December.

When we consider that the regions of space must be occupied by myriads of comets, we are led by analogy, notwithstanding the differences existing between isolated comets and rings filled with asteroids, to regard the frequency of these meteoric streams with less astonishment than the first consideration of the phenomenon would be likely to excite.

* Ferd. v. Wrangle, *Reise längs der Nordküste von Sibirien in den Jahren*, 1820–1824, th. ii., s. 259. Regarding the recurrence of the denser swarm of the November stream after an interval of thirty-three years, see Olbers, in *Jahrb.*, 1837, s. 280. I was informed in Cumana that shortly before the fearful earthquake of 1766, and consequently thirty-three years (the same interval) before the great fall of stars on

was on the Siberian coast of the Polar Sea, he observed, during an Aurora Borealis, certain portions of the vault of heaven, which were not illuminated, light up and continue luminous whenever a shooting star passed over them.

The different meteoric streams, each of which is composed of myriads of small cosmical bodies, probably intersect our Earth's orbit in the same manner as Biela's comet. According to this hypothesis, we may represent to ourselves these asteroid-meteors as composing a closed ring or zone, within which they all pursue one common orbit. The smaller planets between Mars and Jupiter present us, if we except Pallas, with an analogous relation in their constantly intersecting orbits. As yet, however, we have no certain knowledge as to whether changes in the periods at which the stream becomes visible, or the *retardations* of the phenomena of which I have already spoken, indicate a regular precession or oscillation of the nodes—that is to say, of the points of intersection of the Earth's orbit and of that of the ring; or whether this ring or zone attains so considerable a degree of breadth from the irregular grouping and distances apart of the small bodies, that it requires several days for the Earth to traverse it. The system of Saturn's satellites shows us likewise a group of immense width, composed of most intimately-connected cosmical bodies. In this system, the orbit of the outermost (the seventh) satellite has such a vast diameter, that the Earth, in her revolution round the Sun, requires three days to traverse an extent of space equal to this diameter. If, therefore, in one of these rings, which we regard as the orbit of a periodical stream, the asteroids should be so irregularly distributed as to consist of but few groups sufficiently dense to give rise to these phenomena, we may easily understand why we so seldom witness such glorious spectacles as those exhibited in the November months of 1799 and 1833. The acute mind of Olbers led him almost to predict that the next appearance of the phenomenon of shooting stars and fire-balls intermixed, falling like flakes of snow, would not recur until between the 12th and 14th of November, 1867.

the 11th and 12th of November, 1799, a similar fiery manifestation had been observed in the heavens. But it was on the 21st of October, 1766, and not in the beginning of November, that the earthquake occurred. Possibly some traveler in Quito may yet be able to ascertain the day on which the volcano of Cayambe, which is situated there, was for the space of an hour enveloped in falling stars, so that the inhabitants endeavored to appease heaven by religious processions. (*Relat. Hist.*, t i., chap. iv., p 307; chap. x., p. 520 and 527.)

The stream of the November asteroids has occasionally
only been visible in a small section of the Earth. Thus, for
instance, a very splendid *meteoric shower* was seen in England
in the year 1837, while a most attentive and skillful observer
at Braunsberg, in Prussia, only saw, on the same night, which
was there uninterruptedly clear, a few sporadic shooting stars
fall between seven o'clock in the evening and sunrise the next
morning. Bessel* concluded from this "that a dense group
of the bodies composing the great ring may have reached that
part of the Earth in which England is situated, while the
more eastern districts of the Earth might be passing at the
time through a part of the meteoric ring proportionally less
densely studded with bodies." If the hypothesis of a regular
progression or oscillation of the nodes should acquire greater
weight, special interest will be attached to the investigation
of older observations. The Chinese annals, in which great
falls of shooting stars, as well as the phenomena of comets,
are recorded, go back beyond the age of Tyrtæus, or the second Messenian war. They give a description of two streams
in the month of March, one of which is 687 years anterior to
the Christian era. Edward Biot has observed that, among
the fifty-two phenomena which he has collected from the
Chinese annals, those that were of most frequent recurrence
are recorded at periods nearly corresponding with the 20th
and 22d of July, O.S., and might consequently be identical
with the stream of St. Lawrence's day, taking into account
that it has advanced since the epochs† indicated. If the fall
of shooting stars of the 21st of October, 1366, O.S. (a notice
of which was found by the younger Von Boguslawski, in
Benessius de Horowic's *Chronicon Ecclesiæ Pragensis*), be
identical with our November phenomenon, although the occurrence in the fourteenth century was seen in broad daylight, we find by the precession in 477 years that this system
of meteors, or, rather, its common center of gravity, must de-

* From a letter to myself, dated Jan. 24th, 1838. The enormous
swarm of falling stars in November, 1799, was almost exclusively seen
in America, where it was witnessed from New Herrnhut in Greenland
to the equator. The swarms of 1831 and 1832 were visible only in
Europe, and those of 1833 and 1834 only in the United States of North
America.

† Lettre de M. Edouard Biot à M. Quetelet, sur les anciennes apparitions d'Etoiles Filantes en Chine, in the *Bull. de l'Académie de Bruxelles*, 1843, t. x., No. 7, p. 8. On the notice from the *Chronicon Ecclesiæ Pragensis*, see the younger Boguslawski, in Poggend., *Annalen*,
bd. xlviii., s. 612.

scribe a retrograde orbit round the Sun. It also follows, from
the views thus developed, that the non-appearance, during
certain years, in any portion of the Earth, of the two streams
hitherto observed in November and about the time of St.
Lawrence's day, must be ascribed either to an interruption in
the meteoric ring, that is to say, to intervals occurring between the asteroid groups, or, according to Poisson, to the action of the larger planets* on the form and position of this
annulus.

The solid masses which are observed by night to fall to the
earth from fire-balls, and by day, generally when the sky is
clear, from a dark small cloud, are accompanied by much
noise, and although heated, are not in an actual state of incandescence. They undeniably exhibit a great degree of general identity with respect to their external form, the character
of their crust, and the chemical composition of their principal
constituents. These characteristics of identity have been observed at all the different epochs and in the most various parts
of the earth in which these meteoric stones have been found.
This striking and early-observed analogy of physiognomy in
the denser meteoric masses is, however, met by many exceptions regarding individual points. What differences, for instance, do we not find between the malleable masses of iron
of Hradschina in the district of Agram, those from the shores
of the Sisim in the government of Jeniseisk, rendered so celebrated by Pallas, or those which I brought from Mexico,† all
of which contain 96 per cent. of iron, from the aërolites of
Siena, in which the iron scarcely amounts to 2 per cent., or
the earthy aërolite of Alais (in the Department du Gard),
which broke up in water, or, lastly, from those of Jonzac and
Juvenas, which contained no metallic iron, but presented a

* " It appears that an apparently inexhaustible number of bodies, too
small to be observed, are moving in the regions of space, either around
the Sun or the planets, or perhaps even around their satellites. It is
supposed that when these bodies come in contact with our atmosphere,
the difference between their velocity and that of our planet is so great,
that the friction which they experience from their contact with the air
heats them to incandescence, and sometimes causes their explosion. If
the group of falling stars form an annulus around the Sun, its velocity
of circulation may be very different from that of our Earth; and the
displacements it may experience in space, in consequence of the actions
of the various planets, may render the phenomenon of its intersecting
the planes of the ecliptic possible at some epochs, and altogether impossible at others."—Poisson, *Recherches sur la Probabilité des Jugements*, p. 306, 307.

† Humboldt, *Essai Politique sur la Nouv. Espagne* (2de édit.), t. iii.
p. 310.

mixture of oryctognostically distinct crystalline components!
These differences have led mineralogists to separate these cosmical masses into two classes, namely, those containing nickelliferous meteoric iron, and those consisting of fine or coarse-
ly-granular meteoric dust.   The crust or rind of aërolites is
peculiarly characteristic of these bodies, being only a few
tenths of a line in thickness, often glossy and pitch-like, and
occasionally veined.*   There is only one instance on record,
as far as I am aware (the aërolite of Chantonnay, in La Ven-
dée), in which the rind was absent, and this meteor, like that
of Juvenas, presented likewise the peculiarity of having pores
and vesicular cavities.   In all other cases the black crust is
divided from the inner light-gray mass by as sharply-defined
a line of separation as is the black leaden-colored investment
of the white granite blocks† which I brought from the cataracts of the Orinoco, and which are also associated with
many other cataracts, as, for instance, those of the Nile and
of the Congo River.   The greatest heat employed in our
porcelain ovens would be insufficient to produce any thing
similar to the crust of meteoric stones, whose interior remains wholly unchanged.   Here and there, facts have been
observed which would seem to indicate a fusion together of
the meteoric fragments ; but, in general, the character of the
aggregate mass, the absence of compression by the fall, and
the inconsiderable degree of heat possessed by these bodies
when they reach the earth, are all opposed to the hypothesis
of the interior being in a state of fusion during their short
passage from the boundary of the atmosphere to our Earth.
   The chemical elements of which these meteoric masses
consist, and on which Berzelius has thrown so much light,
are the same as those distributed throughout the earth's
crust, and are fifteen in number, namely, iron, nickel, cobalt,
manganese, chromium, copper, arsenic, zinc, potash, soda, sulphur, phosphorus, and carbon, constituting altogether nearly
one third of all the known simple bodies.   Notwithstanding
this similarity with the primary elements into which inorganic
bodies are chemically reducible, the aspect of aërolites, owing
to the mode in which their constituent parts are compounded,
presents, generally, some features foreign to our telluric rocks
and minerals.   The pure native iron, which is almost always

* The peculiar color of their crust was observed even as early as in
the time of Pliny (ii., 56 and 58): "colore adusto."   The phrase "lateri-
bus pluisse" seems also to refer to the burned outer surface of aërolites.
† Humb., *Rel. Hist.*, t. ii., chap xx., p. 299–302.

found incorporated with aërolites, imparts to them a peculiar, but not, consequently, a *selenic* character; for in other regions of space, and in other cosmical bodies besides our Moon, water may be wholly absent, and processes of oxydation of rare occurrence.

Cosmical gelatinous vesicles, similar to the organic *nostoc* (masses which have been supposed since the Middle Ages to be connected with shooting stars), and those pyrites of Sterli tamak, west of the Uralian Mountains, which are said to have constituted the interior of hailstones,* must both be classed among the mythical fables of meteorology. Some few aërolites, as those composed of a finely granular tissue of olivine, augite, and labradorite blended together† (as the meteoric stone found at Juvenas, in the Department de l'Ardèche, which resembled dolorite), are the only ones, as Gustav Rose has remarked, which have a more familiar aspect. These bodies contain, for instance, crystalline substances, perfectly similar to those of our earth's crust; and in the Siberian mass of meteoric iron investigated by Pallas, the olivine only differs from common olivine by the absence of nickel, which is replaced by oxyd of tin.‡ As meteoric olivine, like our basalt, contains from 47 to 49 per cent. of magnesia, constituting, according to Berzelius, almost the half of the earthy components of meteoric stones, we can not be surprised at the great quantity of silicate of magnesia found in these cosmical bodies. If the aërolite of Juvenas contain separable crystals of augite and labradorite, the numerical relation of the constituents

---

* Gustav Rose, *Reise nach dem Ural*, bd. ii., s. 202.

† Gustav Rose, in Poggend., *Ann.*, 1825, bd. iv., s. 173–192. Rammelsberg, *Erstes Suppl. zum chem. Handwörterbuche der Mineralogie*, 1843, s. 102. "It is," says the clear-minded observer Olbers, "a remarkable but hitherto unregarded fact, that while shells are found in secondary and tertiary formations, no *fossil meteoric stones* have as yet been discovered. May we conclude from this circumstance that previous to the present and last modification of the earth's surface no meteoric stones fell on it, although at the present time it appears probable, from the researches of Schreibers, that 700 fall annually?" (Olbers, in Schum., *Jahrb.*, 1838, s. 329.) Problematical nickelliferous masses of native iron have been found in Northern Asia (at the gold-washing establishment at Petropawlowsk, eighty miles southeast of Kusnezk), imbedded thirty-one feet in the ground, and more recently in the Western Carpathians (the mountain chain of Magura, at Szlanicz), both of which are remarkably like meteoric stones. Compare Erman, *Archiv für wissenschaftliche Kunde von Russland*, bd. i., s. 315, and Haidinger, *Bericht über Szlaniczer Schürfe in Ungarn*.

‡ Berzelius. *Jahresber.*, bd. xv., s. 217 und 231. Rammelsberg, *Handwörterb.*, abth. ii., s. 25–28.

render it at least probable that the meteoric masses of Cha-
teau-Renard may be a compound of diorite, consisting of hornblende and albite, and those of Blansko and Chantonnay compounds of hornblende and labradorite.   The proofs of the telluric and atmospheric origin of aërolites, which it is attempted to base upon the oryctognostic analogies presented by these
bodies, do not appear to me to possess any great weight.

Recalling to mind the remarkable interview between Newton and Conduit at Kensington,* I would ask why the elementary substances that compose one group of cosmical bodies,
or one planetary system, may not, in a great measure, be identical ?   Why should we not adopt this view, since we may
conjecture that these planetary bodies, like all the larger or
smaller agglomerated masses revolving round the sun, have
been thrown off from the once far more expanded solar atmosphere, and been formed from vaporous rings describing
their orbits round the central body ?   We are not, it appears
to me, more justified in applying the term telluric to the nickel
and iron, the olivine and pyroxene (augite), found in meteoric
stones, than in indicating the German plants which I found
beyond the Obi as European species of the flora of Northern
Asia.   If the elementary substances composing a group of
cosmical bodies of different magnitudes be identical, why
should they not likewise, in obeying the laws of mutual attraction, blend together under definite relations of mixture,
composing the white glittering snow and ice in the polar zones
of the planet Mars, or constituting in the smaller cosmical
masses mineral bodies inclosing crystals of olivine, augite, and
labradorite ?   Even in the domain of pure conjecture we should
not suffer ourselves to be led away by unphilosophical and arbitrary views devoid of the support of inductive reasoning.

Remarkable obscurations of the sun's disk, during which
the stars have been seen at mid-day (as, for instance, in the
obscuration of 1547, which continued for three days, and occurred about the time of the eventful battle of Mühlberg),
can not be explained as arising from volcanic ashes or mists,
and were regarded by Kepler as owing either to a *materia
cometica*, or to a black cloud formed by the sooty exhalations
of the solar body.   The shorter obscurations of 1090 and
1203, which continued, the one only three, and the other six

* " Sir Isaac Newton said he took all the planets to be composed of
the same matter with the Earth, viz., earth, water, and stone, but variously concocted."—Turner, *Collections for the History of Grantham,
containing authentic Memoirs of Sir Isaac Newton*, p. 172.

hours, were supposed by Chladni and Schnurrer to be occa
sioned by the passage of meteoric masses before the sun's disk.
Since the period that streams of meteoric shooting stars were
first considered with reference to the direction of their orbit
as a closed ring, the epochs of these mysterious celestial phenomena have been observed to present a remarkable connec
tion with the regular recurrence of swarms of shooting stars
Adolph Erman has evinced great acuteness of mind in his ac·
curate investigation of the facts hitherto observed on this subject, and his researches have enabled him to discover the connection of the sun's conjunction with the August asteroids on
the 7th of February, and with the November asteroids on the
12th of May, the latter period corresponding with the days
of St. Mamert (May 11th), St. Pancras (May 12th), and St.
Servatius (May 13th), which, according to popular belief,
were accounted " cold days."*

The Greek natural philosophers, who were but little dis
posed to pursue observations, but evinced inexhaustible fer
tility of imagination in giving the most various interpretation
of half-perceived facts, have, however, left some hypotheses
regarding shooting stars and meteoric stones which strikingly
accord with the views now almost universally admitted of
the cosmical process of these phenomena. " Falling stars,"
says Plutarch, in his life of Lysander,† " are, according to

---

* Adolph Erman, in Poggend., *Annalen*, 1839, bd. xlviii., s. 582–
601. Biot had previously thrown doubt regarding the probability of
the November stream reappearing in the beginning of May (*Comptes
Rendus*, 1836, t. ii., p. 670). Mädler has examined the mean depression of temperature on the three ill-named days of May by Berlin observations for eighty-six years (*Verhandl. des Vereins zur Beförd. des
Gartenbaues*, 1834, s. 377), and found a retrogression of temperature
amounting to 2°·2 Fahr. from the 11th to the 13th of May, a period at
which nearly the most rapid advance of heat takes place. It is much
to be desired that this phenomenon of depressed temperature, which
some have felt inclined to attribute to the melting of the ice in the
northeast of Europe, should be also investigated in very remote spots,
as in America, or in the southern hemisphere. (Comp. *Bull. de l'Acad.
Imp. de St. Pétersbourg*, 1843, t. i., No. 4.)

† Plut., *Vitæ par. in Lysandro*, cap. 22. The statement of Dama-
chos (Daïmachos), that for seventy days continuously there was a fiery
cloud seen in the sky, emitting sparks like falling stars, and which then,
sinking nearer to the earth, let fall the stone of Ægos Potamos, " which,
however, was only a small part of it," is extremely improbable, since
the direction and velocity of the fire-cloud would in that case of necessity have to remain for so many days the same as those of the earth;
and this, in the fire-ball of the 19th of July, 1686, described by Halley
(*Trans.*, vol. xxix., p. 163), lasted only a few minutes. It is not altogether certain whether Daïmachos, the writer, περὶ εὐσεβείας, was the

the opinion of some physicists, not eruptions of the ethereal
fire extinguished in the air immediately after its ignition, nor
yet an inflammatory combustion of the air, which is dissolved
in large quantities in the upper regions of space, but these
meteors are rather a fall of celestial bodies, which, in consequence of a certain intermission in the rotatory force, and by
the impulse of some irregular movement, have been hurled
down not only to the inhabited portions of the Earth, but
also beyond it into the great ocean, where we can not find
them." Diogenes of Apollonia* expresses himself still more
explicitly. According to his views, "Stars that are *invisible*,
and, consequently, have no name, move in space together with
those that are visible. These invisible stars frequently fall
to the earth and are extinguished, as the *stony star* which fell
burning at Ægos Potamos." The Apollonian, who held all
other stellar bodies, when luminous, to be of a pumice-like
nature, probably grounded his opinions regarding shooting
stars and meteoric masses on the doctrine of Anaxagoras the
Clazomenian, who regarded all the bodies in the universe
" as fragments of rocks, which the fiery ether, in the force
of its gyratory motion, had torn from the Earth and converted into stars." In the Ionian school, therefore, according
to the testimony transmitted to us in the views of Diogenes
of Apollonia, aërolites and stars were ranged in one and the
same class; both, when considered with reference to their
primary origin, being equally telluric, this being understood
only so far as the Earth was then regarded as a central body,†

same person as Daïmachos of Platæa, who was sent by Seleucus to
India to the son of Androcottos, and who was charged by Strabo with
being "a speaker of lies" (p. 70, Casaub.). From another passage of
Plutarch (*Compar. Solonis c. Cop.*, cap. 5) we should almost believe
that he was. At all events, we have here only the evidence of a very
late author, who wrote a century and a half after the fall of aërolites
occurred in Thrace, and whose authenticity is also doubted by Plutarch.
* Stob., ed. Heeren, i., 25, p. 508; Plut., *de plac. Philos.*, ii., 13.
† The remarkable passage in Plut., *de plac. Philos.*, ii., 13, runs thus:
"Anaxagoras teaches that the surrounding ether is a fiery substance,
which, by the power of its rotation, tears rocks from the earth, inflames
them, and converts them into stars." Applying an ancient fable to illustrate a physical dogma, the Clazomenian appears to have ascribed
the fall of the Nemæan Lion to the Peloponnesus from the Moon to
such a rotatory or centrifugal force. (Ælian., xii., 7; Plut., *de Facie
in Orbe Lunæ*, c. 24; Schol. ex Cod. Paris., in *Apoll. Argon.*, lib. i.,
p. 498, ed. Schaef., t. ii., p. 40; Meineke, *Annal. Alex.*, 1843, p. 85.)
Here, instead of stones from the Moon, we have an animal from the
Moon! According to an acute remark of Böckh, the ancient mythology of the Nemæan lunar lion has an astronomical origin, and is sym-

forming all things around it in the same manner as we, according to our present views, suppose the planets of our system to have originated in the expanded atmosphere of another central body, the Sun. These views must not, therefore, be confounded with what is commonly termed the telluric or atmospheric origin of meteoric stones, nor yet with the singular opinion of Aristotle, which supposed the enormous mass of Ægos Potamos to have been raised by a hurricane. That arrogant spirit of incredulity, which rejects facts without attempting to investigate them, is in some cases almost more injurious than an unquestioning credulity. Both are alike detrimental to the force of investigation. Notwithstanding that for more than two thousand years the annals of different nations had recorded falls of meteoric stones, many of which had been attested beyond all doubt by the evidence of irreproachable eye-witnesses—notwithstanding the important part enacted by the Bætylia in the meteor-worship of the ancients—notwithstanding the fact of the companions of Cortez having seen an aërolite at Cholula which had fallen on the neighboring pyramid—notwithstanding that califs and Mongolian chiefs had caused swords to be forged from recently-fallen meteoric stones—nay, notwithstanding that several persons had been struck dead by stones falling from heaven, as, for instance, a monk at Crema on the 4th of September, 1511, another monk at Milan in 1650, and two Swedish sailors on board ship in 1674, yet this great cosmical phenomenon remained almost wholly unheeded, and its intimate connection with other planetary systems unknown, until attention was drawn to the subject by Chladni, who had already gained immortal renown by his discovery of the sound-figures. He who is penetrated with a sense of this mysterious connection, and whose mind is open to deep impressions of nature, will feel himself moved by the deepest and most solemn emotion at the sight of every star that shoots across the vault of heaven, no less than at the glorious spectacle of meteoric swarms in the November phenomenon or on St. Lawrence's day. Here motion is suddenly revealed in the midst of nocturnal rest. The still radiance of the vault of heaven is for a moment animated with life and movement. In the mild radiance left on the track of the shooting star, imagination pictures the lengthened path of the meteor through the vault of heaven,

bolically connected in chronology with the cycle of intercalation of the lunar year, with the moon-worship at Nemæa, and the games by which it was accompanied.

while, every where around, the luminous asteroids proclaim the existence of one common material universe.

If we compare the volume of the innermost of Saturn's satellites, or that of Ceres, with the immense volume of the Sun, all relations of magnitude vanish from our minds. The extinction of suddenly resplendent stars in Cassiopeia, Cygnus, and Serpentarius have already led to the assumption of other and non-luminous cosmical bodies. We now know that the meteoric asteroids, spherically agglomerated into small masses, revolve round the Sun, intersect, like comets, the orbits of the luminous larger planets, and become ignited either in the vicinity of our atmosphere or in its upper strata.

The only media by which we are brought in connection with other planetary bodies, and with all portions of the universe beyond our atmosphere, are light and heat (the latter of which can scarcely be separated from the former),* and those mysterious powers of attraction exercised by remote masses, according to the quantity of their constituents, upon our globe, the ocean, and the strata of our atmosphere. Another and different kind of cosmical, or, rather, material mode of contact is, however, opened to us, if we admit falling stars and meteoric stones to be planetary asteroids. They not only act upon us merely from a distance by the excitement of luminous or calorific vibrations, or in obedience to the laws of mutual attraction, but they acquire an actual material existence for us, reaching our atmosphere from the remoter regions of universal space, and remaining on the earth itself. Meteoric stones are the only means by which we can be brought in possible contact with that which is foreign to our own planet. Accustomed to gain our knowledge of what is not telluric solely through measurement, calculations, and the deductions of reason, we experience a sentiment of astonishment at finding that we may examine, weigh, and analyze bodies that ap-

* The following remarkable passage on the radiation of heat from the fixed stars, and on their low combustion and vitality—one of Kepler's many aspirations—occurs in the *Paralipom. in Vitell. Astron. pars Optica*, 1604, Propos. xxxii., p. 25: "Lucis proprium est calor, sydera omnia calefaciunt. De syderum luce claritatis ratio testatur, calorem universorum in minori esse proportione ad calorem unius solis, quam ut ab homine, cujus est certa caloris mensura, uterque simul percipi et judicari possit. De cincindularum lucula tenuissima negare non potes, quin cum calore sit. Vivunt enim et moventur, hoc autem non sine calefactione perficitur. Sic neque putrescentium lignorum lux suo calore destituitur; nam ipsa puetredo quidam lentus ignis est. Inest et stirpibus suus calor." (Compare Kepler, *Epit. Astron. Copernicanæ*, 1618, t. i., lib. i., p. 35.)

pertain to the outer world. This awakens, by the power of the imagination, a meditative, spiritual train of thought, where the untutored mind perceives only scintillations of light in the firmament, and sees in the blackened stone that falls from the exploded cloud nothing beyond the rough product of a powerful natural force.

Although the asteroid-swarms, on which we have been led, from special predilection, to dwell somewhat at length, approximate to a certain degree, in their inconsiderable mass and the diversity of their orbits, to comets, they present this essential difference from the latter bodies, that our knowledge of their existence is almost entirely limited to the moment of their destruction, that is, to the period when, drawn within the sphere of the Earth's attraction, they become luminous and ignite.

In order to complete our view of all that we have learned to consider as appertaining to our solar system, which now, since the discovery of the small planets, of the interior comets of short revolutions, and of the meteoric asteroids, is so rich and complicated in its form, it remains for us to speak of the ring of zodiacal light, to which we have already alluded. Those who have lived for many years in the zone of palms must retain a pleasing impression of the mild radiance with which the zodiacal light, shooting pyramidally upward, illumines a part of the uniform length of tropical nights. I have seen it shine with an intensity of light equal to the milky way in Sagittarius, and that not only in the rare and dry atmosphere of the summits of the Andes, at an elevation of from thirteen to fifteen thousand feet, but even on the boundless grassy plains, the llanos of Venezuela, and on the sea-shore, beneath the ever-clear sky of Cumana. This phenomenon was often rendered especially beautiful by the passage of light, fleecy clouds, which stood out in picturesque and bold relief from the luminous back-ground. A notice of this aërial spectacle is contained in a passage in my journal, while I was on the voyage from Lima to the western coasts of Mexico : " For three or four nights (between 10° and 14° north latitude) the zodiacal light has appeared in greater splendor than I have ever observed it. The transparency of the atmosphere must be remarkably great in this part of the Southern Ocean, to judge by the radiance of the stars and nebulous spots. From the 14th to the 19th of March a regular interval of three quarters of an hour occurred between the disappearance of the sun's disk in the ocean and the first manifestation of the zodiacal light, although the night was already perfectly dark. An hour after sunset it was seen in great brilliancy between Aldebaran and the Pleiades ; and on the 18th of March it attained an altitude of 39° 5′. Narrow elongated clouds are scattered over the beautiful deep azure of the distant horizon, flitting past the zodiacal light as before a golden curtain. Above these, other clouds are from time to time reflecting the most brightly variegated colors. It seems a second sunset. On this side of the vault of heaven the lightness of the night appears to increase almost as much as at the first quarter of the moon. Toward 10 o'clock the zodiacal light generally becomes very faint in this part of the Southern Ocean, and at midnight I have scarcely been able to trace a vestige of it. On the 16th of March, when most strongly luminous, a faint reflection was visible in the east." In our gloomy so-called " temperate" northern zone, the zodiacal light is only distinctly visible in the beginning of Spring, after the evening twilight, in the western part of the sky, and at the close of Autumn, before the dawn of day, above the eastern horizon.

It is difficult to understand how so striking a natural phenomenon should have failed to attract the attention of physicists and astronomers until the middle of the seventeenth century, or how it could have escaped the observation of the Arabian natural philosophers in ancient Bactria, on the Euphrates, and in the south of Spain. Almost equal surprise is excited by the tardiness of observation of the nebulous spots in Andromeda and Orion, first described by Simon Marius and Huygens. The earliest explicit description of the zodiacal light occurs in Childrey's *Britannia Baconica*,* in the year

---

* " There is another thing which I recommend to the observation of mathematical men, which is, that in February, and for a little before and a little after that month (as I have observed several years together), about six in the evening, when the twilight hath almost deserted the horizon, you shall see a plainly discernible way of the twilight striking up toward the Pleiades, and seeming almost to touch them. It is so observed any clear night, but it is best illac nocte. There is no such way to be observed at any other time of the year (that I can perceive), nor any other way at that time to be perceived darting up elsewhere ; and I believe it hath been, and will be constantly visible at that time of the year ; but what the cause of it in nature should be, I can not yet imagine, but leave it to future inquiry." (Childrey, *Britannia Baconica*, 1661, p. 183.) This is the first view and a simple description of the phenomenon. (Cassini, *Découverte de la Lumière Céleste qui paroît dans le Zodiaque*, in the *Mém. de l'Acad.*, t. viii., 1730, p. 276. Mairan, *Traité Phys. de l'Aurore Boréale*, 1754, p. 16.) In this remarkable work by Childrey there are to be found (p. 91) very clear accounts of the epochs of maxima and minima diurnal and annual temperatures.

1661. The first observation of the phenomenon may have been made two or three years prior to this period ; but, notwithstanding, the merit of having (in the spring of 1683) been the first to investigate the phenomenon in all its relations in space is incontestably due to Dominicus Cassini. The light which he saw at Bologna in 1668, and which was observed at the same time in Persia by the celebrated traveler Chardin (the court astrologers of Ispahan called this light, which had never before been observed, *nyzek*, a small lance), was not the zodiacal light, as has often been asserted,* but the

and of the retardation of the extremes of the effects in meteorological processes. It is, however, to be regretted that our Baconian-philosophy-loving author, who was Lord Henry Somerset's chaplain, fell into the same error as Bernardin de St. Pierre, and regarded the Earth as elongated at the poles (see p. 148). At the first, he believes that the Earth was spherical, but supposes that the uninterrupted and increasing addition of layers of ice at both poles has changed its figure ; and that, as the ice is formed from water, the quantity of that liquid is every where diminishing.

* Dominicus Cassini (*Mém. de l'Acad.*, t. viii., 1730, p. 188), and Mairan (*Aurore Bor.*, p. 16), have even maintained that the phenomenon observed in Persia in 1668 was the zodiacal light. Delambre (*Hist. de l'Astron. Moderne*, t. ii., p. 742), in very decided terms, ascribes the discovery of this light to the celebrated traveler Chardin ; but in the *Couronnement de Soliman*, and in several passages of the narrative of his travels (éd. de Langlès, t. iv., p. 326 ; t. x., p. 97), he only applies the term niazouk (nyzek), or " petite lance," to " the great and famous comet which appeared over nearly the whole world in 1668, and whose head was so hidden in the west that it could not be perceived in the horizon of Ispahan" (*Atlas du Voyage de Chardin*, Tab. iv. ; from the observations at Schiraz). The head or nucleus of the comet was, however, visible in the Brazils and in India (Pingré, *Cométogr.*, t. ii., p. 22). Regarding the conjectured identity of the last great comet of March, 1843, with this, which Cassini mistook for the zodiacal light, see Schum., *Astr. Nachr.*, 1843, No. 476 and 480. In Persian, the term "nîzehi âteschîn" (fiery spears or lances) is also applied to the rays of the rising or setting sun, in the same way as "nayâzik," according to Freytag's Arabic Lexicon, signifies " stellæ cadentes." The comparison of comets to lances and swords was, however, in the Middle Ages, very common in all languages. The great comet of 1500, which was visible from April to June, was always termed by the Italian writers of that time *il Signor Astone* (see my *Examen Critique de l'Hist. de la Géographie*, t. v., p. 80). All the hypotheses that have been advanced to show that Descartes (Cassini, p. 230 ; Mairan, p. 16), and even Kepler (Delambre, t. i., p. 601), were acquainted with the zodiacal light, appear to me altogether untenable. Descartes (*Principes*, iii., art. 136, 137) is very obscure in his remarks on comets, observing that their tails are formed " by oblique rays, which, falling on different parts of the planetary orbs, strike the eye laterally by extraordinary refraction," and that they might be seen morning and evening, " like a long beam," when the Sun is between the comet and the Earth. This passage no more refers to the zodiacal light than those in which Kepler (*Epit. A*

enormous tail of a comet, whose head was concealed in the
vapory mist of the horizon, and which, from its length and
appearance, presented much similarity to the great comet of
1843. We may conjecture, with much probability, that the
remarkable light on the elevated plains of Mexico, seen for
forty nights consecutively in 1509, and observed in the eastern
horizon rising pyramidally from the earth, was the zodiacal
light. I found a notice of this phenomenon in an ancient Aztec MS., the *Codex Telleriano-Remensis,** preserved in the
Royal Library at Paris.

This phenomenon, whose primordial antiquity can scarcely
be doubted, and which was first noticed in Europe by Childrey
and Dominicus Cassini, is not the luminous solar atmosphere
itself, since this can not, in accordance with mechanical laws,
be more compressed than in the relation of 2 to 3, and consequently can not be diffused beyond $\frac{9}{20}$ths of Mercury's heliocentric distance. These same laws teach us that the altitude
of the extreme boundaries of the atmosphere of a cosmical

tron. Copernicanæ, t. i., p. 57, and t. ii., p. 893) speaks of the existence
of a solar atmosphere (limbus circa solem, coma lucida), which, in
eclipses of the Sun, prevents it "from being quite night;" and even
more uncertain, or indeed erroneous, is the assumption that the "trabes
quas δοκοὺς vocant" (Plin., ii., 26 and 27) had reference to the tongue-
shaped rising zodiacal light, as Cassini (p. 231, art. xxxi.) and Mairan
(p. 15) have maintained. Every where among the ancients the trabes
are associated with the bolides (ardores et faces) and other fiery meteors, and even with long-barbed comets. (Regarding δοκὸς, δοκίας,
δοκίτης, see Schäfer, *Schol. Par. ad Apoll. Rhod.*, 1813, t. ii., p. 206;
Pseudo-Aristot., *de Mundo*, 2, 9 ; *Comment. Alex. Joh. Philop. et Olymp.
in Aristot. Meteor.*, lib. i., cap. vii., 3, p. 195, Ideler; Seneca, *Nat.
Quæst.*, i., 1.)

* Humboldt, *Monumens des Peuples Indigènes de l'Amérique*, t. ii..
p. 301. The rare manuscript which belonged to the Archbishop of
Rheims, Le Tellier, contains various kinds of extracts from an Aztec
ritual, an astrological calendar, and historical annals, extending from
1197 to 1549, and embracing a notice of different natural phenomena,
epochs of earthquakes and comets (as, for instance, those of 1490 and
1529), and of (which are important in relation to Mexican chronology)
solar eclipses. In Camargo's manuscript *Historia de Tlascala*, the light
rising in the east almost to the zenith is, singularly enough, described
as "sparkling, and as if sown with stars." The description of this
phenomenon, which lasted forty days, can not in any way apply to volcanic eruptions of Popocatepetl, which lies very near, in the southeast-
ern direction. (Prescott, *History of the Conquest of Mexico*, vol. i., p.
284.) Later commentators have confounded this phenomenon, which
Montezuma regarded as a warning of his misfortunes, with the "estrella
que humeava" (literally, *which spring forth;* Mexican *choloa, to leap or
spring forth*). With respect to the connection of this vapor with the
star Citlal Choloha (Venus) and with "the mountain of the star" (Cit-
laltepetl, the volcano of Orizaba), see my *Monumens*, t. ii., p. 303.

body above its equator, that is to say, the point at which
gravity and centrifugal force are in equilibrium, must be the
same as the altitude at which a satellite would rotate round
the central body simultaneously with the diurnal revolution
of the latter.* This limitation of the solar atmosphere in its
present concentrated condition is especially remarkable when
we compare the central body of our system with the nucleus
of other nebulous stars. Herschel has discovered several, in
which the radius of the nebulous matter surrounding the star
appeared at an angle of 150″. On the assumption that the
parallax is not fully equal to 1″, we find that the outermost
nebulous layer of such a star must be 150 times further from
the central body than our Earth is from the Sun. If, therefore, the nebulous star were to occupy the place of our Sun,
its atmosphere would not only include the orbit of Uranus,
but even extend eight times beyond it.†

Considering the narrow limitation of the Sun's atmosphere,
which we have just described, we may with much probability
regard the existence of a very compressed annulus of nebulous
matter,‡ revolving freely in space between the orbits of Venus
and Mars, as the material cause of the zodiacal light. As

* Laplace, *Expos. du Syst. du Monde*, p. 270 ; *Mécanique Céleste*,
t. ii., p. 169 and 171 ; Schubert, *Astr.*, bd. iii., § 206.

† Arago, in the *Annuaire*, 1842, p. 408. Compare Sir John Herschel's considerations on the volume and faintness of light of planetary
nebulæ, in Mary Somerville's *Connection of the Physical Sciences*, 1835,
p. 108. The opinion that the Sun is a nebulous star, whose atmosphere presents the phenomenon of zodiacal light, did not originate with
Dominicus Cassini, but was first promulgated by Mairan in 1730 (*Traité
de l'Aurore Bor.*, p. 47 and 263 ; Arago, in the *Annuaire*, 1842, p.
412). It is a renewal of Kepler's views.

‡ Dominicus Cassini was the first to assume, as did subsequently
Laplace, Schubert, and Poisson, the hypothesis of a separate ring to
explain the form of the zodiacal light. He says distinctly, " If the
orbits of Mercury and Venus were visible (throughout their whole extent), we should invariably observe them with the same figure and in
the same position with regard to the Sun, and at the same time of the
year with the zodiacal light." (*Mém. de l'Acad.*, t. viii., 1730, p. 218,
and Biot, in the *Comptes Rendus*, 1836, t. iii., p. 666.) Cassini believed that the nebulous ring of zodiacal light consisted of innumerable
small planetary bodies revolving round the Sun. He even went so
far as to believe that the fall of fire-balls might be connected with the
passage of the Earth through the zodiacal nebulous ring. Olmsted,
and especially Biot (op. cit., p. 673), have attempted to establish its
connection with the November phenomenon—a connection which Ol
bers doubts. (Schum., *Jahrb.*, 1837, s. 281.) Regarding the question
whether the place of the zodiacal light perfectly coincides with that
of the Sun's equator, see Houzeau, in Schum., *Astr. Nachr.*, 1843, No
492, s. 190.

yet we certainly know nothing definite regarding its actual material dimensions ; its augmentation* by emanations from the tails of myriads of comets that come within the Sun's vicinity ; the singular changes affecting its expansion, since it sometimes does not appear to extend beyond our Earth's orbit ; or, lastly, regarding its conjectural intimate connection with the more condensed cosmical vapor in the vicinity of the Sun. The nebulous particles composing this ring, and revolving round the Sun in accordance with planetary laws, may either be self-luminous or receive light from that luminary. Even in the case of a terrestrial mist (and this fact is very remarkable), which occurred at the time of the new moon at midnight in 1743, the phosphorescence was so intense that objects could be distinctly recognized at a distance of more than 600 feet.

I have occasionally been astonished, in the tropical climates of South America, to observe the variable intensity of the zodiacal light. As I passed the nights, during many months, in the open air, on the shores of rivers and on llanos, I enjoyed ample opportunities of carefully examining this phenomenon. When the zodiacal light had been most intense, I have observed that it would be perceptibly weakened for a few minutes, until it again suddenly shone forth in full brilliancy. In some few instances I have thought that I could perceive—not exactly a reddish coloration, nor the lower portion darkened in an arc-like form, nor even a scintillation, as Mairan affirms he has observed—but a kind of flickering and wavering of the light.† Must we suppose that changes are actually in progress in the nebulous ring ? or is it not more probable that, although I could not, by my meteorological instruments, detect any change of heat or moisture near the ground, and small stars of the fifth and sixth magnitudes appeared to shine with equally undiminished intensity of light, processes of condensation may be going on in the uppermost strata of the air, by means of which the transparency, or, rather, the reflection of light, may be modified in some peculiar and unknown man-

* Sir John Herschel, *Astron.*, § 487.

† Arago, in the *Annuaire*, 1832, p. 246. Several physical facts appear to indicate that, in a mechanical separation of matter into its smallest particles, if the mass be very small in relation to the surface, the electrical tension may increase sufficiently for the production of light and heat. Experiments with a large concave mirror have not hitherto given any positive evidence of the presence of radiant heat in the zodiacal light. (Lettre de M. Matthiessen à M. Arago, in the *Comptes Rendus*, t. xvi., 1843, Avril, p. 687.)

ner ? An assumption of the existence of such meteorological causes on the confines of our atmosphere is strengthened by the " sudden flash and pulsation of light," which, according to the acute observations of Olbers, vibrated for several sec-onds through the tail of a comet, which appeared during the continuance of the pulsations of light to be lengthened by sev-eral degrees, and then again contracted.* As, however, the separate particles of a comet's tail, measuring millions of miles,

---

* "What you tell me of the changes of light in the zodiacal light, and of the causes to which you ascribe such changes within the trop-ics, is of the greater interest to me, since I have been for a long time past particularly attentive, every spring, to this phenomenon in our northern latitudes. I, too, have always believed that the zodiacal light rotated; but I assumed (contrary to Poisson's opinion, which you have communicated to me) that it completely extended to the Sun, with considerably augmenting brightness. The light circle which, in total solar eclipses, is seen surrounding the darkened Sun, I have regarded as the brightest portion of the zodiacal light. I have convinced my self that this light is very different in different years, often for several successive years being very bright and diffused, while in other years it is scarcely perceptible. I think that I find the first trace of an allu-sion to the zodiacal light in a letter from Rothmann to Tycho, in which he mentions that in spring he has observed the twilight did not close until the sun was 24° below the horizon. Rothmann must certainly have confounded the disappearance of the setting zodiacal light in the vapors of the western horizon with the actual cessation of twilight. I have failed to observe the pulsations of the light, probably on account of the faintness with which it appears in these countries. You are, however, certainly right in ascribing those rapid variations in the light of the heavenly bodies, which you have perceived in tropical climates, to our own atmosphere, and especially to its higher regions. This is most strikingly seen in the tails of large comets. We often observe, especially in the clearest weather, that these tails exhibit pulsations, commencing from the head, as being the lowest part, and vibrating in one or two seconds through the entire tail, which thus appears rapidly to become some degrees longer, but again as rapidly contracts. That these undulations, which were formerly noticed with attention by Robert Hooke, and in more recent times by Schröter and Chladni, *do not actually occur in the tails of the comets*, but are produced by our at-mosphere, is obvious when we recollect that the individual parts of those tails (which are many millions of miles in length) lie *at very dif-ferent distances* from us, and that the light from their extreme points can only reach us at intervals of time which differ several minutes from one another. Whether what you saw on the Orinoco, not at intervals of seconds, but of minutes, were actual coruscations of the zodiacal light, or whether they belonged exclusively to the upper strata of our atmosphere, I will not attempt to decide; neither can I explain the remarkable *lightness of whole nights*, nor the anomalous augmentation and prolongation of the twilight in the year 1831, particularly if, as has been remarked, the lightest part of these singular twilights did not coin-cide with the Sun's place below the horizon." (From a letter written by Dr. Olbers to myself, and dated Bremen, March 26th, 1833.)

are very unequally distant from the earth, it is not possible, according to the laws of the velocity and transmission of light, that we should be able, in so short a period of time, to perceive any actual changes in a cosmical body of such vast extent. These considerations in no way exclude the reality of the changes that have been observed in the emanations from the more condensed envelopes around the nucleus of a comet, nor that of the sudden irradiation of the zodiacal light from internal molecular motion, nor of the increased or diminished reflection of light in the cosmical vapor of the luminous ring, but should simply be the means of drawing our attention to the differences existing between that which appertains to the air of heaven (the realms of universal space) and that which belongs to the strata of our terrestrial atmosphere. It is not possible, as well-attested facts prove, perfectly to explain the operations at work in the much-contested upper boundaries of our atmosphere. The extraordinary lightness of whole nights in the year 1831, during which small print might be read at midnight in the latitudes of Italy and the north of Germany, is a fact directly at variance with all that we know, according to the most recent and acute researches on the crepuscular theory, and of the height of the atmosphere.* The phenom ena of light depend upon conditions still less understood, and their variability at twilight, as well as in the zodiacal light, excite our astonishment.

We have hitherto considered that which belongs to our solar system—that world of material forms governed by the Sun—which includes the primary and secondary planets, comets of short and long periods of revolution, meteoric asteroids, which move thronged together in streams, either sporadically or in closed rings, and finally a luminous nebulous ring, that revolves round the Sun in the vicinity of the Earth, and for which, owing to its position, we may retain the name of zodiacal light. Every where the law of periodicity governs the motions of these bodies, however different may be the amount of tangential velocity, or the quantity of their agglomerated material parts ; the meteoric asteroids which enter our atmosphere from the external regions of universal space are alone arrested in the course of their planetary revolution, and retained within the sphere of a larger planet. In the solar system, whose boundaries determine the attractive force of the central body, comets are made to revolve in their elliptical

* Biot, *Traité d'Astron. Physique*, 3ème éd., 1841, t. i., p. 171, 238 and 312.

orbits at a distance 44 times greater than that of Uranus; nay, in those comets whose nucleus appears to us, from its inconsiderable mass, like a mere passing cosmical cloud, the Sun exercises its attractive force on the outermost parts of the emanations radiating from the tail over a space of many millions of miles. Central forces, therefore, at once constitute and maintain the system.

Our Sun may be considered as at rest when compared to all the large and small, dense and almost vaporous cosmical bodies that appertain to and revolve around it; but it actually rotates round the common center of gravity of the whole system, which occasionally falls within itself, that is to say, remains within the material circumference of the Sun, whatever changes may be assumed by the positions of the planets. A very different phenomenon is that presented by the translatory motion of the Sun, that is, the progressive motion of the center of gravity of the whole solar system in universal space. Its velocity is such* that, according to Bessel, the relative motion of the Sun, and that of 61 Cygni, is not less in one day than 3,336,000 geographical miles. This change of the entire solar system would remain unknown to us, if the admirable exactness of our astronomical instruments of measurement, and the advancement recently made in the art of observing, did not cause our advance toward remote stars to be perceptible, like an approximation to the objects of a distant shore in apparent motion. The proper motion of the star 61 Cygni, for instance, is so considerable, that it has amounted to a whole degree in the course of 700 years.

The amount or quantity of these alterations in the fixed stars (that is to say, the changes in the relative position of self-luminous stars toward each other), can be determined with a greater degree of certainty than we are able to attach to the genetic explanation of the phenomenon. After taking into consideration what is due to the precession of the equinoxes, and the nutation of the earth's axis produced by the action of the Sun and Moon on the spheroidal figure of our globe, and what may be ascribed to the transmission of light, that is to say, to its aberration, and to the parallax formed by the diametrically opposite position of the Earth in its course round the Sun, we still find that there is a residual portion

* Bessel, in Schum., *Jahrb. für* 1839, s. 51; probably four millions of miles daily, in a *relative* velocity of at the least 3,336,000 miles, or more than double the velocity of revolution of the Earth in her orbit round the Sun.

of the annual motion of the fixed stars due to the translation
of the whole solar system in universal space, and to the true
proper motion of the stars.   The difficult problem of numerically separating these two elements, the true and the apparent motion, has been effected by the careful study of the direction of the motion of certain individual stars, and by the
consideration of the fact that, if all the stars were in a state
of absolute rest, they would appear perspectively to recede
from the point in space toward which the Sun was directing
its course.   But the ultimate result of this investigation, confirmed by the calculus of probabilities, is, that our solar system and the stars both change their places in space.   According to the admirable researches of Argelander at Abo, who
has extended and more perfectly developed the work begun by
William Herschel and Prevost, the Sun moves in the direction of the constellation Hercules, and probably, from the
combination of the observations made of 537 stars, toward a
point lying (at the equinox of 1792·5) at 257° 49'·7 R.A., and
28° 49'·7 N.D.   It is extremely difficult, in investigations of
this nature, to separate the absolute from the relative motion,
and to determine what is alone owing to the solar system.*

If we consider the proper, and not the perspective motions
of the stars, we shall find many that appear to be distributed
in groups, having an opposite direction ; and facts hitherto
observed do not, at any rate, render it a necessary assumption
that all parts of our starry stratum, or the whole of the stellar
islands filling space, should move round one large unknown
luminous or non-luminous central body.   The tendency of the
human mind to investigate ultimate and highest causes certainly inclines the intellectual activity, no less than the imagination of mankind, to adopt such an hypothesis.   Even the
Stagirite proclaimed that " every thing which is moved must
be referable to a motor, and that there would be no end to

---

* Regarding the motion of the solar system, according to Bradley,
Tobias Mayer, Lambert, Lalande, and William Herschel, see Arago, in
the *Annuaire*, 1842, p. 388–399 ; Argelander, in Schum., *Astron. Nachr.*,
No. 363, 364, 398, and in the treatise *Von der eigenen Bewegung des
Sonnensystems* (On the proper Motion of the Solar System), 1837, s. 43,
respecting Perseus as the central body of the whole stellar stratum,
likewise Otho Struve, in the *Bull. de l'Acad. de St. Pétersb.*, 1842, t. x.,
No. 9, p. 137–139.   The last-named astronomer has found, by a more
recent combination, 261° 23' R.A. + 37° 36' Decl. for the direction of
the Sun's motion ; and, taking the mean of his own results with that of
Argelander, we have, by a combination of 797 stars, the formula 259°
9' R.A. + 34° 36' Decl.

the concatenation of causes if there were not one primordial
immovable motor."[*]

The manifold translatory changes of the stars, not those
produced by the parallaxes at which they are seen from the
changing position of the spectator, but the true changes constantly going on in the regions of space, afford us incontrovertible evidence of the *dominion of the laws of attraction* in
the remotest regions of space, beyond the limits of our solar
system. The existence of these laws is revealed to us by
many phenomena, as, for instance, by the motion of double
stars, and by the amount of retarded or accelerated motion in
different parts of their elliptic orbits. Human inquiry need
no longer pursue this subject in the domain of vague conjecture, or amid the undefined analogies of the ideal world ; for
even here the progress made in the method of astronomical
observations and calculations has enabled astronomy to take
up its position on a firm basis. It is not only the discovery
of the astounding numbers of double and multiple stars revolving round a center of gravity lying *without* their system
(2800 such systems having been discovered up to 1837), but
rather the extension of our knowledge regarding the fundamental forces of the whole material world, and the proofs we
have obtained of the universal empire of the laws of attraction, that must be ranked among the most brilliant discoveries
of the age. The periods of revolution of colored stars present
the greatest differences ; thus, in some instances, the period
extends to 43 years, as in η of Corona, and in others to several thousands, as in 66 of Cetus, 38 of Gemini, and 100 of
Pisces. Since Herschel's measurements in 1782, the satellite
of the nearest star in the triple system of ζ of Cancer has completed more than one entire revolution. By a skillful combination of the altered distances and angles of position,[†] the
elements of these orbits may be found, conclusions drawn regarding the absolute distance of the double stars from the
Earth, and comparisons made between their mass and that
of the Sun. Whether, however, here and in our solar system, quantity of matter is the only standard of the amount
of attractive force, or whether *specific* forces of attraction proportionate to the mass may not at the same time come into
operation, as Bessel was the first to conjecture, are questions

[*] Aristot., *de Cœlo*, iii., 2, p. 301, Bekker ; *Phys.*, viii., 5, p. 256.
[†] Savary, in the *Connaissance des Tems*, 1830, p. 56 and 163. Encke,
Berl. Jahrb., 1832, s. 253, &c. Arago, in the *Annuaire*, 1834, p. 260,
295. John Herschel, in the *Memoirs of the Astronom. Soc.*, vol. v., p. 171.

whose practical solution must be left to future ages.* When we compare our Sun with the other fixed stars, that is, with oth er self-luminous Suns in the lenticular starry stratum of which our system forms a part, we find, at least in the case of some, that channels are opened to us, which may lead, at all events, to an *approximate* and limited knowledge of their relative distances, volumes, and masses, and of the velocities of their translatory motion. If we assume the distance of Uranus from the Sun to be nineteen times that of the Earth, that is to say, nineteen times as great as that of the Sun from the Earth, the central body of our planetary system will be 11,900 times the distance of Uranus from the star a in the constellation Centaur, almost 31,300 from 61 Cygni, and 41,600 from Vega in the constellation Lyra. The comparison of the volume of the Sun with that of the fixed stars of the first magnitude is dependent upon the apparent diameter of the latter bodies—an extremely uncertain optical element. If even we assume, with Herschel, that the apparent diameter of Arcturus is only a tenth part of a second, it still follows that the true diameter of this star is eleven times greater than that of the Sun.† The distance of the star 61 Cygni, made known by Bessel, has led approximately to a knowledge of the quantity of matter contained in this body as a double star. Notwithstanding that, since Bradley's observations, the portion of the apparent orbit traversed by this star is not sufficiently great to admit of our arriving with perfect exactness at the true orbit and the major axis of this star, it has been conjectured with much probability by the great Königsberg astronomer,‡ " that the mass of this double star can not be very considerably larger or smaller than half of the mass of the Sun." This result is from actual measurement. The analogies deduced from the relatively larger mass of those planets in our solar system that are attended by satellites, and from the fact that Struve has discovered six times more double stars among

* Bessel, *Untersuchung des Theils der planetarischen Störungen, welche aus der Bewegung der Sonne entstehen* (An Investigation of the portion of the Planetary Disturbances depending on the Motion of the Sun) in *Abh. der Berl. Akad. der Wissensch.*, 1824 (Mathem. Classe), s. 2–6. The question has been raised by John Tobias Mayer, in *Comment. Soc. Reg. Götting.*, 1804–1808, vol. xvi., p. 31–68.

† *Philos. Trans.* for 1803, p. 225. Arago, in the *Annuaire*, 1842, p. 375. In order to obtain a clearer idea of the distances ascribed in a rather earlier part of the text to the fixed stars, let us assume that the Earth is a distance of one foot from the Sun; Uranus is then 19 feet, and Vega Lyræ is 158 geographical miles from it.

‡ Bessel, in Schum., *Jahrb.*, 1839, s. 53.

the brighter than among the telescopic fixed stars, have led other astronomers to conjecture that the average mass of the larger number of the binary stars exceeds the mass of the Sun.* We are, however, far from having arrived at general results regarding this subject. Our Sun, according to Arge-lander, belongs, with reference to proper motion in space, to the class of rapidly-moving fixed stars.

The aspect of the starry heavens, the relative position of stars and nebulæ, the distribution of their luminous masses, the picturesque beauty, if I may so express myself, of the whole firmament, depend in the course of ages conjointly upon the proper motion of the stars and nebulæ, the translation of our solar system in space, the appearance of new stars, and the disappearance or sudden diminution in the intensity of the light of others, and, lastly and specially, on the changes which the Earth's axis experiences from the attraction of the Sun and Moon. The beautiful stars in the constellation of the Centaur and the Southern Cross will at some future time be visible in our northern latitudes, while other stars, as Sirius and the stars in the Belt of Orion, will in their turn disappear below the horizon. The places of the North Pole will suc-cessively be indicated by the stars β and α Cephei, and δ Cygni, until after a period of 12,000 years, Vega in Lyra will shine forth as the brightest of all possible pole stars. These data give us some idea of the extent of the motions which, divided into infinitely small portions of time, proceed without inter-mission in the great chronometer of the universe. If for a moment we could yield to the power of fancy, and imagine the acuteness of our visual organs to be made equal with the extremest bounds of telescopic vision, and bring together that which is now divided by long periods of time, the apparent rest that reigns in space would suddenly disappear. We should see the countless host of fixed stars moving in thronged groups in different directions; nebulæ wandering through space, and becoming condensed and dissolved like cosmical clouds; the vail of the Milky Way separated and broken up in many parts, and *motion* ruling supreme in every portion of the vault of heaven, even as on the Earth's surface, where we see it unfolded in the germ, the leaf, and the blossom, the or-ganisms of the vegetable world. The celebrated Spanish bot-anist Cavanilles was the first who entertained the idea of "seeing grass grow," and he directed the horizontal microme-ter threads of a powerfully magnifying glass at one time to

* Mädler, *Astron.*, s. 476; also in Schum., *Jahrb.*, 1839, s. 95.

the apex of the shoot of a bambusa, and at another on the
rapidly-growing stem of an American aloe (*Agave Americana*),
precisely as the astronomer places his cross of net-work against
a culminating star.   In the collective life of physical nature,
in the organic as in the sidereal world, all things that have
been, that are, and will be, are alike dependent on motion.

The breaking up of the Milky Way, of which I have just
spoken, requires special notice.   William Herschel, our safe
and admirable guide to this portion of the regions of space,
has discovered by his star-guagings that the telescopic breadth
of the Miiky Way extends from six to seven degrees beyond
what is indicated by our astronomical maps and by the extent
of the sidereal radiance visible to the naked eye.*   The two
brilliant nodes in which the branches of the zone unite, in the
region of Cepheus and Cassiopeia, and in the vicinity of Scorpio and Sagittarius, appear to exercise a powerful attraction
on the contiguous stars ; in the most brilliant part, however,
between β and γ Cygni, one half of the 330,000 stars that
have been discovered in a breadth of 5° are directed toward
one side, and the remainder to the other.   It is in this part
that Herschel supposes the layer to be broken up.†   The number of telescopic stars in the Milky Way uninterrupted by any
nebulæ is estimated at 18 millions.   In order, I will not say,
to realize the greatness of this number, but, at any rate, to
compare it with something analogous, I will call attention to
the fact that there are not in the whole heavens more than
about 8000 stars, between the first and the sixth magnitudes,
visible to the naked eye.   The barren astonishment excited
by numbers and dimensions in space, when not considered
with reference to applications engaging the mental and perceptive powers of man, is awakened in both extremes of the
universe, in the celestial bodies as in the minutest animalcules.‡   A cubic inch of the polishing slate of Bilin contains,
according to Ehrenberg, 40,000 millions of the silicious shells
of Galionellæ.

The stellar Milky Way, in the region of which, according to
Argelander's admirable observations, the brightest stars of the
firmament appear to be congregated, is almost at right angles

* Sir William Herschel, in the *Philos. Transact.* for 1817, Part ii
p. 328.   † Arago, in the *Annuaire*, 1842, p. 459.
‡ Sir John Herschel, in a letter from Feldhuysen, dated Jan. 13th,
1836.   Nicholl, *Architecture of the Heavens*, 1838, p. 22.   (See, also,
some separate notices by Sir William Herschel on the starless space
which separates us by a great distance from the Milky Way, in the
Philos. Transact. for 1817, Part ii., p. 328.)

with another Milky Way, composed of nebulæ. The former constitutes, according to Sir John Herschel's views, an annulus, that is to say, an independent zone, somewhat remote from our lenticular-shaped starry stratum, and similar to Saturn's ring. Our planetary system lies in an eccentric direction, nearer to the region of the Cross than to the diametrically opposite point, Cassiopeia.* An imperfectly seen nebulous spot, discovered by Messier in 1774, appeared to present a remarkable similarity to the form of our starry stratum and the divided ring of our Milky Way.† The Milky Way composed of nebulæ does not belong to our starry stratum, but surrounds it at a great distance without being physically connected with it, passing almost in the form of a large cross through the dense nebulæ of Virgo, especially in the northern wing, through Comæ Berenicis, Ursa Major, Andromeda's girdle, and Pisces Boreales. It probably intersects the stellar Milky Way in Cassiopeia, and connects its dreary poles (rendered starless from the attractive forces by which stellar bodies are made to agglomerate into groups) in the least dense portion of the starry stratum.

We see from these considerations that our starry cluster, which bears traces in its projecting branches of having been subject in the course of time to various metamorphoses, and evinces a tendency to dissolve and separate, owing to secondary centers of attraction—is surrounded by two rings, one of which, the nebulous zone, is very remote, while the other is nearer, and composed of stars alone. The latter, which we generally term the Milky Way, is composed of nebulous stars, averaging from the tenth to the eleventh degree of magnitude,‡ but appearing, when considered individually, of very different magnitudes, while isolated starry clusters (starry swarms) almost always exhibit throughout a character of great uniformity in magnitude and brilliancy.

In whatever part the vault of heaven has been pierced by powerful and far-penetrating telescopic instruments, stars or luminous nebulæ are every where discoverable, the former, in

* Sir John Herschel, *Astronom.*, § 624; likewise in his *Observations on Nebulæ and Clusters of Stars* (*Phil. Transact.*, 1833, Part ii., p. 479, fig. 25): "We have here a brother system, bearing a real physical resemblance and strong analogy of structure to our own."

† Sir William Herschel, in the *Phil. Trans.* for 1785, Part i., p. 257. Sir John Herschel, *Astron.*, § 616. ("The *nebulous* region of the heavens forms *a nebulous Milky Way*, composed of distinct nebulæ, as the other of stars." The same observation was made in a letter he addressed to me in March, 1829.) ‡ Sir John Herschel, *Astron.*, § 585.

some cases, not exceeding the twentieth or twenty-fourth de‐
gree of telescopic magnitude. A portion of the nebulous vapor
would probably be found resolvable into stars by more power‐
ful optical instruments. As the retina retains a less vivid im‐
pression of separate than of infinitely near luminous points,
less strongly marked photometric relations are excited in the
latter case, as Arago has recently shown.* The definite or
amorphous cosmical vapor so universally diffused, and which
generates heat through condensation, probably modifies the
transparency of the universal atmosphere, and diminishes that
uniform intensity of light which, according to Halley and Ol‐
bers, should arise, if every point throughout the depths of space
were filled by an infinite series of stars.† The assumption of
such a distribution in space is, however, at variance with ob‐
servation, which shows us large starless regions of space, *open‐
ings* in the heavens, as William Herschel terms them—one,
four degrees in width, in Scorpio, and another in Serpentari‐
us. In the vicinity of both, near their margin, we find un‐
resolvable nebulæ, of which that on the western edge of the
opening in Scorpio is one of the most richly thronged of the
clusters of small stars by which the firmament is adorned.
Herschel ascribes these openings or starless regions to the at‐
tractive and agglomerative forces of the marginal groups.‡
" They are parts of our starry stratum," says he, with his
usual graceful animation of style, " that have experienced
great devastation from time." If we picture to ourselves the
telescopic stars lying behind one another as a starry canopy
spread over the vault of heaven, these starless regions in Scor‐
pio and Serpentarius may, I think, be regarded as tubes
through which we may look into the remotest depths of space.
Other stars may certainly lie in those parts where the strata
forming the canopy are interrupted, but these are unattainable
by our instruments. The aspect of fiery meteors had led the
ancients likewise to the idea of clefts or openings (*chasmata*)
in the vault of heaven. These openings were, however, only
regarded as transient, while the reason of their being luminous
and fiery, instead of obscure, was supposed to be owing to the

* Arago, in the *Annuaire*, 1842, p. 282–285, 409–411, and 439–442.
† Olbers, on the transparency of celestial space, in Bode's *Jahrb.,*
1826, s. 110–121.
‡ "An opening in the heavens." William Herschel, in the *Phil. Trans.*
for 1785, vol. lxxv., Part i., p. 256. Le Français Lalande, in the *Con‐
naiss. des Tems pour l'An.* VIII., p. 383. Arago, in the *Annuaire*,
1842, p. 425.

translucent illuminated ether which lay beyond them.* Derham, and even Huygens, did not appear disinclined to explain in a similar manner the mild radiance of the nebulæ.†

When we compare the stars of the first magnitude, which, on an average, are certainly the nearest to us, with the non-nebulous telescopic stars, and further, when we compare the nebulous stars with unresolvable nebulæ, for instance, with the nebula in Andromeda, or even with the so-called planetary nebulous vapor, a fact is made manifest to us by the consideration of the varying distances and the boundlessness of space, which shows the world of phenomena, and that which constitutes its causal reality, to be dependent upon the *propagation of light*. The velocity of this propagation is, according to Struve's most recent investigations, 166,072 geographical miles in a second, consequently almost a million of times greater than the velocity of sound. According to the measurements of Maclear, Bessel, and Struve, of the parallaxes and distances of three fixed stars of very unequal magnitudes (*a* Centauri, 16 Cygni, and *a* Lyræ), a ray of light requires respectively 3, 9¼, and 12 years to reach us from these three bodies. In the short but memorable period between 1572 and 1604, from the time of Cornelius Gemma and Tycho Brahe to that of Kepler, three new stars suddenly appeared in Cassiopeia and Cygnus, and in the foot of Serpentarius. A similar phenomenon exhibited itself at intervals in 1670, in the constellation Vulpis. In recent times, even since 1837, Sir John Herschel has observed, at the Cape of Good Hope, the brilliant star η in Argo increase in splendor from the second to the first magnitude.‡ These events in the universe belong, however, with reference to their historical reality, to other periods of time than those in which the phenomena of light are first revealed to the inhabitants of the Earth : they reach us like the voices of the past. It has been truly said, that with our large and powerful telescopic instruments we penetrate alike through the boundaries of time and space : we measure the former through the latter, for in the course of an

* Aristot., *Meteor.*, ii., 5, 1. Seneca, *Natur. Quæst.*, i., 14, 2. "Cœlum discessisse," in Cic., *de Divin.*, i., 43.

† Arago, in the *Annuaire*, 1842, p. 429.

‡ In December, 1837, Sir John Herschel saw the star η Argo, which till that time appeared as of the second magnitude, and liable to no change, rapidly increase till it became of the first magnitude. In January, 1838, the intensity of its light was equal to that of *a* Centauri. According to our latest information, Maclear, in March, 1843, found it as bright as Canopus; and even *a* Crucis looked faint by η Argo.

hour a ray of light traverses over a space of 592 millions of miles. While, according to the theogony of Hesiod, the dimensions of the universe were supposed to be expressed by the time occupied by bodies in falling to the ground ("the brazen anvil was not more than nine days and nine nights in falling from heaven to earth"), the elder Herschel was of opinion* that light required almost two millions of years to pass to the Earth from the remotest luminous vapor reached by his forty-foot reflector. Much, therefore, has vanished long before it is rendered visible to us—much that we see was once differently arranged from what it now appears. The aspect of the starry heavens presents us with the spectacle of that which is only apparently simultaneous, and however much we may endeavor, by the aid of optical instruments, to bring the mildly-radiant vapor of nebulous masses or the faintly-glimmering starry clusters nearer, and diminish the thousands of years interposed between us and them, that serve as a criterion of their distance, it still remains more than probable, from the knowledge we possess of the velocity of the transmission of luminous rays, that the light of remote heavenly bodies presents us with the most ancient perceptible evidence of the existence of matter. It is thus that the reflective mind of man is led from simple premises to rise to those exalted heights of nature, where, in the light-illumined realms of space, "myriads of worlds are bursting into life like the grass of the night."†

From the regions of celestial forms, the domain of Uranus, we will now descend to the more contracted sphere of terrestrial forces—to the interior of the Earth itself. A mysterious chain links together both classes of phenomena. According to the ancient signification of the Titanic myth,‡ the powers of organic life, that is to say, the great order of nature, depend upon the combined action of heaven and earth. If we suppose that the Earth, like all the other planets, primordially belonged, according to its origin, to the central body, the Sun, and to the solar atmosphere that has been separated into neb-

---

* "Hence it follows that the rays of light of the remotest nebulæ must have been almost two millions of years on their way, and that consequently, so many years ago, this object must already have had an existence in the sidereal heaven, in order to send out those rays by which we now perceive it." William Herschel, in the *Phil. Trans.* for 1802, p. 498. John Herschel, *Astron.*, § 590. Arago, in the *Annuaire*, 1842. p. 334, 359, and 382–385.

† From my brother's beautiful sonnet "Freiheit und Gesetz." (Wilhelm von Humboldt, *Gesammelte Werke*, bd. iv., s. 358, No. 25.)

‡ Otfried Müller, *Prolegomena*, s. 373.

ulous rings, the same connection with this contiguous Sun, as well as with all the remote suns that shine in the firmament, is still revealed through the phenomena of light and radiating heat. The difference in the degree of these actions must not lead the physicist, in his delineation of nature, to forget the connection and the common empire of similar forces in the universe. A small fraction of telluric heat is derived from the regions of universal space in which our planetary system is moving, whose temperature (which, according to Fourier, is almost equal to our mean icy polar heat) is the result of the combined radiation of all the stars. The causes that more powerfully excite the light of the Sun in the atmosphere and in the upper strata of our air, that give rise to heat-engendering electric and magnetic currents, and awaken and genially vivify the vital spark in organic structures on the earth's surface, must be reserved for the subject of our future consideration.

As we purpose for the present to confine ourselves exclusively within the telluric sphere of nature, it will be expedient to cast a preliminary glance over the relations in space of solids and fluids, the form of the Earth, its mean density, and the partial distribution of this density in the interior of our planet, its temperature and its electro-magnetic tension. From the consideration of these relations in space, and of the forces inherent in matter, we shall pass to the reaction of the interior on the exterior of our globe ; and to the special consideration of a universally distributed natural power—subterranean heat ; to the phenomena of earthquakes, exhibited in unequally expanded circles of commotion, which are not referable to the action of dynamic laws alone ; to the springing forth of hot wells ; and, lastly, to the more powerful actions of volcanic processes. The crust of the Earth, which may scarcely have been perceptibly elevated by the sudden and repeated, or almost uninterrupted shocks by which it has been moved from below, undergoes, nevertheless, great changes in the course of centuries in the relations of the elevation of solid portions, when compared with the surface of the liquid parts, and even in the form of the bottom of the sea. In this manner simultaneous temporary or permanent fissures are opened, by which the interior of the Earth is brought in contact with the external atmosphere. Molten masses, rising from an unknown depth, flow in narrow streams along the declivity of mountains, rushing impetuously onward, or moving slowly and gently, until the fiery source is quenched in the midst of exhalations, and the lava becomes incrusted, as it were, by the solidification of its outer surface. New masses of rocks are thus formed before our eyes, while the older ones are in their turn converted into other forms by the greater or lesser agency of Plutonic forces. Even where no disruption takes place the crystalline molecules are displaced, combining to form bodies of denser texture. The water presents structures of a totally different nature, as, for instance, concretions of animal and vegetable remains, of earthy, calcareous, or aluminous precipitates, agglomerations of finely-pulverized mineral bodies, covered with layers of the silicious shields of infusoria, and with transported soils containing the bones of fossil animal forms of a more ancient world. The study of the strata which are so differently formed and arranged before our eyes, and of all that has been so variously dislocated, contorted, and upheaved, by mutual compression and volcanic force, leads the reflective observer, by simple analogies, to draw a com parison between the present and an age that has long passed. It is by a combination of actual phenomena, by an ideal enlargement of relations in space, and of the amount of active forces, that we are able to advance into the long sought and indefinitely anticipated domain of geognosy, which has only within the last half century been based on the solid foundation of scientific deduction.

It has been acutely remarked, " that, notwithstanding our continual employment of large telescopes, we are less acquainted with the exterior than with the interior of other planets, excepting, perhaps, our own satellite." They have been weighed, and their volume measured ; and their mass and density are becoming known with constantly-increasing exactness ; thanks to the progress made in astronomical observation and calculation. Their physical character is, however, hidden in obscurity, for it is only in our own globe that we can be brought in immediate contact with all the elements of organic and inorganic creation. The diversity of the most heterogeneous substances, their admixtures and metamorphoses, and the ever-changing play of the forces called into action, afford to the human mind both nourishment and enjoyment, and open an immeasurable field of observation, from which the intellectual activity of man derives a great portion of its grandeur and power. The world of perceptive phenomena is reflected in the depths of the ideal world, and the richness of nature and the mass of all that admits of classification gradually become the objects of inductive reasoning.

I would here allude to the advantage, of which I have already spoken, possessed by that portion of physical science whose origin is familiar to us, and is connected with our earthly existence. The physical description of celestial bodies, from the remotely-glimmering nebulæ with their suns, to the central body of our own system, is limited, as we have seen, to general conceptions of the volume and quantity of matter. No manifestation of vital activity is there presented to our senses. It is only from analogies, frequently from purely ideal combinations, that we hazard conjectures on the specific elements of matter, or on their various modifications in the different planetary bodies. But the physical knowledge of the heterogeneous nature of matter, its chemical differences, the regular forms in which its molecules combine together, whether in crystals or granules; its relations to the deflected or decomposed waves of light by which it is penetrated; to radiating, transmitted, or polarized heat; and to the brilliant or invisible, but not, on that account, less active phenomena of electro-magnetism—all this inexhaustible treasure, by which the enjoyment of the contemplation of nature is so much heightened, is dependent on the surface of the planet which we inhabit, and more on its solid than on its liquid parts. I have already remarked how greatly the study of natural objects and forces, and the infinite diversity of the sources they open for our consideration, strengthen the mental activity, and call into action every manifestation of intellectual progress. These relations require, however, as little comment as that concatenation of causes by which particular nations are permitted to enjoy a superiority over others in the exercise of a material power derived from their command of a portion of these elementary forces of nature.

If, on the one hand, it were necessary to indicate the difference existing between the nature of our knowledge of the Earth and of that of the celestial regions and their contents, I am no less desirous, on the other hand, to draw attention to the limited boundaries of that portion of space from which we derive all our knowledge of the heterogeneous character of matter. This has been somewhat inappropriately termed the Earth's crust; it includes the strata most contiguous to the upper surface of our planet, and which have been laid open before us by deep fissure-like valleys, or by the labors of man, in the bores and shafts formed by miners. These labors*

* In speaking of the greatest depths within the Earth reached by human labor, we must recollect that there is a difference between the *absolute depth* (that is to say, the depth below the Earth's surface at that

do not extend beyond a vertical depth of somewhat more than
2000 feet (about one third of a geographical mile) below the

point) and the *relative depth* (or that beneath the level of the sea). The
greatest relative depth that man has hitherto reached is probably the
bore at the new salt-works at Minden, in Prussia: in June, 1844, it
was exactly 1993 feet, the absolute depth being 2231 feet. The tem
perature of the water at the bottom was 91° F., which, assuming the
mean temperature of the air at 49°·3, gives an augmentation of temperature of 1° for every 54 feet. The absolute depth of the Artesian
well of Grenelle, near Paris, is only 1795 feet. According to the account of the missionary Imbert, the fire-springs, " Ho-tsing," of the Chinese, which are sunk to obtain [carbureted] hydrogen gas for salt-boil-
ing, far exceed our Artesian springs in depth. In the Chinese province
of Szü-tschuan these fire-springs are very commonly of the depth of
more than 2000 feet; indeed, at Tseu-lieu-tsing (the place of continual
flow) there is a Ho-tsing which, in the year 1812, was found to be 3197
feet deep. (Humboldt, *Asie Centrale*, t. ii., p. 521 and 525. *Annales
de l'Association de la Propagation de la Foi*, 1829, No. 16, p. 369.)

The relative depth reached at Mount Massi, in Tuscany, south ·of
Volterra, amounts, according to Matteuci, to only 1253 feet. The boring at the new salt-works near Minden is probably of about the same
relative depth as the coal-mine at Apendale, near Newcastle-under-
Lyme, in Staffordshire, where men work 725 yards below the surface
of the earth. (Thomas Smith, *Miner's Guide*, 1836, p. 160.) Unfortunately, I do not know the exact height of its mouth above the level
of the sea. The relative depth of the Monk-wearmouth mine, near
Newcastle, is only 1496 feet. (Phillips, in the *Philos. Mag.*, vol. v.,
1834, p. 446.) That of the Liege coal-mine, *l'Espérançe*, at Seraing,
is 1355 feet, according to M. von Dechen, the director; and the old
mine of Marihaye, near Val-St.-Lambert, in the valley of the Maes,
is, according to M. Gernaert, Ingénieur des Mines, 1233 feet in depth.
The works of greatest absolute depth that have ever been formed
are for the most part situated in such elevated plains or valleys that
they either do not descend so low as the level of the sea, or at most
reach very little below it. Thus the Eselschacht, at Kuttenberg, in Bohemia, a mine which can not now be worked, had the enormous absolute depth of 3778 feet. (Fr. A. Schmidt, *Berggesetze der öster Mon.*,
abth. i., bd. i., s. xxxii.) Also, at St. Daniel and at Geish, on the Rörerbühel, in the *Landgericht* (or provincial district) of Kitzbühl, there
were, in the sixteenth century, excavations of 3107 feet. The plans
of the works of the Rörerbühel are still preserved. (See Joseph von
Sperges, *Tyroler Bergwerksgeschichte*, s. 121. Compare, also, Humboldt, *Gutachten über Herantreibung des Meissner Stollens in die Frei-
berger Erzrevier*, printed in Herder, *über den jetz begonnenen Erbstol-
len*, 1838, s. cxxiv.) We may presume that the knowledge of the extraordinary depth of the Rörerbühel reached England at an early period,
for I find it remarked in Gilbert, *de Magnete*, that men have penetrated
2400 or even 3000 feet into the crust of the Earth. (" Exigua videtur
terræ portio, quæ unquam hominibus spectanda emerget aut eruitur;
cum profundius in ejus viscera, ultra · fflorescentis extremitatis corrupte-
lam, aut propter aquas in magnis fodin. tanquam per venas scaturientes
aut propter aeris salubrioris ad vitam o erariorum sustinendam neces-
sarii defectum, aut propter ingentes sumṭtus ad tantos labores exant-
landos, multasque difficultates, ad profundi ᵎres terræ partes penetrare

level of the sea, and consequently only about $\frac{1}{9800}$th of the Earth's radius. The crystalline masses that have been erupted from active volcanoes, and are generally similar to the rocks on the upper surface, have come from depths which, although not accurately determined, must certainly be sixty times greater than those to which human labor has been enabled to penetrate. We are able to give in numbers the depth of the shaft where the strata of coal, after penetrating a certain way, rise again at a distance that admits of being accurately defined by measurements. These dips show that the carboniferous strata, together with the fossil organic remains which they contain, must lie, as, for instance, in Belgium, more than five or six thousand feet* below the present level

non possumus; adeo ut quadringentas aut [quod rarissime] quingentas orgyas in quibusdam metallis descendisse, stupendus omnibus videatur conatus." — Gulielmi Gilberti, Colcestrensis, *de Magnete Physiologia nova.* Lond., 1600, p. 40.)

The absolute depth of the mines in the Saxon Erzgebirge, near Frei burg, are : in the Thurmhofer mines, 1944 feet; in the Honenbirker mines, 1827 feet; the relative depths are only 677 and 277 feet, if, in order to calculate the elevation of the mine's mouth above the level of the sea, we regard the elevation of Freiburg as determined by Reich's recent observations to be 1209 feet. The absolute depth of the celebrated mine of Joachimsthal, in Bohemia (Verkreuzung des Jung Hauer Zechen-und Andreasganges), is full 2120 feet; so that, as Von Dechen's measurements show that its surface is about 2388 feet above the level of the sea, it follows that the excavations have not as yet reached that point. In the Harz, the Samson mine at Andreasberg has an absolute depth of 2197 feet. In what was formerly Spanish America, I know of no mine deeper than the Valenciana, near Guanaxuato (Mexico), where I found the absolute depth of the Planes de San Bernardo to be 1686 feet; but these planes are 5960 feet above the level of the sea. If we compare the depth of the old Kuttenberger mine (a depth greater than the height of our Brocken, and only 200 feet less than that of Vesuvius) with the loftiest structures that the hands of man have erected (with the Pyramid of Cheops and with the Cathedral of Strasburg), we find that they stand in the ratio of eight to one. In this note I have collected all the certain information I could find regarding the greatest absolute and relative depths of mines and borings. In descending eastward from Jerusalem toward the Dead Sea, a view presents itself to the eye, which, according to our present hypsometrical knowledge of the surface of our planet, is unrivaled in any country; as we approach the open ravine through which the Jordan takes its course, we tread, with the open sky above us, on rocks which, according to the barometric measurements of Berton and Russegger, are 1385 feet below the level of the Mediterranean. (Humboldt, *Asie Centrale*, th. ii., p. 323.)

* Basin-shaped curved strata, which dip and reappear at measurable distances, although their deepest portions are beyond the reach of the miner, afford sensible evidence of the nature of the earth's crust at great depths below its surface. Testimony of this kind possesses, consequently, a great geognostic interest. I am indebted to that excellent geog-

ɔf the sea, and that the calcareous and the curved strata of the Devonian basin penetrate twice that depth. If we compare these subterranean basins with the summits of mountains that have hitherto been considered as the most elevated portions of the raised crust of the Earth, we obtain a distance of 37,000 feet (about seven miles), that is, about the $\frac{1}{524}$th of the Earth's radius. These, therefore, would be the limits of vertical depth and of the superposition of mineral strata to which geognostical inquiry could penetrate, even if the general elevation of the upper surface of the earth were equal to the height of the Dhawalagiri in the Himalaya, or of the Sorata in Bolivia. All that lies at a greater depth below the level of the sea than the shafts or the basins of which I have spoken, the limits to which man's labors have penetrated, or than the depths to which the sea has in some few instances been sounded (Sir James Ross was unable to find bottom with 27,600 feet of line), is as much unknown to us as the interior of the other planets of our solar system. We only know the mass of the whole Earth and its mean density by comparing it with the open strata, which alone are accessible to us. In the interior of the Earth, where all knowledge of its chemical and mineralogical character fails, we are again limited to as pure conjecture, as in the remotest bodies that revolve round the Sun. We can determine nothing with certainty regarding the depth at which the geological strata must be supposed to be in state of softening or of liquid fusion, of the cavities occupied by elastic vapor, of the condition of fluids when heated under an enormous pressure, or of the law of the in-

nosist, Von Dechen, for the following observations. " The depth of the coal basin of Liège, at Mont St. Gilles, which I, in conjunction with our friend Von Oeynhausen, have ascertained to be 3890 feet below the surface, extends 3464 feet below the surface of the sea, for the absolute height of Mont St. Gilles certainly does not much exceed 400 feet; the coal basin of Mons is fully 1865 feet deeper. But all these depths are trifling compared with those which are presented by the coal strata of Saar-Revier (Saarbrücken). I have found, after repeated examinations, that the lowest coal stratum which is known in the neighborhood of Duttweiler, near Bettingen, northeast of Saarlouis, must descend to depths of 20,682 and 22,015 feet (or 3·6 geographical miles) below the level of the sea." This result exceeds, by more than 8000 feet, the assumption made in the text regarding the basin of the Devonian strata. This coal-field is therefore sunk as far below the surface of the sea as Chimborazo is elevated above it—at a depth at which the Earth's temperature must be as high as 435° F. Hence, from the highest pinnacles of the Himalaya to the lowest basins containing the vegetation of an earlier world, there is a vertical distance of about 48,000 feet, or of the 435th part of the Earth's radius.

crease of density from the upper surface to the center of the Earth.

The consideration of the increase of heat with the increase of depth toward the interior of our planet, and of the reaction of the interior on the external crust, leads us to the long series of volcanic phenomena. These elastic forces are manifested in earthquakes, eruptions of gas, hot wells, mud volcanoes and lava currents from craters of eruptions, and even in producing alterations in the level of the sea.* Large plains and variously indented continents are raised or sunk, lands are sep arated from seas, and the ocean itself, which is permeated by hot and cold currents, coagulates at both poles, converting water into dense masses of rock, which are either stratified and fixed, or broken up into floating banks. The boundaries of sea and land, of fluids and solids, are thus variously and frequently changed. Plains have undergone oscillatory movements, being alternately elevated and depressed. After the elevation of continents, mountain chains were raised upon long fissures, mostly parallel, and, in that case, probably cotemporaneous ; and salt lakes and inland seas, long inhabited by the same creatures, were forcibly separated, the fossil remains of shells and zoophytes still giving evidence of their original connection. Thus, in following phenomena in their mutual dependence, we are led from the consideration of the forces acting in the interior of the Earth to those which cause eruptions on its surface, and by the pressure of elastic vapors give rise to burning streams of lava that flow from open fissures.

The same powers that raised the chains of the Andes and the Himalaya to the regions of perpetual snow, have occasioned new compositions and new textures in the rocky masses, and have altered the strata which had been previously deposited from fluids impregnated with organic substances. We here trace the series of formations, divided and superposed according to their age, and depending upon the changes of configuration of the surface, the dynamic relations of upheaving forces, and the chemical action of vapors issuing from the fissures.

The form and distribution of continents, that is to say, of that solid portion of the Earth's surface which is suited to the luxurious development of vegetable life, are associated by intimate connection and reciprocal action with the encircling

---

* [See Daubeney *On Volcanoes*, 2d edit., 1848, p. 539, &c., on the so-called *mud volcanoes*, and the reasons advanced in favor of adopting the term " salses" to designate these phenomena.]—*Tr.*

sea, in which organic life is almost entirely limited to the animal world.   The liquid element is again covered by the atmosphere, an aërial ocean in which the mountain chains and
high plains of the dry land rise like shoals, occasioning a variety of currents and changes of temperature, collecting vapor
from the region of clouds, and distributing life and motion by
the action of the streams of water which flow from their declivities.

While the geography of plants and animals depends on
these intricate relations of the distribution of sea and land, the
configuration of the surface. and the direction of isothermal
lines (or zones of equal mean annual heat), we find that the
case is totally different when we consider the human race—
the last and noblest subject in a physical description of the
globe.   The characteristic differences in races, and their relative numerical distribution over the Earth's surface, are conditions affected not by natural relations alone, but at the same
time and specially, by the progress of civilization, and by moral
and intellectual cultivation, on which depends the political
superiority that distinguishes national progress.   Some few
races, clinging, as it were, to the soil, are supplanted and ruined
by the dangerous vicinity of others more civilized than themselves, until scarce a trace of their existence remains.   Other
races, again, not the strongest in numbers, traverse the liquid
element, and thus become the first to acquire, although late,
a geographical knowledge of at least the maritime lands of the
whole surface of our globe, from pole to pole.

I have thus, before we enter on the individual characters
of that portion of the delineation of nature which includes the
sphere of telluric phenomena, shown generally in what manner the consideration of the form of the Earth and the incessant action of electro-magnetism and subterranean heat may
enable us to embrace in one view the relations of horizontal
expansion and elevation on the Earth's surface, the geognostic
type of formations, the domain of the ocean (of the liquid portions of the Earth), the atmosphere with its meteorological
processes, the geographical distribution of plants and animals,
and, finally, the physical gradations of the human race, which
is, exclusively and every where, susceptible of intellectual culture.   This unity of contemplation presupposes a connection
of phenomena according to their internal combination.   A
mere tabular arrangement of these facts would not fulfill the
object I have proposed to myself, and would not satisfy that
requirement for cosmical presentation awakened in me by the

aspect of nature in my journeyings by sea and land, by the
careful study of forms and forces, and by a vivid impression
of the unity of nature in the midst of the most varied portions
of the Earth. In the rapid advance of all branches of physical
science, much that is deficient in this attempt will, perhaps,
at no remote period, be corrected, and rendered more perfect,
for it belongs to the history of the development of knowledge
that portions which have long stood isolated become gradually
connected, and subject to higher laws. I only indicate the
empirical path in which I and many others of similar pursuits
with myself are advancing, full of expectation that, as Plato
tells us Socrates once desired, "Nature may be interpreted by
reason alone."[*]

The delineation of the principal characteristics of telluric
phenomena must begin with the form of our planet and its
relations in space. Here, too, we may say that it is not only
the mineralogical character of rocks, whether they are crystalline, granular, or densely fossiliferous, but the geometrical
form of the Earth itself, which indicates the mode of its origin,
and is, in fact, its history. An elliptical spheroid of revolution gives evidence of having once been a soft or fluid mass.
Thus the Earth's compression constitutes one of the most ancient geognostic events, as every attentive reader of the book
of nature can easily discern; and an analogous fact is presented in the case of the Moon, the perpetual direction of whose
axes toward the Earth, that is to say, the increased accumulation of matter on that half of the Moon which is turned toward us, determines the relations of the periods of rotation and
revolution, and is probably cotemporaneous with the earliest
epoch in the formative history of this satellite. The mathematical figure of the Earth is that which it would have were
its surface covered entirely by water in a state of rest; and it
is this assumed form to which all geodesical measurements of
degrees refer. This mathematical surface is different from
that true physical surface which is affected by all the acci-
dents and inequalities of the solid parts.[†] The whole figure
of the Earth is determined when we know the amount of the

* Plato, *Phædo*, p. 97. (Arist., *Metaph.*, p. 985.) Compare Hegel,
Philosophie der Geschichte, 1840, s. 16.

† Bessel, *Allgemeine Betrachtungen über Gradmessungen nach astro-
nomisch-geodätischen Arbeiten*, at the conclusion of Bessel and Baeyer,
Gradmessung in Ostpreussen, s. 427. Regarding the accumulation of
matter on the side of the Moon turned toward us (a subject noticed
in an earlier part of the text), see Laplace, *Expos. du Syst. du Monde*,
p. 308.

compression at the poles and the equatorial diameter; in order, however, to obtain a perfect representation of its form it
is necessary to have measurements in two directions, perpendicular to one another.

Eleven measurements of degrees (or determinations of the
curvature of the Earth's surface in different parts), of which
nine only belong to the present century, have made us acquainted with the size of our globe, which Pliny named " a
point in the immeasurable universe."* If these measurements
do not always accord in the curvatures of different meridians
under the same degree of latitude, this very circumstance
speaks in favor of the exactness of the instruments and the
methods employed, and of the accuracy and the fidelity to
nature of these partial results. The conclusion to be drawn
from the increase of forces of attraction (in the direction from
the equator to the poles) with respect to the figure of a planet
is dependent on the distribution of density in its interior.
Newton, from theoretical principles, and perhaps likewise
prompted by Cassini's discovery, previously to 1666, of the
compression of Jupiter,† determined, in his immortal work,
Philosophiæ Naturalis Principia, that the compression of the
Earth, as a homogeneous mass, was $\frac{1}{230}$th. Actual meas-

---

* Plin., ii., 68. Seneca, *Nat. Quæst., Præf.*, c. ii. "El mundo es
poco" (the Earth is small and narrow), writes Columbus from Jamaica
to Queen Isabella on the 7th of July, 1503: not because he entertained
the philosophic views of the aforesaid Romans, but because it appeared
advantageous to him to maintain that the journey from Spain was not
long, if, as he observes, "we seek the east from the west." Compare
my *Examen Crit. de l'Hist. de la Géogr. du 15me Siècle*, t. i., p. 83, and
t. ii., p. 327, where I have shown that the opinion maintained by De-
lisle, Fréret, and Gosselin, that the excessive differences in the statements regarding the Earth's circumference, found in the writings of
the Greeks, are only apparent, and dependent on different values being
attached to the stadia, was put forward as early as 1495 by Jaime Fer-
rer, in a proposition regarding the determination of the line of demarkation of the papal dominions.

† Brewster, *Life of Sir Isaac Newton*, 1831, p. 162. "The discovery
of the spheroidal form of Jupiter by Cassini had probably directed the
attention of Newton to the determination of its cause, and, consequently, to the investigation of the true figure of the Earth." Although Cassini did not announce the amount of the compression of Jupiter ($\frac{1}{15}$th)
till 1691 (*Anciens Mémoires de l'Acad. des Sciences*, t. ii., p. 108), yet
we know from Lalande (*Astron.*, 3me éd., t. iii., p. 335) that Moraldi
possessed some printed sheets of a Latin work, "On the Spots of the
Planets," commenced by Cassini, from which it was obvious that he
was aware of the compression of Jupiter before the year 1666, and
therefore at least twenty-one years before the publication of Newton's
Principia.

urements, made by the aid of new and more perfect analysis, have, however, shown that the compression of the poles of the terrestrial spheroid, when the density of the strata is regarded as increasing toward the center, is very nearly $\frac{1}{300}$th.

Three methods have been employed to investigate the curvature of the Earth's surface, viz., measurements of degrees, oscillations of the pendulum, and observations of the inequalities in the Moon's orbit. The first is a direct geometrical and astronomical method, while in the other two we determine from accurately observed movements the amount of the forces which occasion those movements, and from these forces we arrive at the cause from whence they have originated, viz., the compression of our terrestrial spheroid. In this part of my delineation of nature, contrary to my usual practice, I have instanced methods because their accuracy affords a striking illustration of the intimate connection existing among the forms and forces of natural phenomena, and also because their application has given occasion to improvements in the exactness of instruments (as those employed in the measurements of space) in optical and chronological observations; to greater perfection in the fundamental branches of astronomy and mechanics in respect to lunar motion and to the resistance experienced by the oscillations of the pendulum; and to the discovery of new and hitherto untrodden paths of analysis. With the exception of the investigations of the parallax of stars, which led to the discovery of aberration and nutation, the history of science presents no problem in which the object attained—the knowledge of the compression and of the irregular form of our planet—is so far exceeded in importance by the incidental gain which has accrued, through a long and weary course of investigation, in the general furtherance and improvement of the mathematical and astronomical sciences. The comparison of eleven measurements of degrees (in which are included three extra-European, namely, the old Peruvian and two East Indian) gives, according to the most strictly theoretical requirements allowed for by Bessel,* a compression

* According to Bessel's examination of ten measurements of degrees, in which the error discovered by Puissant in the calculation of the French measurements is taken into consideration (Schumacher, *Astron. Nachr.*, 1841, No. 438, s. 116), the semi-axis major of the elliptical spheroid of revolution to which the irregular figure of the Earth most closely approximates is 3,272,077·14 toises, or 20,924,774 feet; the semi-axis minor, 3,261,159·83 toises, or 20,854,821 feet; and the amount of compression or eccentricity $\frac{1}{299.1528}$d; the length of a mean degree of the meridian, 57,013·109 toises, or 364,596 feet, with an error of $+$

of $\frac{1}{299}$th. In accordance with this, the polar radius is 10,938 toises (69,944 feet), or about $11\frac{1}{2}$ miles, shorter than the equatorial radius of our terrestrial spheroid. The excess at the equator in consequence of the curvature of the upper surface of the globe amounts, consequently, in the direction of gravitation, to somewhat more than $4\frac{3}{7}$th times the height of Mont Blanc, or only $2\frac{1}{2}$ times the probable height of the summit of the Dhawalagiri, in the Himalaya chain. The lunar inequalities (perturbation in the moon's latitude and longitude) give, according to the last investigations of Laplace, almost the same result for the ellipticity as the measurements of degrees, viz., $\frac{1}{299}$th. The results yielded by the oscillation of the pendulum give, on the whole, a much greater amount of compression, viz., $\frac{1}{288}$th.[*]

2·8403 toises, or 18·16 feet, whence the length of a geographical mile is 3807·23 toises, or 6086·7 feet. Previous combinations of measurements of degrees varied between $\frac{1}{302}$d and $\frac{1}{297}$th; thus Walbeck (*De Forma et Magnitudine telluris in demensis arcubus Meridiani definiendis,* 1819) gives $\frac{1}{302\cdot78}$th : Ed. Schmidt (*Lehrbuch der Mathem. und Phys. Geographie,* 1829, s. 5) gives $\frac{1}{20\cdot742}$d, as the mean of seven measures. Respecting the influence of great differences of longitude on the polar compression, see *Bibliothèque Universelle,* t. xxxiii., p. 181, and t. xxxv., p. 56; likewise *Connaissance des Tems,* 1829, p. 290. From the lunar inequalities alone, Laplace (*Exposition du Syst. du Monde,* p. 229) found it, by the older tables of Bürg, to be $\frac{1}{324\cdot7}$th; and subsequently, from the lunar observations of Burckhardt and Bouvard, he fixed it at $\frac{1}{299\cdot1}$th (*Mécanique Céleste,* t. v., p. 13 and 43).

 * The oscillations of the pendulum give $\frac{1}{288\cdot7}$th as the general result of Sabine's great expedition (1822 and 1823, from the equator to 80° north latitude); according to Freycinet, $\frac{1}{286\cdot2}$d, exclusive of the experiments instituted at the Isle of France, Guam, and Mowi (Mawi); according to Forster, $\frac{1}{289\cdot3}$th; according to Duperrey, $\frac{1}{286\cdot4}$th; and according to Lütke (*Partie Nautique,* 1836, p. 232), $\frac{1}{269}$th, calculated from eleven stations. On the other hand, Mathieu (*Connaiss. des Temps,* 1816, p. 330) fixed the amount at $\frac{1}{299\cdot2}$d, from observations made between Formentera and Dunkirk; and Biot, at $\frac{1}{304}$th, from observations between Formentera and the island of Unst. Compare Baily, *Report on Pendulum Experiments,* in the *Memoirs of the Royal Astronomical Society,* vol. vii., p. 96; also Borenius, in the *Bulletin de l'Acad. de St. Pétersbourg,* 1843, t. i., p. 25. The first proposal to apply the length of the pendulum as a standard of measure, and to establish the third part of the seconds pendulum (then supposed to be every where of equal length) as a *pes horarius,* or general measure, that might be recovered at any age and by all nations, is to be found in Huygens's *Horologium Oscillatorium,* 1673, Prop. 25. A similar wish was afterward publicly expressed, in 1742, on a monument erected at the equator by Bouguer, La Condamine, and Godin. On the beautiful marble tablet which exists, as yet uninjured, in the old Jesuits' College at Quito, I have myself read the inscription, *Penduli simplicis æquinoctialis unius minuti secundi*

Galileo, who first observed when a boy (having, probably, suffered his thoughts to wander from the service) that the height of the vaulted roof of a church might be measured by the time of the vibration of the chandeliers suspended at different altitudes, could hardly have anticipated that the pendulum would one day be carried from pole to pole, in order to determine the form of the Earth, or, rather, that the unequal density of the strata of the Earth affects the length of the seconds pendulum by means of intricate forces of local attraction, which are, however, almost regular in large tracts of land. These geognostic relations of an instrument intended for the measurement of time—this property of the pendulum, by which, like a sounding line, it searches unknown depths, and reveals in volcanic islands,* or in the declivity of elevated continental mountain chains,† dense masses of basalt and mela-

archetypus, mensuræ naturalis exemplar, utinam universalis! From an observation made by La Condamine, in his *Journal du Voyage à l'Equateur,* 1751, p. 163, regarding parts of the inscription that were not filled up, and a slight difference between Bouguer and himself respecting the numbers, I was led to expect that I should find considerable discrepancies between the marble tablet and the inscription as it had been described in Paris; but, after a careful comparison, I merely found two perfectly unimportant differences: "ex arcu graduum 3½" instead of "ex arcu graduum plusquam trium," and the date of 1745 instead of 1742. The latter circumstance is singular, because La Condamine returned to Europe in November, 1744, Bouguer in June of the same year, and Godin had left South America in July, 1744. The most necessary and useful amendment to the numbers on this inscription would have been the astronomical longitude of Quito. (Humboldt, *Recueil d'Observ. Astron.,* t. ii., p. 319–354.) Nouet's latitudes, engraved on Egyptian monuments, offer a more recent example of the danger presented by the grave perpetuation of false or careless results.

* Respecting the augmented intensity of the attraction of gravitation in volcanic islands (St. Helena, Ualan, Fernando de Noronha, Isle of France, Guam, Mowi, and Galapagos), Rawak (Lütke, p. 240) being an exception, probably in consequence of its proximity to the high land of New Guinea, see Mathieu, in Delambre, *Hist. de l'Astronomie, au* 18*me Siècle,* p. 701.

† Numerous observations also show great irregularities in the length of the pendulum in the midst of continents, and which are ascribed to local attractions. (Delambre, *Mesure de la Méridienne,* t. iii., p. 548; Biot, in the *Mém. de l'Académie des Sciences,* t. viii., 1829, p. 18 and 23.) In passing over the South of France and Lombardy from west to east, we find the minimum intensity of gravitation at Bordeaux; from thence it increases rapidly as we advance eastward, through Figeac, Clermont-Ferrand, Milan, and Padua; and in the last town we find that the intensity has attained its maximum. The influence of the southern declivities of the Alps is not merely dependent on the general size of their mass, but (much more), in the opinion of Elie de Beaumont (*Rech. sur les Révol. de la Surface du Globe,* 1830, p. 729), on the rocks of melaphyre and serpentine, which have elevated the chain. On the

phyre instead of cavities, render it difficult, notwithstanding
the admirable simplicity of the method, to arrive at any great
result regarding the figure of the Earth from observation of
the oscillations of the pendulum.   In the astronomical part of
the determination of degrees of latitude, mountain chains, or
the denser strata of the Earth, likewise exercise, although in a
less degree, an unfavorable influence on the measurement.

As the form of the Earth exerts a powerful influence on the
motions of other cosmical bodies, and especially on that of its
own neighboring satellite, a more perfect knowledge of the motion of the latter will enable us reciprocally to draw an inference regarding the figure of the Earth.   Thus, as Laplace ably
remarks,* "An astronomer, without leaving his observatory,
may, by a comparison of lunar theory with true observations,
not only be enabled to determine the form and size of the
Earth, but also its distance from the Sun and Moon—results
that otherwise could only be arrived at by long and arduous
expeditions to the most remote parts of both hemispheres."

declivity of Ararat, which with Caucasus may be said to lie in the center of gravity of the old continent formed by Europe, Asia, and Africa,
the very exact pendulum experiments of Fedorow give indications, not
of subterranean cavities, but of dense volcanic masses. (Parrot, *Reise
zum Ararat*, bd. ii., s. 143.)   In the geodesic operations of Carlini and
Plana, in Lombardy, differences ranging from $20''$ to $47''\cdot8$ have been
found between direct observations of latitude and the results of these
operations.   (See the instances of Andrate and Mondovi, and those of
Milan and Padua, in the *Opérations Geodes. et Astron. pour la Mesure
d'un Arc du Parallèle Moyen*, t. ii., p. 347 ; *Effemeridi Astron. di Milano*, 1842, p. 57.)   The latitude of Milan, deduced from that of Berne,
according to the French triangulation, is $45^\circ\ 27'\ 52''$, while, according
to direct astronomical observations, it is $45^\circ\ 27'\ 35''$.   As the perturbations extend in the plain of Lombardy to Parma, which is far south of
the Po (Plana, *Opérat. Geod.*, t. ii., p. 347), it is probable that there are
deflecting causes *concealed beneath the soil of the plain itself.*   Struve
has made similar experiments [with corresponding results] in the most
level parts of eastern Europe.   (Schumacher, *Astron. Nachrichten*, 1830,
No. 164, s. 399.)   Regarding the influence of dense masses supposed to
lie at a small depth, equal to the mean height of the Alps, see the analytical expressions given by Hossard and Rozet, in the *Comptes Rendus*,
t. xviii.; 1844, p. 292, and compare them with Poisson, *Traité de Mécanique* (2me éd.), t. i., p. 482.   The earliest observations on the influence which different kinds of rocks exercise on the vibration of the
pendulum are those of Thomas Young, in the *Philos. Transactions* for
1819, p. 70–96.   In drawing conclusions regarding the Earth's curvature from the length of the pendulum, we ought not to overlook the
possibility that its crust may have undergone a process of hardening
previously to metallic and dense basaltic masses having penetrated from
great depths, through open clefts, and approached near the surface.

* Laplace, *Expos. du Syst. du Monde*, p. 231.

The compression which may be inferred from lunar inequalities affords an advantage not yielded by individual measurements of degrees or experiments with the pendulum, since it gives a mean amount which is referable to the whole planet. The comparison of the Earth's compression with the velocity of rotation shows, further, the increase of density from the strata from the surface toward the center—an increase which a comparison of the ratios of the axes of Jupiter and Saturn with their times of rotation likewise shows to exist in these two large planets. Thus the knowledge of the external form of planetary bodies leads us to draw conclusions regarding their internal character.

The northern and southern hemispheres appear to present nearly the same curvature under equal degrees of latitude, but, as has already been observed, pendulum experiments and measurements of degrees yield such different results for individual portions of the Earth's surface that no regular figure can be given which would reconcile all the results hitherto obtained by this method. The true figure of the Earth is to a regular figure as the uneven surfaces of water in motion are to the even surface of water at rest.

When the Earth had been measured, it still had to be weighed. The oscillations of the pendulum* and the plummet have here likewise served to determine the mean density of the Earth, either in connection with astronomical and geodetic operations, with the view of finding the deflection of the plummet from a vertical line in the vicinity of a mountain, or by a comparison of the length of the pendulum in a plain and on the summit of an elevation, or, finally, by the employment of a torsion balance, which may be considered as a horizontally vibrating pendulum for the measurement of the relative density of neighboring strata. Of these three methods† the

* La Caille's pendulum measurements at the Cape of Good Hope, which have been calculated with much care by Mathieu (Delambre, *Hist. de l'Astron. au* 18*me Siècle*, p. 479), give a compression of $\frac{1}{285\cdot4}$th; but, from several comparisons of observations made in equal latitudes n the two hemispheres (New Holland and the Malouines (Falkland Islands), compared with Barcelona, New York, and Dunkirk), there is as yet no reason for supposing that the mean compression of the southern hemisphere is greater than that of the northern. (Biot, in the *Mém. de l'Acad. des Sciences*, t. viii., 1829, p. 39–41.)

† The three methods of observation give the following results: (1.) by the deflection of the plumb-line in the proximity of the Shehallien Mountain (Gaelic, Thichallin) in Perthshire, 4·713, as determined by Maskelyne, Hutton, and Playfair (1774–1776 and 1810), according to a method that had been proposed by Newton; (2.) by pendulum vibra

last is the most certain, since it is independent of the difficult
determination of the density of the mineral masses of which
the spherical segment of the mountain consists near which the
observations are made.   According to the most recent experiments of Reich, the result obtained is 5·44 ; that is to say, the
mean density of the whole Earth is 5·44 times greater than
that of pure water.   As, according to the nature of the mineralogical strata constituting the dry continental part of the
Earth's surface, the mean density of this portion scarcely
amounts to 2·7, and the density of the dry and liquid surface
conjointly to scarcely 1·6, it follows that the elliptical unequally compressed layers of the interior must greatly increase
in density toward the center, either through pressure or owing
to the heterogeneous nature of the substances.   Here again
we see that the vertical, as well as the horizontally vibrating
pendulum, may justly be termed a geognostical instrument.

The results obtained by the employment of an instrument
of this kind have led celebrated physicists, according to the
difference of the hypothesis from which they started, to adopt

tions on mountains, 4·837 (Carlini's observations on Mount Cenis com
pared with Biot's observations at Bordeaux, *Effemer. Astron. di Milano*,
1824, p. 184); (3.) by the torsion balance used by Cavendish, with an
apparatus originally devised by Mitchell, 5·48 (according to Hutton's
revision of the calculation, 5·32, and according to that of Eduard
Schmidt, 5·52 ; *Lehrbuch der Math. Geographie*, bd. i., s. 487); by the
torsion balance, according to Reich, 5·44.   In the calculation of these
experiments of Professor Reich, which have been made with masterly
accuracy, the original mean result was 5·43 (with a probable error of
only 0·0233), a result which, being increased by the quantity by which
the Earth's centrifugal force diminishes the force of gravity for the latitude of Freiberg (50° 55'), becomes changed to 5·44.   The employment of cast iron instead of lead has not presented any sensible difference, or none exceeding the limits of errors of observation, hence dis-
closing no traces of magnetic influences. (Reich, *Versuche über die mitt-
lere Dichtigkeit der Erde*, 1838, s. 60, 62, and 66.)   By the assumption
of too slight a degree of ellipticity of the Earth, and by the uncertainty
of the estimations regarding the density of rocks on its surface, the
mean density of the Earth, as deduced from experiments on and near
mountains, was found about one sixth smaller than it really is, namely, 4·761 (Laplace, *Mécan. Céleste*, t. v., p. 46), or 4·785.   (Eduard
Schmidt, *Lehrb. der Math. Geogr.*, bd. i., § 387 und 418.)   On Halley's
hypothesis of the Earth being a hollow sphere (noticed in page 171),
which was the germ of Franklin's ideas concerning earthquakes, see
Philos. Trans. for the year 1693, vol. xvii., p. 563 (*On the Structure of
the Internal Parts of the Earth, and the concave habited Arch of the
Shell*).   Halley regarded it as more worthy of the Creator "that the
Earth, like a house of several stories, should be inhabited both without
and within.   For light in the hollow sphere (p. 576) provision might in
some manner be contrived."

entirely opposite views regarding the nature of the interior of the globe. It has been computed at what depths liquid or even gaseous substances would, from the pressure of their own superimposed strata, attain a density exceeding that of platinum or even iridium ; and in order that the compression which has been determined within such narrow limits migh⁺ be brought into harmony with the assumption of simple and infinitely compressible matter, Leslie has ingeniously conceived the nucleus of the world to be a hollow sphere, filled with an assumed " imponderable matter, having an enormous force of expansion." These venturesome and arbitrary conjectures have given rise, in wholly unscientific circles, to still more fantastic notions. The hollow sphere has by degrees been peopled with plants and animals, and two small subterranean revolving planets—Pluto and Proserpine—were imaginatively supposed to shed over it their mild light ; as, however, it was further imagined that an ever-uniform temperature reigned in these internal regions, the air, which was made self-luminous by compression, might well render the planets of this lower world unnecessary. Near the north pole, at 82° latitude, whence the polar light emanates, was an enormous opening, through which a descent might be made into the hollow sphere, and Sir Humphrey Davy and myself were even publicly and frequently invited by Captain Symmes to enter upon this subterranean expedition : so powerful is the morbid inclination of men to fill unknown spaces with shapes of wonder, totally unmindful of the counter evidence furnished by well-attested facts and universally acknowledged natural laws. Even the celebrated Halley, at the end of the seventeenth century, hollowed out the Earth in his magnetic speculations Men were invited to believe that a subterranean freely-ro tating nucleus occasions by its position the diurnal and an nual changes of magnetic declination. It has thus been attempted in our own day, with tedious solemnity, to clothe in a scientific garb the quaintly-devised fiction of the humorous Holberg.*

* [The work referred to, one of the wittiest productions of the learned Norwegian satirist and dramatist Holberg, was written in Latin, and first appeared under the following title : *Nicolai Klimii iter subterraneum novam telluris theoriam ac historiam quintæ monarchiæ adhuc nobis incognitæ exhibens e bibliotheca b. Abelini. Hafniæ et Lipsiæ sumt. Jac. Preuss,* 1741. An admirable Danish translation of this learned but severe satire on the institutions, morals, and manners of the inhabitants of the upper Earth, appeared at Copenhagen in 1789, and was entitled *Niels Klim's underjordiske reise ved Ludwig Holberg, oversat*

The figure of the Earth and the amount of solidification (density) which it has acquired are intimately connected with the forces by which it is animated, in so far, at least, as they have been excited or awakened from without, through its planetary position with reference to a luminous central body. Compression, when considered as a consequence of centrifugal force acting on a rotating mass, explains the earlier condition of fluidity of our planet. During the solidification of this fluid, which is commonly conjectured to have been gaseous and primordially heated to a very high temperature, an enormous quantity of latent heat must have been liberated. If the process of solidification began, as Fourier conjectures, by radiation from the cooling surface exposed to the atmosphere, the particles near the center would have continued fluid and hot. As, after long emanation of heat from the center toward the exterior, a stable condition of the temperature of the Earth would at length be established, it has been assumed that with increasing depth the subterranean heat likewise uninterruptedly increases. The heat of the water which flows from deep borings (Artesian wells), direct experiments regarding the temperature of rocks in mines, but, above all, the volcanic activity of the Earth, shown by the flow of molten masses from open fissures, afford unquestionable evidence of this increase for very considerable depths from the upper strata. According to conclusions based certainly upon mere analogies, this increase is probably much greater toward the center.

That which has been learned by an ingenious analytic calculation, expressly perfected for this class of investigations,*

efter den Latinske original af Jens Baggesen. Holberg, who studied for a time at Oxford, was born at Bergen in 1685, and died in 1754 as Rector of the University of Copenhagen.]— Tr.

* Here we must notice the admirable analytical labors of Fourier, Biot, Laplace, Poisson, Duhamel, and Lamé. In his Théorie Mathématique de la Chaleur, 1835, p. 3, 428–430, 436, and 521–524 (see, also, De la Rive's abstract in the Bibliothèque Universelle de Genève), Poisson has developed an hypothesis totally different from Fourier's view (Théorie Analytique de la Chaleur.) He denies the present fluid state of the Earth's center; he believes that "in cooling by radiation to the medium surrounding the Earth, the parts which were first solidified sunk, and that by a double descending and ascending current, the great inequality was lessened which would have taken place in a solid body cooling from the surface." It seems more probable to this great geometer that the solidification began in the parts lying nearest to the center: "the phenomenon of the increase of heat with the depth does not extend to the whole mass of the Earth, and is merely a consequence of the motion of our planetary system in space, of which some parts

regaiding the motion of heat in homogeneous metallic sphe-roids, must be applied with much caution to the actual char-acter of our planet, considering our present imperfect knowl-edge of the substances of which the Earth is composed, the difference in the capacity of heat and in the conducting power of different superimposed masses, and the chemical changes experienced by solid and liquid masses from any enormous compression. It is with the greatest difficulty that our pow-ers of comprehension can conceive the boundary line which di-vides the fluid mass of the interior from the hardened mineral masses of the external surface, or the gradual increase of the solid strata, and the condition of semi-fluidity of the earthy substances, these being conditions to which known laws of hydraulics can only apply under considerable modifications. The Sun and Moon, which cause the sea to ebb and flow, most probably also affect these subterranean depths.   We may suppose that the periodic elevations and depressions of the molten mass under the already solidified strata must have caused inequalities in the vaulted surface from the force of pressure.   The amount and action of such oscillations must, however, be small ; and if the relative position of the attract-ing cosmical bodies may here also excite " spring tides," it is certainly not to these, but to more powerful internal forces, that we must ascribe the movements that shake the Earth's surface.   There are groups of phenomena to whose existence it is necessary to draw attention, in order to indicate the universality of the influence of the attraction of the Sun and Moon on the external and internal conditions of the Earth, however little we may be able to determine the quantity of this influence.

According to tolerably accordant experiments in Artesian wells, it has been shown that the heat increases on an average about $1°$ foi every $54·5$ feet.   If this increase can be reduced

are of a very different temperature from others, in consequence of stel-lar heat (chaleur stellaire)."   Thus, according to Poisson, the warmth of the water of our Artesian wells is merely that which has penetrated into the Earth from without; and the Earth itself " might be regarded as in the same circumstances as a mass of rock conveyed from the equator to the pole in so short a time as not to have entirely cooled. The increase of temperature in such a block would not extend to the central strata."   The physical doubts which have reasonably been entertained against this extraordinary cosmical view (which attributes to the regions of space that which probably is more dependent on the first transition of matter condensing from the gaseo-fluid into the solid state) will be found collected in Poggendorf's *Annalen*, bd. xxxix., s. 93–100.

to arithmetical relations, it will follow, as I have already observed,* that a stratum of granite would be in a state of fusion
at a depth of nearly twenty-one geographical miles, or between
four and five times the elevation of the highest summit of the
Himalaya.

We must distinguish in our globe three different modes for
the transmission of heat. The first is periodic, and affects
the temperature of the terrestrial strata according as the heat
penetrates from above downward or from below upward, being
influenced by the different positions of the Sun and the seasons of the year. The second is likewise an effect of the Sun,
although extremely slow : a portion of the heat that has penetrated into the equatorial regions moves in the interior of the
globe toward the poles, where it escapes into the atmosphere
and the remoter regions of space. The third mode of transmission is the slowest of all, and is derived from the secular
cooling of the globe, and from the small portion of the primitive heat which is still being disengaged from the surface.

---

* See the Introduction. This increase of temperature has been found
in the Puits de Grenelle, at Paris, at 58·3 feet; in the boring at the new
salt-works at Minden, almost 53·6 ; at Pregny, near Geneva, according
to Auguste de la Rive and Marcet, notwithstanding that the mouth of
the boring is 1609 feet above the level of the sea, it is also 53·6 feet.
This coincidence between the results of a method first proposed by
Arago in the year 1821 (*Annuaire du Bureau des Longitudes*, 1835, p.
234), for three different mines, of the absolute depths of 1794, 2231,
and 725 feet respectively, is remarkable. The two points on the Earth,
lying at a small vertical distance from each other, whose annual mean
temperatures are most accurately known, are probably at the spot on
which the Paris Observatory stands, and the Caves de l'Observatoire
beneath it: the mean temperature of the former is $51°·5$, and of the
latter $53°·3$, the difference being $1°·8$ for 92 feet, or $1°$ for 51·77 feet.
(Poisson, *Théorie Math. de la Chaleur*, p. 415 and 462.) In the course
of the last seventeen years, from causes not yet perfectly understood,
but probably not connected with the actual temperature of the caves,
the thermometer standing there has risen very nearly $0°·4$. Although
in Artesian wells there are sometimes slight errors from the lateral
permeation of water, these errors are less injurious to the accuracy of
conclusions than those resulting from currents of cold air, which are
almost always present in mines. The general result of Reich's great
work on the temperature of the mines in the Saxony mining districts
gives a somewhat slower increase of the terrestrial heat, or $1°$ to 76·3
feet. (Reich, *Beob. über die Temperatur des Gesteins in verschieden en
Tiefen*, 1834, s. 134.) Phillips, however, found (Pogg., *Annalen*, bd.
xxxiv., s. 191), in a shaft of the coal-mine of Monk-wearmouth, near
Newcastle, in which, as I have already remarked, excavations are going
on at a depth of about 1500 feet below the level of the sea, an increase
of $1°$ to 59·06 feet, a result almost identical with that found by Arago
in the Puits de Grenell.

This loss experienced by the central heat must have been very considerable in the earliest epochs of the Earth's revolutions, but within historical periods it has hardly been appreciable by our instruments. The surface of the Earth is therefore situated between the glowing heat of the inferior strata and the universal regions of space, whose temperature is probably below the freezing-point of mercury.

The periodic changes of temperature which have been occasioned on the Earth's surface by the Sun's position and by meteorological processes, are continued in its interior, although to a very inconsiderable depth. The slow conducting power of the ground diminishes this loss of heat in the winter, and is very favorable to deep-rooted trees. Points that lie at very different depths on the same vertical line attain the maximum and minimum of the imparted temperature at very different periods of time. The further they are removed from the surface, the smaller is this difference between the extremes. In the latitudes of our temperate zone (between 48° and 52°), the stratum of invariable temperature is at a depth of from 59 to 64 feet, and at half that depth the oscillations of the thermometer, from the influence of the seasons, scarcely amount to half a degree. In tropical climates this invariable stratum is only one foot below the surface, and this fact has been ingeniously made use of by Boussingault to obtain a convenient, and, as he believes, certain determination of the mean temperature of the air of different places.* This mean temperature of the air at a fixed point, or at a group of contiguous points on the surface, is to a certain degree the fundamental element of the climate and agricultural relations of a district ; but the mean temperature of the whole surface is very different from that of the globe itself. The questions so often agitated, whether the mean temperature has experienced any considerable differences in the course of centuries, whether the climate of a country has deteriorated, and whether the winters have not become milder and the summers cooler, can only be answered by means of the thermometer ; this instrument has, however, scarcely been invented more than two centuries and a half, and its scientific application hardly dates back 120 years. The nature and novelty of the means interpose, therefore, very narrow limits to our investigation regarding the temperature

* Boussingault, *Sur la Profondeur à laquelle se trouve la Couche de Température invariable entre les Tropiques*, in the *Annales de Chimie et de Physique*, t. liii., 1833, p. 225–247.

of the air. It is quite otherwise, however, with the solution
of the great problem of the internal heat of the whole Earth.
As we may judge of uniformity of temperature from the unaltered time of vibration of a pendulum, so we may also learn,
from the unaltered rotatory velocity of the Earth, the amount
of stability in the mean temperature of our globe. This
insight into the relations between the *length of the day* and
the *heat of the Earth* is the result of one of the most brilliant
applications of the knowledge we had long possessed of the
movement of the heavens to the thermic condition of our
planet. The rotatory velocity of the Earth depends on its
volume ; and since, by the gradual cooling of the mass by
radiation, the axis of rotation would become shorter, the rotatory velocity would necessarily increase, and the length of the
day diminish, with a decrease of the temperature. From the
comparison of the secular inequalities in the motions of the
Moon with the eclipses observed in ancient times, it follows
that, since the time of Hipparchus, that is, for full 2000
years, the length of the day has certainly not diminished by
the hundredth part of a second. The decrease of the mean
heat of the globe during a period of 2000 years has not, therefore, taking the extremest limits, diminished as much as $\frac{1}{306}$th
of a degree of Fahrenheit.*

This invariability of form presupposes also a great invariability in the distribution of relations of density in the interior
of the globe. The translatory movements, which occasion
the eruptions of our present volcanoes and of ferruginous lava,
and the filling up of previously empty fissures and cavities
with dense masses of stone, are consequently only to be regarded as slight superficial phenomena affecting merely one
portion of the Earth's crust, which, from their smallness
when compared to the Earth's radius, become wholly insignificant.

I have described the internal heat of our planet, both with
reference to its cause and distribution, almost solely from the
results of Fourier's admirable investigations. Poisson doubts
the fact of the uninterrupted increase of the Earth's heat

* Laplace, *Exp. du Syst. du Monde*, p. 229 and 263 ; *Mécanique
Céleste*, t. v., p. 18 and 72. It should be remarked that the fraction
$\frac{1}{306}$th of a degree of Fahrenheit of the mercurial thermometer, given in
the text as the limit of the stability of the Earth's temperature since
the days of Hipparchus, rests on the assumption that the dilatation of
the substances of which the Earth is composed is equal to that of glass,
that is to say, $\frac{1}{18.000}$th for 1°. Regarding this hypothesis, see Arago
in the *Annuaire* for 1834, p. 177–190.

from the surface to the center, and is of opinion that all heat has penetrated from without inward, and that the temperature of the globe depends upon the very high or very low temperature of the regions of space through which the solar system has moved. This hypothesis, imagined by one of the most acute mathematicians of our time, has not satisfied physicists or geologists, or scarcely, indeed, any one besides its author. But, whatever may be the cause of the internal heat of our planet, and of its limited or unlimited increase in deep strata, it leads us, in this general sketch of nature, through the intimate connection of all primitive phenomena of matter, and through the common bond by which molecular forces are united, into the mysterious domain of magnetism. Changes of temperature call forth magnetic and electric currents. Terrestrial magnetism, whose main character, expressed in the three-fold manifestation of its forces, is incessant periodic variability, is ascribed either to the heated mass of the Earth itself,* or to those galvanic currents which we consider as electricity in motion, that is, electricity moving in a closed circuit.†

The mysterious course of the magnetic needle is equally affected by time and space, by the sun's course, and by changes of place on the Earth's surface. Between the tropics, the hour of the day may be known by the direction of the needle as well as by the oscillations of the barometer. It is affected instantly, but only transiently, by the distant northern light as it shoots from the pole, flashing in beams of colored light across the heavens. When the uniform horary motion of the needle is disturbed by a magnetic storm, the perturbation manifests itself *simultaneously*, in the strictest sense of the word, over hundreds and thousands of miles of sea and land, or propagates itself by degrees, in short intervals of time, in

---

* William Gilbert, of Colchester, whom Galileo pronounced "great to a degree that might be envied," said "magnus magnes ipse est globus terrestris." He ridicules the magnetic mountains of Frascatori, the great cotemporary of Columbus, as being magnetic poles: "rejicienda est vulgaris opinio de montibus magneticis, aut rupe aliqua magnetica, aut polo phantastico a polo mundi distante." He assumes the declination of the magnetic needle at any given point on the surface of the Earth to be invariable (variatio uniuscujusque loci constans est), and refers the curvatures of the isogonic lines to the configuration of continents and the relative positions of sea basins, which possess a weaker magnetic force than the solid masses rising above the ocean. (Gilbert, *de Magnete*, ed. 1633, p. 42, 98, 152, and 155.)

† Gauss, *Allgemeine Theorie des Erdmagnetismus*, in the *Resultate aus den Beob. des Magnet. Vereins*, 1838. s. 41, p. 56.

every direction over the Earth's surface.* In the former case,
the simultaneous manifestation of the storm may serve, within certain limitations, like Jupiter's satellites, fire-signals, and
well-observed falls of shooting stars, for the geographical
determination of degrees of longitude. We here recognize
with astonishment that the perturbations of two small magnetic needles, even if suspended at great depths below the
surface, can measure the distances apart at which they are
placed, teaching us, for instance, how far Kasan is situated
east of Göttingen or of the banks of the Seine. There are
also districts in the earth where the mariner, who has been
enveloped for many days in mist, without seeing either the
sun or stars, and deprived of all means of determining the
time, may know with certainty, from the variations in the
inclination of the magnetic needle, whether he is at the north
or the south of the port he is desirous of entering.†

* There are also perturbations which are of a local character, and
do not extend themselves far, and are probably less deep-seated. Some
years ago I described a rare instance of this kind, in which an extraordinary disturbance was felt in the mines at Freiberg, but was not perceptible at Berlin. (*Lettre de M. de Humboldt à Son Altesse Royale le
Duc de Sussex sur les moyens propres à perfectionner la Connaissance
du Magnétisme Terrestre*, in Becquerel's *Traité Expérimental de l'Electricité*, t. vii., p. 442.) Magnetic storms, which were simultaneously
felt from Sicily to Upsala, did not extend from Upsala to Alten. (Gauss
and Weber, *Resultate des Magnet. Vereins*, 1839, § 128; Lloyd, in the
Comptes Rendus de l'Acad. des Sciences, t. xiii., 1843, Sém. ii., p. 725
and 827.) Among the numerous examples that have been recently
observed, of perturbations occurring simultaneously and extending over
wide portions of the Earth's surface, and which are collected in Sabine's
important work (*Observ. on Days of unusual Magnetic Disturbance*,
1843), one of the most remarkable is that of the 25th of September,
1841, which was observed at Toronto in Canada, at the Cape of Good
Hope, at Prague, and partially in Van Diemen's Land. The English
Sunday, on which it is deemed sinful, after midnight on Saturday, to
register an observation, and to follow out the great phenomena of creation in their perfect development, interrupted the observations in Van
Diemen's Land, where, in consequence of the difference of the longitude, the magnetic storm fell on the Sunday. (*Observ.*, p. xiv., 78, 85,
and 87.)

† I have described, in Lamétherie's *Journal de Physique*, 1804, t.
lix., p. 449, the application (alluded to in the text) of the magnetic inclination to the determination of latitude along a coast running north
and south, and which, like that of Chili and Peru, is for a part of the
year enveloped in mist (*garua*). In the locality I have just mentioned,
this application is of the greater importance, because, in consequence
of the strong current running northward as far as to Cape Pareña, navigators incur a great loss of time if they approach the coast to the north
of the haven they are seeking. In the South Sea, from Callao de Lima
harbor to Truxillo, which differ from each other in latitude by 3° 57′

When the needle, by its sudden disturbance in its horary course, indicates the presence of a magnetic storm, we are still unfortunately ignorant whether the seat of the disturbing cause is to be sought in the Earth itself or in the upper regions of the atmosphere. If we regard the Earth as a true magnet, we are obliged, according to the views entertained by Friedrich Gauss (the acute propounder of a general theory of terrestrial magnetism), to ascribe to every portion of the globe measuring one eighth of a cubic meter (or $3\frac{7}{10}$ths of a French cubic foot) in volume, an average amount of magnetism equal to that contained in a magnetic rod of 1 lb. weight.[*]
If iron and nickel, and probably, also, cobalt (but not chrome, as has long been believed),[†] are the only substances which become permanently magnetic, and retain polarity from a certain coercive force, the phenomena of Arago's magnetism of rotation and of Faraday's induced currents show, on the other hand, that all telluric substances may possibly be made transitorily magnetic. According to the experiments of the

I have observed a variation of the magnetic inclination amounting to 9° (centesimal division); and from Callao to Guayaquil, which differ in latitude by 0° 50′, a variation of 23°·5. (See my *Relat. Hist.*, t. iii., p. 622.) At Guarmey (10° 4′ south lat.), Huaura (11° 3′ south lat.), and Chancay (11° 32′ south lat.), the inclinations are 6°·80, 9°, and 10°·35 of the centesimal division. The determination of position by means of the magnetic inclination has this remarkable feature connected with it, that where the ship's course cuts the isoclinal line almost perpendicularly, it is the only one that is independent of all determination of time, and, consequently, of observations of the sun or stars. It is only lately that I discovered, for the first time, that as early as at the close of the sixteenth century, and consequently hardly twenty years after Robert Norman had invented the inclinatorium, William Gilbert, in his great work *De Magnete*, proposed to determine the latitude by the inclination of the magnetic needle. Gilbert (*Physiologia Nova de Magnete*, lib. v., cap. 8, p. 200) commends the method as applicable "aëre caliginoso." Edward Wright, in the introduction which he added to his master's great work, describes this proposal as "worth much gold." As he fell into the same error with Gilbert, of presuming that the isoclinal lines coincided with the geographical parallel circles, and that the magnetic and geographical equators were identical, he did not perceive that the proposed method had only a local and very limited application.

[*] Gauss and Weber, *Resultate des Magnet. Vereins*, 1838, § 31, s. 146.

[†] According to Faraday (*London and Edinburgh Philosophical Magazine*, 1836, vol. viii., p. 178), pure cobalt is totally devoid of magnetic power. I know, however, that other celebrated chemists (Heinrich Rose and Wöhler) do not admit this as absolutely certain. If out of two carefully-purified masses of cobalt totally free from nickel, one appears altogether non-magnetic (in a state of equilibrium), I think it probable that the other owes its magnetic property to a want of purity; and this opinion coincides with Faraday's view.

first-mentioned of these great physicists, water, ice, glass, and
carbon affect the vibrations of the needle entirely in the same
manner as mercury in the rotation experiments.*   Almost all
substances show themselves to be, in a certain degree, magnetic when they are conductors, that is to say, when a current
of electricity is passing through them.

Although the knowledge of the attracting power of native
iron magnets or loadstones appears to be of very ancient date
among the nations of the West, there is strong historical evidence in proof of the striking fact that the knowledge of the
directive power of a magnetic needle and of its relation to
terrestrial magnetism was peculiar to the Chinese, a people
living in the extremest eastern portions of Asia.   More than
a thousand years before our era, in the obscure age of Codrus,
and about the time of the return of the Heraclidæ to the Peloponnesus, the Chinese had already magnetic carriages, on
which the movable arm of the figure of a man continually
pointed to the south, as a guide by which to find the way
across the boundless grass plains of Tartary ; nay, even in the
third century of our era, therefore at least 700 years before
the use of the mariner's compass in European seas, Chinese
vessels navigated the Indian Ocean† under the direction of
magnetic needles pointing to the south.   I have shown, in
another work, what advantages this means of topographical direction, and the early knowledge and application of the magnetic needle gave the Chinese geographers over the Greeks
and Romans, to whom, for instance, even the true direction
of the Apennines and Pyrenees always remained unknown.‡

The magnetic power of our globe is manifested on the terrestrial surface in three classes of phenomena, one of which
exhibits itself in the varying intensity of the force, and the
two others in the varying direction of the inclination, and in

* Arago, in the *Annales de Chimie*, t. xxxii., p. 214 ; Brewster, *Treatise on Magnetism*, 1837, p. 111; Baumgartner, in the *Zeitschrift für
Phys. und Mathem.*, bd. ii., s. 419.

† Humboldt, *Examen Critique de l'Hist. de la Géographie*, t. iii., p. 36.

‡ *Asie Centrale*, t. i., Introduction, p. xxxviii.–xlii.   The Western
nations, the Greeks and the Romans, knew that magnetism could be
communicated to iron, *and that that metal would retain it for a length of
time.*   (" Sola hæc materia ferri vires, a magnete lapide accipit, *retinet-
que longo tempore.*"   Plin., xxxiv., 14.)   The great discovery of the terrestrial directive force depended, therefore, alone on this, that no one
in the West had happened to observe an elongated fragment of magnetic iron stone, or a magnetic iron rod, floating, by the aid of a piece of
wood, in water, or suspended in the air by a thread, in such a position
as to admit of free motion.

the horizontal deviation from the terrestrial meridian of the spot. Their combined action may therefore be graphically represented by three systems of lines, the *isodynamic, isoclinic,* and *isogonic* (or those of equal force, equal inclination, and equal declination). The distances apart, and the relative positions of these moving, oscillating, and advancing curves, do not always remain the same. The total deviation (variation or declination of the magnetic needle) has not at all changed, or, at any rate, not in any appreciable degree, during a whole century, at any particular point on the Earth's surface,* as, for instance, the western part of the Antilles, or Spitzbergen. In like manner, we observe that the isogonic curves, when they pass in their secular motion from the surface of the sea to a continent or an island of considerable extent, continue for a long time in the same position, and become inflected as they advance.

These gradual changes in the forms assumed by the lines in their translatory motions, and which so unequally modify the amount of eastern and western declination, in the course of time render it difficult to trace the transitions and analogies of forms in the graphic representations belonging to different centuries. Each branch of a curve has its history, but this history does not reach further back among the nations of the West than the memorable epoch of the 13th of September, 1492, when the re-discoverer of the New World found a line of no variation 3° west of the meridian of the island of Flores, one of the Azores.† The whole of Europe, excepting a small

* A very slow secular progression, or a local invariability of the magnetic declination, prevents the confusion which might arise from terrestrial influences in the boundaries of land, when, with an utter disregard for the correction of declination, estates are, after long intervals, measured by the mere application of the compass. "The whole mass of West Indian property," says Sir John Herschel, "has been saved from the bottomless pit of endless litigation by the invariability of the magnetic declination in Jamaica and the surrounding Archipelago during the whole of the last century, all surveys of property there having been conducted solely by the compass." See Robertson, in the *Philosophical Transactions* for 1806, Part ii., p. 348, *On the Permanency of the Compass in Jamaica since* 1660. In the mother country (England) the magnetic declination has varied by fully 14° during that period.

† I have elsewhere shown that, from the documents which have come down to us regarding the voyages of Columbus, we can, with much certainty, fix upon three places *in the Atlantic line of no declination* for the 13th of September, 1492, the 21st of May, 1496, and the 16th of August, 1498. The Atlantic line of no declination at that period ran from northeast to southwest. It then touched the South American continent a little east of Cape Codera, while it is now observed to reach that continent on the northern coast of the Brazils. (Humboldt, *Examen Critique de l'Hist. de la Géogr.*, t. iii., p. 44-48.)

part of Russia, has now a western declination, while at the
close of the seventeenth century the needle first pointed due
north, in London in 1657, and in Paris in 1669, there being
thus a difference of twelve years, notwithstanding the small
distance between these two places.   In Eastern Russia, to
the east of the mouth of the Volga, of Saratow, Nischni-Now-
gorod, and Archangel, the easterly declination of Asia is advancing toward us.   Two admirable observers, Hansteen and
Adolphus Erman, have made us acquainted with the remarkable double curvature of the lines of declination in the vast
region of Northern Asia ; these being concave toward the
pole between Obdorsk, on the Oby, and Turuchansk, and convex between the Lake of Baikal and the Gulf of Ochotsk.   In
this portion of the earth, in northern Asia, between the mountains of Werchojansk, Jakutsk, and the northern Korea, the
isogonic lines form a remarkable closed system.   This oval
configuration* recurs regularly, and over a great extent of the
South Sea, almost as far as the meridian of Pitcairn and the
group of the Marquesas Islands, between 20° north and 45°

From Gilbert's *Physiologia Nova de Magnete*, we see plainly (and the
fact is very remarkable) that in 1600 the declination was still null in
the region of the Azores, just as it had been in the time of Columbus
(lib. 4, cap. 1).   I believe that in my *Examen Critique* (t. iii., p. 54)
I have proved from documents that the celebrated line of demarkation
by which Pope Alexander VI. divided the Western hemisphere between
Portugal and Spain was not drawn through the most western point of
the Azores, because Columbus wished to convert a physical into a political division.   He attached great importance to the zone (raya) " in
which the compass shows no variation, where air and ocean, the latter
covered with pastures of sea-weed, exhibit a peculiar constitution,
where cooling winds begin to blow, and where [as erroneous observations of the polar star led him to imagine] the form (sphericity) of the
Earth is no longer the same."

 * To determine whether the two oval systems of isogonic lines, so
singularly included each within itself, will continue to advance for centuries in the same inclosed form, or will unfold and expand themselves,
is a question of the highest interest in the problem of the physical
causes of terrestrial magnetism.   In the Eastern Asiatic nodes the declination increases from without inward, while in the node or oval system of the South Sea the opposite holds good ; in fact, at the present
time, in the whole South Sea to the east of the meridian of Kamtschatka, there is no line where the declination is null, or, indeed, in
which it is less than 2° (Erman, in Pogg., *Annal.*, bd. xxxi., § 129).
Yet Cornelius Schouten, on Easter Sunday, 1616, appears to have found
the declination null somewhere to the southeast of Nukahiva, in 15°
south lat. and 132° west long., and consequently in the middle of the
present closed isogonal system.   (Hansteen, *Magnet. der Erde*, 1819, §
28.)   It must not be forgotten, in the midst of all these considerations,
that we can only follow the direction of the magnetic lines in their
progress as they are projected upon the surface of the Earth.

south lat. One would almost be inclined to regard this singular configuration of closed, almost concentric, lines of declination as the effect of a local character of that portion of the globe ; but if, in the course of centuries, these apparently isolated systems should also advance, we must suppose, as in the case of all great natural forces, that the phenomenon arises from some general cause.

The horary variations of the declination, which, although dependent upon true time, are apparently governed by the Sun, as long as it remains above the horizon, diminish in angular value with the magnetic latitude of place. Near the equator, for instance, in the island of Rawak, they scarcely amount to three or four minutes, while they are from thirteen to fourteen minutes in the middle of Europe. As in the whole northern hemisphere the north point of the needle moves from east to west on an average from $8\frac{1}{2}$ in the morning until $1\frac{1}{2}$ at mid-day, while in the southern hemisphere the same north point moves from west to east,[*] attention has recently been drawn, with much justice, to the fact that there must be a region of the Earth between the terrestrial and the magnetic equator where no horary deviations in the declination are to be observed. This fourth curve, which might be called the *curve of no motion,* or, rather, *the line of no variation of horary declination,* has not yet been discovered.

The term *magnetic poles* has been applied to those points of the Earth's surface where the horizontal power disappears, and more importance has been attached to these points than properly appertains to them ;[†] and in like manner, the curve, where the inclination of the needle is null, has been termed the *magnetic equator.* The position of this line and its secular change of configuration have been made an object of careful investigation in modern times. According to the admirable work of Duperrey,[‡] who crossed the magnetic equator six times between 1822 and 1825, the nodes of the two equators, that is to say, the two points at which the line without inclination intersects the terrestrial equator, and consequently passes from one hemisphere into the other, are so unequally placed, that in 1825 the node near the island of St. Thomas, on the west-

---

[*] Arago, in the *Annuaire,* 1836, p. 284, and 1840, p. 330–338.

[†] Gauss, *Allg. Theorie des Erdmagnet.,* § 31.

[‡] Duperrey, *De la Configuration de l'Équateur Magnétique,* in the *Annales de Chimie,* t. xlv., p. 371 and 379. (See, also, Morlet, in the *Mémoires présentés par divers Savans à l'Acad. Roy. des Sciences,* t. iii., p. 132.)

ern coast of Africa, was 188½° distant from the node in the South Sea, close to the little islands of Gilbert, nearly in the meridian of the Viti group. In the beginning of the present century, at an elevation of 11,936 feet above the level of the sea, I made an astronomical determination of the point (7° 1′ south lat., 48° 40′ west long. from Paris), where, in the interior of the New Continent, the chain of the Andes is intersected by the magnetic equator between Quito and Lima. To the west of this point, the magnetic equator continues to traverse the South Sea in the southern hemisphere, at the same time slowly drawing near the terrestrial equator. It first passes into the northern hemisphere a little before it approaches the Indian Archipelago, just touches the southern points of Asia, and enters the African continent to the west of Socotora, almost in the Straits of Bab-el-Mandeb, where it is most distant from the terrestrial equator. After intersecting the unknown regions of the interior of Africa in a southwest direction, the magnetic equator re-enters the south tropical zone in the Gulf of Guinea, and retreats so far from the terrestrial equator that it touches the Brazilian coast near Os Ilheos, north of Porto Seguro, in 15° south lat. From thence to the elevated plateaux of the Cordilleras, between the silver mines of Micuipampa and Caxamarca, the ancient seat of the Incas, where I observed the inclination, the line traverses the whole of South America, which in these latitudes is as much a magnetic *terra incognita* as the interior of Africa.

The recent observations of Sabine* have shown that the node near the island of St. Thomas has moved 4° from east to west between 1825 and 1837. It would be extremely important to know whether the opposite pole, near the Gilbert Islands, in the South Sea, has approached the meridian of the Carolinas in a westerly direction. These general remarks will be sufficient to connect the different systems of isoclinic nonparallel lines with the great phenomenon of equilibrium which is manifested in the magnetic equator. It is no small advantage, in the exposition of the laws of terrestrial magnetism, that the magnetic equator (whose oscillatory change of form and whose nodal motion exercise an influence on the inclination of the needle in the remotest districts of the world, in consequence of the altered magnetic latitudes)† should traverse the

---

* See the remarkable chart of isoclinic lines in the Atlantic Ocean for the years 1825 and 1837, in Sabine's *Contributions to Terrestrial Magnetism*, 1840, p. 134.

† Humboldt, *Ueber die seculäre Veränderung der Magnetischen In-*

ocean throughout its whole course, excepting about one fifth, and consequently be made so much more accessible, owing to the remarkable relations in space between the sea and land, and to the means of which we are now possessed for determining with much exactness both the declination and the inclination at sea.

We have described the distribution of magnetism on the surface of our planet according to the two forms of *declination* and *inclination;* it now, therefore, remains for us to speak of the *intensity of the force* which is graphically expressed by isodynamic curves (or lines of equal intensity). The investigation and measurement of this force by the oscillations of a vertical or horizontal needle have only excited a general and lively interest in its telluric relations since the beginning of the nineteenth century. The application of delicate optical and chronometrical instruments has rendered the measurement of this horizontal power susceptible of a degree of accuracy far surpassing that attained in any other magnetic determinations. The isogonic lines are the more important in their immediate application to navigation, while we find from the most recent views that isodynamic lines, especially those which indicate the horizontal force, are the most valuable elements in the theory of terrestrial magnetism.* One of the earliest facts yielded by observation is, that the intensity of the total force increases from the equator toward the pole.†

clination (On the secular Change in the Magnetic Inclination), in Pogg. *Annal.*, bd. xv., s. 322.

* Gauss, *Resultate der Beob. des Magn. Vereins*, 1838, § 21; Sabine, *Report on the Variations of the Magnetic Intensity*, p. 63.

† The following is the history of the discovery of the law that the intensity of the force increases (in general) with the magnetic latitude. When I was anxious to attach myself, in 1798, to the expedition of Captain Baudin, who intended to circumnavigate the globe, I was requested by Borda, who took a warm interest in the success of my project, to examine the oscillations of a vertical needle in the magnetic meridian in different latitudes in each hemisphere, in order to determine whether the intensity of the force was the same, or whether it varied in different places. During my travels in the tropical regions of America, I paid much attention to this subject. I observed that the same needle, which in the space of ten minutes made 245 oscillations in Paris, 246 in the Havana. and 242 in Mexico, performed only 216 oscillations during the same period at St. Carlos del Rio Negro (1° 53′ north lat. and 80° 40) west long. from Paris), on the magnetic equator, *i. e.*, the line in which the inclination =0; in Peru (7° 1′ south lat. and 80° 40′ west long. from Paris) only 211; while at Lima (12° 2′ south lat.) the number rose to 219. I found, in the years intervening between 1799 and 1803, that the whole force, if we assume it at 1·0000 on the magnetic equator in the Peruvian Andes, between Micuipampa and Caxamarca,

The knowledge which we possess of the quantity of this increase, and of all the numerical relations of the law of in-

may be expressed at Paris by 1·3482, in Mexico by 1·3155, in San Carlos
del Rio Negro by 1·0480, and in Lima by 1·0773. When I developed
this law of the variable intensity of terrestrial magnetic force, and supported it by the numerical value of observations instituted in 104 different places, in a Memoir read before the Paris Institute on the 26th
Frimaire, An. XIII. (of which the mathematical portion was contributed
by M. Biot), the facts were regarded as altogether new. It was only
after the reading of the paper, as Biot expressly states (Lamétherie,
Journal de Physique, t. lix., p. 446, note 2), and as I have repeated in
the Relation Historique, t. i., p. 262, note 1, that M. de Rossel communicated to Biot his oscillation experiments made six years earlier (between 1791 and 1794) in Van Diemen's Land, in Java, and in Amboyna.
These experiments gave evidence of the same law of decreasing force
in the Indian Archipelago. It must, I think, be supposed, that this excellent man, when he wrote his work, was not aware of the regularity
of the augmentation and diminution of the intensity, as before the reading of my paper he never mentioned this (certainly not unimportant)
physical law to any of our mutual friends, La Place, Delambre, Prony,
or Biot. It was not till 1808, four years after my return from America,
that the observations made by M. de Rossel were published in the *Voyage de l'Entrecasteaux*, t. ii., p. 287, 291, 321, 480, and 644. Up to the
present day it is still usual, in all the tables of magnetic intensity which
have been published in Germany (Hansteen, *Magnet. der Erde*, 1819,
s. 71; Gauss, *Beob. des Magnct. Vereins*, 1838, s. 36–39; Erman, *Physikal. Beob.*, 1841, s. 529–579), in England (Sabine, *Report on Magnet.
Intensity*, 1838, p. 43–62; *Contributions to Terrestrial Magnetism*, 1843),
and in France (Becquerel, *Traité de Electr. et de Magnét.*, t. vii., p.
354–367), to reduce the oscillations observed in any part of the Earth
to the standard of force which I found on the magnetic equator in
Northern Peru, so that, according to the unit thus arbitrarily assumed,
the intensity of the magnetic force at Paris is put down as 1·348. The
observations made by Lamanon in the unfortunate expedition of La
Perouse, during the stay at Teneriffe (1785), and on the voyage to
Macao (1787), are still older than those of Admiral Rossel. They were
sent to the Academy of Sciences, and it is known that they were in the
possession of Condorcet in the July of 1787 (Becquerel, t. vii., p. 320);
but, notwithstanding the most careful search, they are not now to be
found. From a copy of a very important letter of Lamanon, now in the
possession of Captain Duperrey, which was addressed to the then perpetual secretary of the Academy of Sciences, but was omitted in the
narrative of the *Voyage de La Perouse*, it is stated " that the attractive
force of the magnet is less in the tropics than when we approach the
poles, and that the magnetic intensity deduced from the number of oscillations of the needle of the inclination-compass varies and increases
with the latitude." If the Academicians, while they continued to expect the return of the unfortunate La Perouse, had felt themselves justified, in the course of 1787, in publishing a truth which had been independently discovered by no less than three different travelers, the theory
of terrestrial magnetism would have been extended by the knowledge
of a new class of observations, dating eighteen years earlier than they
now do. This simple statement of facts may probably justify the observations contained in the third volume of my *Relation Historique* (p

tensity affecting the whole Earth, is especially due, since 1819,
to the unwearied activity of Edward Sabine, who, after having observed the oscillations of the same needles at the American north pole, in Greenland, at Spitzbergen, and on the coasts
of Guinea and Brazil, has continued to collect and arrange
all the facts capable of explaining the direction of the isodynamic lines. I have myself given the first sketch of an isodynamic system in zones for a small part of South America.
These lines are not parallel to lines of equal inclination (isoclinic lines), and the intensity of the force is not at its minimum
at the magnetic equator, as has been supposed, nor is it even
equal at all parts of it. If we compare Erman's observations
in the southern part of the Atlantic Ocean, where a faint zone
(0·706) extends from Angola over the island of St. Helena to
the Brazilian coast, with the most recent investigations of the
celebrated navigator James Clark Ross, we shall find that
on the surface of our planet the force increases almost in the
relation of 1 : 3 toward the magnetic south pole, where Victoria Land extends from Cape Crozier toward the volcano
Erebus, which has been raised to an elevation of 12,600 feet
above the ice.* If the intensity near the magnetic south pole

615): "The observations on the variation of terrestrial magnetism, to
which I have devoted myself for thirty-two years, by means of instruments which admit of comparison with one another, in America, Europe,
and Asia, embrace an area extending over 188 degrees of longitude,
from the frontier of Chinese Dzoungarie to the west of the South Sea
bathing the coasts of Mexico and Peru, and reaching from 60° north
lat. to 12° south lat. I regard the discovery of the law of the decre-
ment of magnetic force from the pole to the equator as the most important result of my American voyage." Although not absolutely certain, it is very probable that Condorcet read Lamanon's letter of July,
1787, at a meeting of the Paris Academy of Sciences; and such a simple reading I regard as a sufficient act of publication. (*Annuaire du
Bureau des Longitudes*, 1842, p. 463.) The first recognition of the law
belongs, therefore, beyond all question, to the companion of La Perouse;
but, long disregarded or forgotten, the knowledge of the law that the
intensity of the magnetic force of the Earth varied with the latitude,
did not, I conceive, acquire an existence in science until the publication of my observations from 1798 to 1804. The object and the length
of this note will not be indifferent to those who are familiar with the
recent history of magnetism, and the doubts that have been started in
connection with it, and who, from their own experience, are aware
that we are apt to attach some value to that which has cost us the uninterrupted labor of five years, under the pressure of a tropical climate,
and of perilous mountain expeditions.

* From the observations hitherto collected, it appears that the maximum of intensity for the whole surface of the Earth is 2·052, and the
minimum 0.706. Both phenomena occur in the southern hemisphere;
the former in 73° 47′ S. lat., and 169° 30′. E. long. from Paris, near

be expressed by 2·052 (the unit still employed being the in‑tensity which I discovered on the magnetic equator in North‑ern Peru), Sabine found it was only 1·624 at the magnetic north pole near Melville Island (74° 27′ north lat.), while it is 1·803 at New York, in the United States, which has al‑most the same latitude as Naples.

The brilliant discoveries of Œrsted, Arago, and Faraday have established a more intimate connection between the elec‑tric tension of the atmosphere and the magnetic tension of our terrestrial globe. While Œrsted has discovered that elec‑tricity excites magnetism in the neighborhood of the conduct‑ing body, Faraday's experiments have elicited electric currents from the liberated magnetism. Magnetism is one of the mani‑fold forms under which electricity reveals itself. The ancient vague presentiment of the identity of electric and magnetic attraction has been verified in our own times. "When elec‑trum (amber)," says Pliny, in the spirit of the Ionic natural philosophy of Thales,* "is *animated* by friction and heat, it will attract bark and dry leaves precisely as the loadstone at‑tracts iron." The same words may be found in the literature of an Asiatic nation, and occur in a eulogium on the load‑stone by the Chinese physicist Kuopho.† I observed with as‑

Mount Crozier, west-northwest of the south magnetic pole, at a place where Captain James Ross found the inclination of the needle to be 87° 11′ (Sabine, *Contributions to Terrestrial Magnetism*, 1843, No. 5, p. 231); the latter, observed by Erman, at 19° 59′ S. lat., and 37° 24′ W. long. from Paris, 320 miles eastward from the Brazilian coast of Espiritu Santo (Erman, *Phys. Beob.*, 1841, s. 570), at a point where the inclina‑tion is only 7° 55′. The actual ratio of the two intensities is therefore as 1 to 2·906. It was long believed that the greatest intensity of the magnetic force was only two and a half times as great as the weakest exhibited on the Earth's surface. (Sabine, *Report on Magnetic In‑tensity*, p. 82.)

* Of amber (succinum, glessum) Pliny observes (xxxvii., 3), "Gen‑era ejus plura. Attritu digitorum accepta caloris anima trahunt in se paleas ac folia arida quæ levia sunt, ac ut magnes lapis ferri ramenta quoque." (Plato, *in Timæo*, p. 80. Martin, *Etude sur le Timée*, t. ii., p. 343–346. Strabo, xv., p. 703, Casaub.; Clemens Alex., *Strom.*, ii., p. 370, where, singularly enough, a difference is made between τὸ σούχιον and τὸ ἤλεκτρον.) When Thales, in Aristot., *de Anima*, 1, 2, and Hippias, in Diog. Laert., i., 24, describe the magnet and amber as possessing a soul, they refer only to a moving principle.

† "The magnet attracts iron as amber does the smallest grain of mus‑tard seed. It is like a breath of wind which mysteriously penetrates through both, and communicates itself with the rapidity of an arrow." These are the words of Kuopho, a Chinese panegyrist on the magnet, who wrote in the beginning of the fourth century. (Klaproth, *Lettre à M. A. de Humboldt, sur l'Invention de la Boussole*, 1834, p. 125.)

tonishment, on the woody banks of the Orinoco, in the sports of the natives, that the excitement of electricity by friction was known to these savage races, who occupy the very lowest place in the scale of humanity. Children may be seen to rub the dry, flat, and shining seeds or husks of a trailing plant (probably a *Negretia*) until they are able to attract threads of cotton and pieces of bamboo cane. That which thus delights the naked copper-colored Indian is calculated to awaken in our minds a deep and earnest impression. What a chasm divides the electric pastime of these savages from the discovery of a metallic conductor discharging its electric shocks, or a pile composed of many chemically-decomposing substances, or a light-engendering magnetic apparatus! In such a chasm lie buried thousands of years that compose the history of the intellectual development of mankind!

The incessant change or oscillatory motion which we discover in all magnetic phenomena, whether in those of the inclination, declination, and intensity of these forces, according to the hours of the day and the night, and the seasons and the course of the whole year, leads us to conjecture the existence of very various and partial systems of electric currents on the surface of the Earth. Are these currents, as in Seebeck's experiments, thermo-magnetic, and excited directly from unequal distribution of heat? or should we not rather regard them as induced by the position of the Sun and by solar heat?* Have the rotation of the planets, and the different degrees of velocity which the individual zones acquire, according to their respective distances from the equator, any influence on the distribution of magnetism? Must we seek the seat of these currents, that is to say, of the disturbed electricity, in the atmosphere, in the regions of planetary space, or in the polarity of the Sun and Moon? Galileo, in his celebrated *Dialogo*, was inclined to ascribe the parallel direction of the axis of the Earth to a magnetic point of attraction seated in universal space.

If we represent to ourselves the interior of the Earth as fused and undergoing an enormous pressure, and at a degree of temperature the amount of which we are unable to assign,

* " The phenomena of periodical variations depend manifestly on the action of solar heat, operating probably through the medium of thermo-electric currents induced on the Earth's surface. Beyond this rude guess, however, nothing is as yet known of their physical cause. It is even still a matter of speculation whether the solar influence be a principal or only a subordinate cause in the phenomena of terrestrial magnetism." (*Observations to be made in the Antarctic Expedition*, 1840, p. 35.)

we must renounce all idea of a magnetic nucleus of the Earth.
All magnetism is certainly not lost until we arrive at a white
heat,[*] and it is manifested when iron is at a dark red heat,
however different, therefore, the modifications may be which
are excited in substances in their molecular state, and in the
coercive force depending upon that condition in experiments
of this nature, there will still remain a considerable thickness
of the terrestrial stratum, which might be assumed to be the
seat of magnetic currents.   The old explanation of the horary
variations of declination by the progressive warming of the
Earth in the apparent revolution of the Sun from east to west
must be limited to the uppermost surface, since thermometers
sunk into the Earth, which are now being accurately observed
at so many different places, show how slowly the solar heat
penetrates even to the inconsiderable depth of a few feet.
Moreover, the thermic condition of the surface of water, by
which two thirds of our planet is covered, is not favorable to
such modes of explanation, when we have reference to an immediate action and not to an effect of induction in the aërial
and aqueous investment of our terrestrial globe.

    In the present condition of our knowledge, it is impossible
to afford a satisfactory reply to all questions regarding the ultimate physical causes of these phenomena.   It is only with reference to that which presents itself in the triple manifestations
of the terrestrial force, as a measurable relation of space and
time, and as a stable element in the midst of change, that
science has recently made such brilliant advances by the aid
of the determination of mean numerical values.   From Toronto in Upper Canada to the Cape of Good Hope and Van Diemen's Land, from Paris to Pekin, the Earth has been covered,
since 1828, with magnetic observatories,[†] in which every regu-

    [*] Barlow, in the *Philos. Trans.* for 1822, Pt. i., p. 117 ; Sir David
Brewster, *Treatise on Magnetism*, p. 129. Long before the times of
Gilbert and Hooke, it was taught in the Chinese work *Ow-thsa-tsou*
that heat diminished the directive force of the magnetic needle. (Klaproth, *Lettre à M. A. de Humboldt, sur l'Invention de la Boussole,* p. 95.)

    [†] As the first demand for the establishment of these observatories (a
net-work of stations, provided with similar instruments) proceeded
from me, I did not dare to cherish the hope that I should live long
enough to see the time when both hemispheres should be uniformly
covered with magnetic houses under the associated activity of able
physicists and astronomers. This has, however, been accomplished,
and chiefly through the liberal and continued support of the Russian and
British governments.

    In the years 1806 and 1807, I and my friend and fellow-laborer, Herr
Oltmanns, while at Berlin, observed the movements of the needle, espe-

lar or irregular manifestation of the terrestrial force is detected
by uninterrupted and simultaneous observations.   A variation

cially at the times of the solstices and equinoxes, from hour to hour,
and often from half hour to half hour, for five or six days and nights
uninterruptedly.   I had persuaded myself that continuous and uninterrupted observations of several days and nights (observatio perpetua)
were preferable to the single observations of many months.   The apparatus, a Prony's magnetic telescope, suspended in a glass case by a
thread devoid of torsion, allowed angles of seven or eight seconds to be
read off on a finely-divided scale, placed at a proper distance, and
lighted at night by lamps.   Magnetic perturbations (storms), which occasionally recurred at the same hour on several successive nights, led
me even then to desire extremely that similar apparatus should be used
to the east and west of Berlin, in order to distinguish general terrestrial phenomena from those which are mere local disturbances, depending on the inequality of heat in different parts of the Earth, or on the
cloudiness of the atmosphere.   My departure to Paris, and the long
period of political disturbance that involved the whole of the west of
Europe, prevented my wish from being then accomplished.   Œrsted's
great discovery (1820) of the intimate connection between electricity
and magnetism again excited a general interest (which had long flag-
ged) in the periodical variations of the electro-magnetic tension of the
Earth.   Arago, who many years previously had commenced in the Observatory at Paris, with a new and excellent declination instrument by
Gambey, the longest uninterrupted series of horary observations which
we possess in Europe, showed, by a comparison with simultaneous observations of perturbation made at Kasan, what advantages might be
obtained from corresponding measurements of declination.   When I
returned to Berlin, after an eighteen years' residence in France, I had
a small magnetic house erected in the autumn of 1828, not only with
the view of carrying on the work commenced in 1806, but more with
the object that simultaneous observations at hours previously determined might be made at Berlin, Paris, and Freiburg, at a depth of 35
fathoms below the surface.   The simultaneous occurrence of the perturbations, and the parallelism of the movements for October and December, 1829, were then graphically represented.   (Pogg., *Annalen,*
bd. xix., s. 357, taf. i.–iii.)   An expedition into Northern Asia, undertaken in 1829, by command of the Emperor of Russia, soon gave me an
opportunity of working out my plan on a larger scale.   This plan was
laid before a select committee of one of the Imperial Academies of
Science, and, under the protection of the Director of the Mining Department, Count von Cancrin, and the excellent superintendence of Professor Kupffer, magnetic stations were appointed over the whole of
Northern Asia, from Nicolajeff, in the line through Catharinenburg, Barnaul, and Nertschinsk, to Pekin.

The year 1832 (*Göttinger gelehrte Anzeigen,* st. 206) is distinguished
as the great epoch in which the profound author of a general theory of
terrestrial magnetism, Friedrich Gauss, erected apparatus, constructed
on a new principle, in the Göttingen Observatory.   The magnetic observatory was finished in 1834, and in the same year Gauss distributed
new instruments, with instructions for their use, in which the celebrated
physicist, Wilhelm Weber, took extreme interest, over a large portion
of Germany and Sweden, and the whole of Italy.   (*Resultate der Beob.
des Magnetischen Vereins im Jahr* 1338, s. 135, and Poggend., *Annalen,*

of $\frac{1}{40000}$th of the magnetic intensity is measured, and, at certain epochs, observations are made at intervals of $2\frac{1}{2}$ minutes,
and continued for twenty-four hours consecutively. A great
English astronomer and physicist has calculated* that the
mass of observations which are in progress will accumulate in
the course of three years to 1,958,000. Never before has so
noble and cheerful a spirit presided over the inquiry into the
quantitative relations of the laws of the phenomena of nature.
We are, therefore, justified in hoping that these laws, when
compared with those which govern the atmosphere and the
remoter regions of space, may, by degrees, lead us to a more
intimate acquaintance with the genetic conditions of magnetic
phenomena. As yet we can only boast of having opened a
greater number of paths which may possibly lead to an explanation of this subject. In the physical science of terres-

bd. xxxiii., s. 426.) In the magnetic association that was now formed
with Göttingen for its center, simultaneous observations have been undertaken four times a year since 1836, and continued uninterruptedly
for twenty-four hours. The periods, however, do not coincide with
those of the equinoxes and solstices, which I had proposed and followed
out in 1830. Up to this period, Great Britain, in possession of the most
extensive commerce and the largest navy in the world, had taken no
part in the movement which since 1828 had begun to yield important
results for the more fixed ground-work of terrestrial magnetism. I had
the good fortune, by a public appeal from Berlin, which I sent in April,
1836, to the Duke of Sussex, at that time President of the Royal Society (Lettre de M. de Humboldt à S.A.R. le Duc de Sussex, sur les
moyens propres à perfectionner la connaissance du magnétisme terrestre
par l'établissement des stations magnétiques et d'observations corre-
spondantes), to excite a friendly interest in the undertaking which it
had so long been the chief object of my wish to carry out. In my letter to the Duke of Sussex I urged the establishment of permanent stations in Canada, St. Helena, the Cape of Good Hope, the Isle of France,
Ceylon, and New Holland, which five years previously I had advanced
as good positions. The Royal Society appointed a joint physical and
meteorological committee, which not only proposed to the government
the establishment of fixed magnetic observatories in both hemispheres,
but also the equipment of a naval expedition for magnetic observations
in the Antarctic Seas. It is needless to proclaim the obligations of
science in this matter to the great activity of Sir John Herschel, Sabine,
Airy, and Lloyd, as well as the powerful support that was afforded by
the British Association for the Advancement of Science at their meeting held at Newcastle in 1838. In June, 1839, the Antarctic magnetic
expedition, under the command of Captain James Clark Ross, was fully
arranged; and now, since its successful return, we reap the double
fruits of highly important geographical discoveries around the south
pole, and a series of simultaneous observations at eight or ten magnetic
stations.

* See the article on *Terrestrial Magnetism*, in the *Quarterly Review*
1840, vol. lxvi., p. 271–312.

trial magnetism, which must not be confounded with the purely mathematical branch of the study, those persons only will obtain perfect satisfaction who, as in the science of the meteorological processes of the atmosphere, conveniently turn aside the practical bearing of all phenomena that can not be explained according to their own views.

Terrestrial magnetism, and the electro-dynamic forces computed by the intellectual Ampère,[*] stand in simultaneous and intimate connection with the terrestrial or polar light, as well as with the internal and external heat of our planet, whose magnetic poles may be considered as the poles of cold.[†] The bold conjecture hazarded one hundred and twenty-eight years since by Halley,[‡] that the Aurora Borealis was a magnetic phenomenon, has acquired empirical certainty from Faraday's brilliant discovery of the evolution of light by magnetic forces. The northern light is preceded by premonitory signs. Thus, in the morning before the occurrence of the phenomenon, the irregular horary course of the magnetic needle generally indicates a disturbance of the equilibrium in the distribution of

[*] Instead of ascribing the internal heat of the Earth to the transition of matter from a vapor-like fluid to a solid condition, which accompanies the formation of the planets, Ampère has propounded the idea, which I regard as highly improbable, that the Earth's temperature may be the consequence of the continuous chemical action of a nucleus of the metals of the earths and alkalies on the oxydizing external crust. "It can not be doubted," he observes in his masterly *Théorie des Phénomènes Electro-dynamiques*, 1826, p. 199, "that electro-magnetic currents exist in the interior of the globe, and that these currents are the cause of its temperature. They arise from the action of a central metallic nucleus, composed of the metals discovered by Sir Humphrey Davy, acting on the surrounding oxydized layer."

[†] The remarkable connection between the curvature of the magnetic lines and that of my isothermal lines was first detected by Sir David Brewster. See the *Transactions of the Royal Society of Edinburgh*, vol. ix., 1821, p. 318, and *Treatise on Magnetism*, 1837, p. 42, 44, 47, and 268. This distinguished physicist admits two cold poles (poles of maximum cold) in the northern hemisphere, an American one near Cape Walker (73° lat., 100° W. long.), and an Asiatic one (73° lat., 80° E. long.); whence arise, according to him, two hot and two cold meridians, *i. e.*, meridians of greatest heat and cold. Even in the sixteenth century, Acosta (*Historia Natural de las Indias*, 1589, lib. i., cap. 17), grounding his opinion on the observations of a very experienced Portuguese pilot, taught that there were four lines without declination. It would seem from the controversy of Henry Bond (the author of *The Longitude Found*, 1676) with Beckborrow, that this view in some measure influenced Halley in his theory of four magnetic poles. See my *Examen Critique de l'Hist. de la Géographie*, t. iii., p. 60.

[‡] Halley, in the *Philosophical Transactions*, vol. xxix. (for 1714–1716), No. 341.

terrestrial magnetism.* When this disturbance attains a great
degree of intensity, the equilibrium of the distribution is restored by a discharge attended by a development of light
· The Aurora† itself is, therefore, not to be regarded as an ex
ternally manifested cause of this disturbance, but rather as a
result of telluric activity, manifested on the one side by the
appearance of the light, and on the other by the vibrations of
the magnetic needle." The splendid appearance of colored
polar light is the act of discharge, the termination of a mag
netic storm, as in an electrical storm a development of light—
the flash of lightning—indicates the restoration of the disturbed equilibrium in the distribution of the electricity. An electric storm is generally confined to a small space, beyond the
limits of which the condition of the atmospheric electricity
remains unchanged. A magnetic storm, on the other hand,

* [The Aurora Borealis of October 24th, 1847, which was one of the
most brilliant ever known in this country, was preceded by great magnetic disturbance. On the 22d of October the maximum of the west
declination was 23° 10′; on the 23d the position of the magnet was
continually changing, and the extreme west declinations were between
22° 44′ and 23° 37′; on the night between the 23d and 24th of October,
the changes of position were very large and very frequent, the magnet
at times moving across the field so rapidly that a difficulty was experienced in following it. During the day of the 24th of October there was
a constant change of position, but after midnight, when the Aurora began perceptibly to decline in brightness, the disturbance entirely ceased.
The changes of position of the horizontal-force magnet were as large and
as frequent as those of the declination magnet, but the vertical-force
magnet was at no time so much affected as the other two instruments.
See *On the Aurora Borealis, as it was seen on Sunday evening, October
24th, 1847, at Blackheath*, by James Glaisher, Esq., of the Royal Observatory, Greenwich, in the *London, Edinburgh, and Dublin Philos. Mag.
and Journal of Science for Nov.*, 1847. See further, *An Account of the
Aurora Borealis of October the 24th*, 1847, by John H. Morgan, Esq.
We must not omit to mention that magnetic disturbance is now regis-
tered by a *photographic* process: the self-registering photographic apparatus used for this purpose in the Observatory at Greenwich was designed by Mr. Brooke, and another ingenious instrument of this kind
has been invented by Mr. F. Ronalds, of the Richmond Observatory.]—
Tr.

† Dove, in Poggend., *Annalen*, bd. xx., s. 341; bd. xix., s. 388.
" The declination needle acts in very nearly the same way as an atmospheric electrometer, whose divergence in like manner shows the increased tension of the electricity before this has become so great as to
yield a spark." See, also, the excellent observations of Professor Kämtz,
in his *Lehrbuch der Meteorologie*, bd. iii., s. 511–519, and Sir David
Brewster, in his *Treatise on Magnetism*, p. 280. Regarding the magnetic properties of the galvanic flame, or luminous arch from a Bunsen's carbon and zinc battery, see Casselmann's *Beobachtungen* (Mar-
burg, 1844), s. 56–62.

shows its influence on the course of the needle over large portions of continents, and, as Arago first discovered, far from the spot where the evolution of light was visible. It is not improbable that, as heavily-charged threatening clouds, owing to frequent transitions of the atmospheric electricity to an opposite condition, are not always discharged, accompanied by lightning, so likewise magnetic storms may occasion far-extending disturbances in the horary course of the needle, without there being any positive necessity that the equilibrium of the distribution should be restored by explosion, or by the passage of luminous effusions from one of the poles to the equator, or from pole to pole.

In collecting all the individual features of the phenomenon in one general picture, we must not omit to describe the origin and course of a perfectly developed Aurora Borealis. Low down in the distant horizon, about the part of the heavens which is intersected by the magnetic meridian, the sky which was previously clear is at once overcast. A dense wall or bank of cloud seems to rise gradually higher and higher, until it attains an elevation of 8 or 10 degrees. The color of the dark segment passes into brown or violet; and stars are visible through the cloudy stratum, as when a dense smoke darkens the sky. A broad, brightly-luminous arch, first white, then yellow, encircles the dark segment; but as the brilliant arch appears subsequently to the smoky gray segment, we can not agree with Argelander in ascribing the latter to the effect of mere contrast with the bright luminous margin.* The highest point of the arch of light is, according to accurate observations made on this subject,† not generally in the magnetic meridian itself, but from $5°$ to $18°$ toward the direction of the magnetic declination of the place.‡ In northern latitudes,

* Argelander, in the important observations on the northern light embodied in the *Vorträgen gehalten in der physikalisch-ökonomischen Gessellschaft zu Königsberg*, bd. i., 1834, s. 257–264.

† For an account of the results of the observations of Lottin, Bravais, and Siljerström, who spent a winter at Bosekop, on the coast of Lapland (70° N. lat.), and in 210 nights saw the northern lights 160 times, see the *Comptes Rendus de l'Acad. des Sciences*, t. x., p. 289, and Martins's *Météorologie*, 1843, p. 453. See, also, Argelander, in the *Vortragen geh. in der Königsberg Gessellschaft*, bd. i., s. 259.

‡ [Professor Challis, of Cambridge, states that in the Aurora of October 24th, 1847, the streamers all converged toward a single point of the heavens, situated in or very near a vertical circle passing through the magnetic pole. Around this point a corona was formed, the rays of which diverged in all directions from the center, leaving a space free from light: its azimuth was 18° 41′ from south to east, and its altitude 69° 54′. See Professor Challis, in the *Athenæum*, Oct. 31, 1847.]—*Tr*

in the immediate vicinity of the magnetic pole, the smoke-like conical segment appears less dark, and sometimes is not even seen. Where the horizontal force is the weakest, the middle of the luminous arch deviates the most from the magnetic meridian.

The luminous arch remains sometimes for hours together flashing and kindling in ever-varying undulations, before rays and streamers emanate from it, and shoot up to the zenith. The more intense the discharges of the northern light, the more bright is the play of colors, through all the varying gradations from violet and bluish white to green and crimson. Even in ordinary electricity excited by friction, the sparks are only colored in cases where the explosion is very violent after great tension. The magnetic columns of flame rise either singly from the luminous arch, blended with black rays similar to thick smoke, or simultaneously in many opposite points of the horizon, uniting together to form a flickering sea of flame, whose brilliant beauty admits of no adequate description, as the luminous waves are every moment assuming new and varying forms. The intensity of this light is at times so great, that Lowenörn (on the 29th of June, 1786) recognized the coruscation of the polar light in bright sunshine. Motion renders the phenomenon more visible. Round the point in the vault of heaven which corresponds to the direction of the inclination of the needle, the beams unite together to form the so-called corona, the crown of the northern light, which encircles the summit of the heavenly canopy with a milder radiance and unflickering emanations of light. It is only in rare instances that a perfect crown or circle is formed, but on its completion the phenomenon has invariably reached its maximum, and the radiations become less frequent, shorter, and more colorless. The crown and the luminous arches break up, and the whole vault of heaven becomes covered with irregularly-scattered, broad, faint, almost ashy-gray luminous immovable patches, which in their turn disappear, leaving nothing but a trace of the dark, smoke-like segment on the horizon. There often remains nothing of the whole spectacle but a white, delicate cloud with feathery edges, or divided at equal distances into small roundish groups like cirro-cumuli.

This connection of the polar light with the most delicate cirrous clouds deserves special attention, because it shows that the electro-magnetic evolution of light is a part of a meteorological process. Terrestrial magnetism here manifests its in

fluence on the atmosphere and on the condensation of aqueous
vapor. The fleecy clouds seen in Iceland by Thienemann,
and which he considered to be the northern light, have been
seen in recent times by Franklin and Richardson near the
American north pole, and by Admiral Wrangel on the Siberian coast of the Polar Sea. All remarked " that the Aurora
flashed forth in the most vivid beams when masses of cirrous
strata were hovering in the upper regions of the air, and when
these were so thin that their presence could only be recognized
by the formation of a halo round the moon." These clouds
sometimes range themselves, even by day, in a similar manner
to the beams of the Aurora, and then disturb the course of
the magnetic needle in the same manner as the latter. On
the morning after every distinct nocturnal Aurora, the same
superimposed strata of clouds have still been observed that
had previously been luminous.* The apparently converging
polar zones (streaks of clouds in the direction of the magnetic
meridian), which constantly occupied my attention during my
journeys on the elevated plateaux of Mexico and in Northern
Asia, belong probably to the same group of diurnal phenomena.†

* John Franklin, *Narrative of a Journey to the Shores of the Polar
Sea, in the Years* 1819–1822, p. 552 and 597; Thienemann, in the
Edinburgh Philosophical Journal, vol. xx., p. 336; Farquharson, in vol.
vi., p. 392, of the same journal; Wrangel, *Phys. Beob.*, s. 59. Parry
even saw the great arch of the northern light continue throughout the
day. (*Journal of a Second Voyage, performed in* 1821–1823, p. 156.)
Something of the same nature was seen in England on the 9th of September, 1827. A luminous arch, 20° high, with columns proceeding
from it, was seen at noon in a part of the sky that had been clear after
rain. (*Journal of the Royal Institution of Great Britain*, 1828, Jan.,
p. 429.)

† On my return from my American travels, I described the delicate
cirro-cumulus cloud, which appears uniformly divided, as if by the
action of repulsive forces, under the name of polar bands (*bandes po-
laires*), because their perspective point of convergence is mostly at first
in the magnetic pole, so that the parallel rows of fleecy clouds follow
the magnetic meridian. One peculiarity of this mysterious phenomenon
is the oscillation, or occasionally the gradually progressive motion, of
the point of convergence. It is usually observed that the bands are
only fully developed in one region of the heavens, and they are seen
to move first from south to north, and then gradually from east to west.
I could not trace any connection between the advancing motion of the
bands and alterations of the currents of air in the higher regions of the
atmosphere. They occur when the air is extremely calm and the
heavens are quite serene, and are much more common under the
tropics than in the temperate and frigid zones. I have seen this phenomenon on the Andes, almost under the equator, at an elevation of
15,920 feet, and in Northern Asia, in the plains of Krasnojarski, south

Southern lights have often been seen in England by the intelligent and indefatigable observer Dalton, and northern lights
have been observed in the southern hemisphere as far as 45°
latitude (as on the 14th of January, 1831). On occasions
that are by no means of rare occurrence, the equilibrium at
both poles has been simultaneously disturbed. I have discovered with certainty that northern polar lights have been seen
within the tropics in Mexico and Peru. We must distinguish
between the sphere of simultaneous visibility of the phenomenon and the zones of the Earth where it is seen almost night-
ly. Every observer no doubt sees a separate Aurora of his
own, as he sees a separate rainbow. A great portion of the
Earth simultaneously engenders these phenomena of emanations of light. Many nights may be instanced in which the
phenomenon has been simultaneously observed in England
and in Pennsylvania, in Rome and in Pekin. When it is
stated that Auroras diminish with the decrease of latitude,
the latitude must be understood to be magnetic, and as measured by its distance from the magnetic pole. In Iceland, in
Greenland, Newfoundland, on the shores of the Slave Lake,
and at Fort Enterprise in Northern Canada, these lights appear almost every night at certain seasons of the year, cele-
brating with their flashing beams, according to the mode of
expression common to the inhabitants of the Shetland Isles,
" a merry dance in heaven."* While the Aurora is a phenomenon of rare occurrence in Italy, it is frequently seen in
the latitude of Philadelphia (39° 57′), owing to the southern
position of the American magnetic pole. In the districts
which are remarkable, in the New Continent and the Siberian coasts, for the frequent occurrence of this phenomenon,
there are special regions or zones of longitude in which the
polar light is particularly bright and brilliant.† The exist-

of Buchtarminsk, so similarly developed, that we must regard the in
fluences producing it as very widely distributed, and as depending on
general natural forces. See the important observations of Kämtz (*Vorlesungen über Meteorologie*, 1840, s. 146), and the more recent ones of
Martins and Bravais (*Météorologie*, 1843, p. 117). In south polar bands,
composed of very delicate clouds, observed by Arago at Paris on the
23d of June, 1844, dark rays shot upward from an arch running east
and west. We have already made mention of black rays, resembling
dark smoke, as occurring in brilliant nocturnal northern lights.
    * The northern lights are called by the Shetland Islanders " the
merry dancers." (Kendal, in the *Quarterly Journal of Science*, new
series, vol. iv., p. 395.)
    † See Muncke's excellent work in the new edition of Gehler's *Physik
Wörterbuch*, bd. vii., i., s 113–268, and especially s. 158.

ence of local influences can not, therefore, be denied in these cases. Wrangel saw the brilliancy diminish as he left the shores of the Polar Sea, about Nischne-Kolymsk. The observations made in the North Polar expedition appear to prove that in the immediate vicinity of the magnetic pole the development of light is not in the least degree more intense or frequent than at some distance from it.

The knowledge which we at present possess of the altitude of the polar light is based on measurements which, from their nature, the constant oscillation of the phenomenon of light, and the consequent uncertainty of the angle of parallax, are not deserving of much confidence. The results obtained, setting aside the older data, fluctuate between several miles and an elevation of 3000 or 4000 feet ; and, in all probability, the northern lights at different times occur at very different elevations.* The most recent observers are disposed to place the phenomenon in the region of clouds, and not on the confines of the atmosphere ; and they even believe that the rays of the Aurora may be affected by winds and currents of air, if the phenomenon of light, by which alone the existence of an electro-magnetic current is appreciable, be actually connected with material groups of vesicles of vapor in motion, or, more correctly speaking, if light penetrate them, passing from one vesicle to another. Franklin saw near Great Bear Lake a beaming northern light, the lower side of which he thought illuminated a stratum of clouds, while, at a distance of only eighteen geographical miles, Kendal, who was on watch throughout the whole night, and never lost sight of the sky, perceived no phenomenon of light. The assertion, so frequently maintained of late, that the rays of the Aurora have been seen to shoot down to the ground between the spectator and some neighboring hill, is open to the charge of optical delusion, as in the cases of strokes of lightning or of the fall of fire-balls.

Whether the magnetic storms, whose local character we have illustrated by such remarkable examples, share noise as well as light in common with electric storms, is a question

* Farquharson in the *Edinburgh Philos. Journal*, vol. xvi., p. 304 ; *Philos. Transact.* for 1829, p. 113.

[The height of the bow of light of the Aurora seen at the Cambridge Observatory, March 19, 1847, was determined by Professors Challis, of Cambridge, and Chevallier, of Durham, to be 177 miles above the surface of the Earth. See the notice of this meteor in *An Account of the Aurora Borealis of Oct.* 24, 1847, by John H. Morgan, Esq., 1848.]— *Tr.*

that has become difficult to answer, since implicit confidence
is no longer yielded to the relations of Greenland whale-fish-
ers and Siberian fox-hunters. Northern lights appear to have
become less noisy since their occurrences have been more accurately recorded. Parry, Franklin, and Richardson, near
the north pole ; Thienemann in Iceland ; Gieseke in Greenland ; Lottin and Bravais, near the North Cape ; Wrangel
and Anjou, on the coast of the Polar Sea, have together seen
the Aurora thousands of times, but never heard any sound
attending the phenomenon. If this negative testimony should
not be deemed equivalent to the positive counter-evidence of
Hearne on the mouth of the Copper River and of Henderson
in Iceland, it must be remembered that, although Hood heard
a noise as of quickly-moved musket-balls and a slight crack-
ing sound during an Aurora, he also noticed the same noise
on the following day, when there was no northern light to be
seen ; and it must not be forgotten that Wrangel and Gieseke
were fully convinced that the sound they had heard was to
be ascribed to the contraction of the ice and the crust of the
snow on the sudden cooling of the atmosphere. The belief
in a crackling sound has arisen, not among the people generally, but rather among learned travelers, because in earlier
times the northern light was declared to be an effect of atmospheric electricity, on account of the luminous manifestation
of the electricity in rarefied space, and the observers found it
easy to hear what they wished to hear. Recent experiments
with very sensitive electrometers have hitherto, contrary to
the expectation generally entertained, yielded only negative
results. The condition of the electricity in the atmosphere[*]

* [Mr. James Glaisher, of the Royal Observatory, Greenwich, in his
interesting *Remarks on the Weather during the Quarter ending December 31st*, 1847, says, " It is a fact well worthy of notice, that from the
beginning of this quarter till the 20th of December, the electricity of
the atmosphere was almost always in a neutral state, so that no signs of
electricity were shown for several days together by any of the electrical instruments." During this period there were *eight* exhibitions of
the Aurora Borealis, of which one was the peculiarly bright display of
the meteor on the 24th of October. These frequent exhibitions of brilliant Auroræ seem to depend upon many remarkable meteorological relations, for we find, according to Mr. Glaisher's statement in the paper
to which we have already alluded, that the previous fifty years afford
no parallel season to the closing one of 1847. The mean temperature
of evaporation and of the dew point, the mean elastic force of vapor,
the mean reading of the barometer, and the mean daily range of the
readings of the thermometers in air, were all greater at Greenwich
during that season of 1847 than the average range of many preceding
years.]—*Tr.*

is not found to be changed during the most intense Aurora ;
but, on the other hand, the three expressions of the power of
terrestrial magnetism, declination, inclination, and intensity,
are all affected by polar light, so that in the same night, and
at different periods of the magnetic development, the same
end of the needle is both attracted and repelled. The asser
tion made by Parry, on the strength of the data yielded by
his observations in the neighborhood of the magnetic pole at
Melville Island, that the Aurora did not disturb, but rathei
exercised a calming influence on the magnetic needle, has been
satisfactorily refuted by Parry's own more exact researches,*
detailed in his journal, and by the admirable observations of
Richardson, Hood, and Franklin in Northern Canada, and
lastly by Bravais and Lottin in Lapland. The process of the
Aurora is, as has already been observed, the restoration of a
disturbed condition of equilibrium. The effect on the needle
is different according to the degree of intensity of the explosion. It was only unappreciable at the gloomy winter station
of Bosekop when the phenomenon of light was very faint and
low in the horizon. The shooting cylinders of rays have been
aptly compared to the flame which rises in the closed circuit
of a voltaic pile between two points of carbon at a considerable distance apart, or, according to Fizeau, to the flame rising
between a silver and a carbon point, and attracted or repelled
by the magnet. This analogy certainly sets aside the necessity of assuming the existence of metallic vapors in the atmosphere, which some celebrated physicists have regarded as the
substratum of the northern light.

When we apply the indefinite term *polar light* to the luminous phenomenon which we ascribe to a galvanic current, that
is to say, to the motion of electricity in a closed circuit, we
merely indicate the local direction in which the evolution of
light is most frequently, although by no means invariably,
seen. This phenomenon derives the greater part of its importance from the fact that the Earth becomes *self-luminous*,
and that as a planet, besides the light which it receives from
the central body, the Sun, it shows itself capable in itself of
developing light. The intensity of the terrestrial light, or,
rather, the luminosity which is diffused, exceeds, in cases of
the brightest colored radiation toward the zenith, the light
of the Moon in its first quarter. Occasionally, as on the 7th
of January, 1831, printed characters could be read without
difficulty. This almost uninterrupted development of light

* Kämtz, *Lehrbuch der Meteorologie*, bd. iii., s. 498 und 501.

in the Earth leads us by analogy to the remarkable process exhibited in Venus. The portion of this planet which is not illumined by the Sun often shines with a phosphorescent light of its own. It is not improbable that the Moon, Jupiter, and the comets shine with an independent light, besides the reflected solar light visible through the polariscope. Without speaking of the problematical but yet ordinary mode in which the sky is illuminated, when a low cloud may be seen to shine with an uninterrupted flickering light for many minutes together, we still meet with other instances of terrestrial development of light in our atmosphere. In this category we may reckon the celebrated luminous mists seen in 1783 and 1831 ; the steady luminous appearance exhibited without any flickering in great clouds observed by Rozier and Beccaria ; and lastly, as Arago* well remarks, the faint diffused light which guides the steps of the traveler in cloudy, starless, and moonless nights in autumn and winter, even when there is no snow on the ground. As in polar light or the electro-magnetic storm, a current of brilliant and often colored light streams through the atmosphere in high latitudes, so also in the torrid zones between the tropics, the ocean simultaneously develops light over a space of many thousand square miles. Here the magical effect of light is owing to the forces of organic nature. Foaming with light, the eddying waves flash in phosphorescent sparks over the wide expanse of waters, where every scintillation is the vital manifestation of an invisible animal world. So varied are the sources of terrestrial light ! Must we still suppose this light to be latent, and combined in vapors, in order to explain *Moser's images produced at a distance*—a discovery in which reality has hitherto manifested itself like a mere phantom of the imagination.

As the internal heat of our planet is connected on the one hand with the generation of electro-magnetic currents and the process of terrestrial light (a consequence of the magnetic storm), it, on the other hand, discloses to us the chief source of geognostic phenomena. We shall consider these in their connection with and their transition from merely dynamic disturbances, from the elevation of whole continents and mountain chains to the development and effusion of gaseous and

* Arago, on the dry fogs of 1783 and 1831, which illuminated the night, in the *Annuaire du Bureau des Longitudes,* 1832, p. 246 and 250 ; and, regarding extraordinary luminous appearances in clouds without storms, see *Notices sur la Tonnerre,* in the *Annuaire pour l'an.* 1838, p. 279–285.

liquid fluids, of hot mud, and of those heated and molten earths which become solidified into crystalline mineral masses. Modern geognosy, the mineral portion of terrestrial physics, has made no slight advance in having investigated this con nection of phenomena. This investigation has led us away from the delusive hypothesis, by which it was customary formerly to endeavor to explain, individually, every expression of force in the terrestrial globe : it shows us the connection of the occurrence of heterogeneous substances with that which only appertains to changes in space (disturbances or elevations), and groups together phenomena which at first sight appeared most heterogeneous, as thermal springs, effusion of carbonic acid and sulphurous vapor, innocuous salses (mud eruptions), and the dreadful devastations of volcanic mountains.* In a general view of nature, all these phenomena are fused together in one sole idea of the reaction of the interior of a planet on its external surface. We thus recognize in the depths of the earth, and in the increase of temperature with the increase of depth from the surface, not only the germ of disturbing movements, but also of the gradual elevation of whole continents (as mountain chains on long fissures), of volcanic eruptions, and of the manifold production of mountains and mineral masses. The influence of this reaction of the interior on the exterior is not, however, limited to inorganic nature alone. It is highly probable that, in an earlier world, more powerful emanations of carbonic acid gas, blended with the atmosphere, must have increased the assimilation of carbon in vegetables, and that an inexhaustible supply of combustible matter (lignites and carboniferous formations) must have been thus buried in the upper strata of the earth by the revolutions attending the destruction of vast tracts of forest. We likewise perceive that the destiny of mankind is in part dependent on the formation of the external surface of the earth, the direction of mountain tracts and high lands, and on the distribution of elevated continents. It is thus granted to the inquiring mind to pass from link to link along the chain of phenomena until it reaches the period when, in the solidifying process of our planet, and in its first transition from the gaseous form to the agglomeration of matter, that portion of the inner heat of the Earth was developed, which does not belong to the action of the Sun.

* [See Mantell's *Wonders of Geology*, 1848, vol. i., p. 34, 36, 105 ; also Lyell's *Principles of Geology*, vol. ii., and Daubeney *On Volcanoes*, 2d ed., 1848. Part ii., ch. xxxii., xxxiii.] — *Tr.*

In order to give a general delineation of the causal connection of geognostical phenomena, we will begin with those
whose chief characteristic is dynamic, consisting in motion
and in change in space. Earthquakes manifest themselves
by quick and successive vertical, or horizontal, or rotatory vibrations.* In the very considerable number of earthquakes
which I have experienced in both hemispheres, alike on land
and at sea, the two first-named kinds of motion have often appeared to me to occur simultaneously. The mine-like explosion—the vertical action from below upward—was most strikingly manifested in the overthrow of the town of Riobamba
in 1797, when the bodies of many of the inhabitants were
found to have been hurled to Cullca, a hill several hundred
feet in height, and on the opposite side of the River Lican.
The propagation is most generally effected by undulations in
a linear direction,† with a velocity of from twenty to twenty-
eight miles in a minute, but partly in circles of commotion or
large ellipses, in which the vibrations are propagated with
decreasing intensity from a center toward the circumference.
There are districts exposed to the action of two intersecting
circles of commotion. In Northern Asia, where the Father
of History,‡ and subsequently Theophylactus Simocatta,§ described the districts of Scythia as free from earthquakes, I
have observed the metalliferous portion of the Altai Mountains under the influence of a two-fold focus of commotion, the
Lake of Baikal, and the volcano of the Celestial Mountain
(Thianschan).‖ When the circles of commotion intersect one
another—when, for instance, an elevated plain lies between
two volcanoes simultaneously in a state of eruption, several
wave-systems may exist together, as in fluids, and not mutually disturb one another. We may even suppose *interfer-*

* [See Daubeney *On Volcanoes*, 2d ed., 1848, p. 509.]—*Tr.*

† [On the linear direction of earthquakes, see Daubeney *On Volcanoes*, p. 515.]—*Tr.*

‡ Herod, iv., 28. The prostration of the colossal statue of Memnon,
which has been again restored (Letronne, *La Statue Vocale de Memnon*,
1835, p. 25, 26), presents a fact in opposition to the ancient prejudice
that Egypt is free from earthquakes (Pliny, ii., 80); but the valley of
the Nile does lie external to the circle of commotion of Byzantium, the
Archipelago, and Syria (Ideler ad Aristot., *Meteor.*, p. 584).

§ Saint-Martin, in the learned notes to Lebeau, *Hist. du Bas Empire*,
t. ix., p. 401.

‖ Humboldt, *Asie Centrale*, t. ii., p. 110–118. In regard to the difference between agitation of the surface and of the strata lying beneath
it, see Gay-Lussac, in the *Annales de Chimie et de Physique*, t. xxii., p.
429.

ence to exist here, as in the intersecting waves of sound. The extent of the propagated waves of commotion will be increased on the upper surface of the earth, according to the general law of mechanics, by which, on the transmission of motion in elastic bodies, the stratum lying free on the one side endeavors to separate itself from the other strata.

Waves of commotion have been investigated by means of the pendulum and the seismometer* with tolerable accuracy in respect to their direction and total intensity, but by no means with reference to the internal nature of their alternations and their periodic intumescence. In the city of Quito, which lies at the foot of a still active volcano (the Rucu Pichincha), and at an elevation of 9540 feet above the level of the sea, which has beautiful cupolas, high vaulted churches, and massive edifices of several stories, I have often been astonished that the violence of the nocturnal earthquakes so seldom causes fissures in the walls, while in the Peruvian plains oscillations apparently much less intense injure low reed cottages. The natives, who have experienced many hundred earthquakes, believe that the difference depends less upon the length or shortness of the waves, and the slowness or rapidity of the horizontal vibrations,† than on the uniformity of the motion in opposite directions. The circling rotatory commotions are the most uncommon, but, at the same time, the most dangerous. Walls were observed to be twisted, but not thrown down ; rows of trees turned from their previous parallel direc-

---

* [This instrument, in its simplest form, consists merely of a basin filled with some viscid liquid, which, on the occurrence of a shock of an earthquake of sufficient force to disturb the equilibrium of the building in which it is placed, is tilted on one side, and the liquid made to rise in the same direction, thus showing by its height the degree of the disturbance. Professor J. Forbes has invented an instrument of this nature, although on a greatly improved plan. It consists of a vertical metal rod, having a ball of lead movable upon it. It is supported upon a cylindrical steel wire, which may be compressed at pleasure by means of a screw. A lateral movement, such as that of an earthquake, which carries forward the base of the instrument, can only act upon the ball through the medium of the elasticity of the wire, and the direction of the displacement will be indicated by the plane of vibration of the pendulum. A self-registering apparatus is attached to the machine. See Professor J. Forbes's account of his invention in *Edinb. Phil. Trans.*, vol. xv., Part i.]—*Tr.*

† " Tutissimum est cum vibrat crispante ædificiorum crepitu ; et cum intumescit assurgens alternoque motu residet, innoxium et cum concurrentia tecta contrario ictu arietant ; quoniam alter motus alteri renititur. Undantis inclinatio et fluctus more quædam volutatio infesta est, aut cum in unam partem totus se motus impellit."—Plin., ii., 82.

tion ; and fields covered with different kinds of plants found to be displaced in the great earthquake of Riobamba, in the province of Quito, on the 4th of February, 1797, and in that of Calabria, between the 5th of February and the 28th of March, 1783 The phenomenon of the inversion or displacement of fields and pieces of land, by which one is made to occupy the place of another, is connected with a translatory motion or penetration of separate terrestrial strata. When I made the plan of the ruined town of Riobamba, one particular spot was pointed out to me, where all the furniture of one house had been found under the ruins of another. The loose earth had evidently moved like a fluid in currents, which must be assumed to have been directed first downward, then horizontally, and lastly upward. It was found necessary to appeal to the *Audiencia*, or Council of Justice, to decide upon the contentions that arose regarding the proprietorship of objects that had been removed to a distance of many hundred toises.

In countries where earthquakes are comparatively of much less frequent occurrence (as, for instance, in Southern Europe), a very general belief prevails, although unsupported by the authority of inductive reasoning,* that a calm, an oppressive

---

* Even in Italy they have begun to observe that earthquakes are unconnected with the state of the weather, that is to say, with the appearance of the heavens immediately before the shock.   The numerical results of Friedrich Hoffmann (*Hinterlassene Werke*, bd. ii., 366–375) exactly correspond with the experience of the Abbate Scina of Palermo. I have myself several times observed reddish clouds on the day of an earthquake, and shortly before it; on the 4th of November, 1799, I experienced two sharp shocks at the moment of a loud clap of thunder. (*Relat. Hist.*, liv. iv., chap. 10.) The Turin physicist, Vassalli Eandi, observed Volta's electrometer to be strongly agitated during the protracted earthquake of Pignerol, which lasted from the 2d of April to the 17th of May, 1808; *Journal de Physique*, t. lxvii., p. 291. But these indications presented by clouds, by modifications of atmospheric electricity, or by calms, can not be regarded as *generally* or *necessarily* connected with earthquakes, since in Quito, Peru, and Chili, as well as in Canada and Italy, many earthquakes are observed along with the purest and clearest skies, and with the freshest land and sea breezes. But if no meteorological phenomenon indicates the coming earthquake either on the morning of the shock or a few days previously, the influence of certain periods of the year (the vernal and autumnal equinoxes), the commencement of the rainy season in the tropics after long drought, and the change of the monsoons (according to general belief), can not be overlooked, even though the genetic connection of meteorological processes with those going on in the interior of our globe is still enveloped in obscurity.   Numerical inquiries on the distribution of earthquakes throughout the course of the year, such as those of Von Hoff, Peter Merian, and Friedrich Hoffmann, bear testimony to their frequency

ιeat, and a misty horizon, are always the forerunners of this phenomenon. The fallacy of this popular opinion is not only refuted by my own experience, but likewise by the observations of all those who have lived many years in districts where, as in Cumana, Quito, Peru, and Chili, the earth is frequently and violently agitated. I have felt earthquakes in clear air and a fresh east wind, as well as in rain and thunder storms. The regularity of the horary changes in the declination of the magnetic needle and in the atmospheric pressure remained un disturbed between the tropics on the days when earthquakes occurred.* These facts agree with the observations made by Adolph Erman (in the temperate zone, on the 8th of March, 1829) on the occasion of an earthquake at Irkutsk, near the Lake of Baikal. During the violent earthquake of Cumana, on the 4th of November, 1799, I found the declination and the intensity of the magnetic force alike unchanged, but, to my surprise, the inclination of the needle was diminished about 48'.† There was no ground to suspect an error in the calcu- lation, and yet, in the many other earthquakes which I have experienced on the elevated plateaux of Quito and Lima, the inclination as well as the other elements of terrestrial mag- netism remained always unchanged. Although, in general, the processes at work within the interior of the earth may not be announced by any meteorological phenomena or any special appearance of the sky, it is, on the contrary, not improbable, as we shall soon see, that in cases of violent earthquakes some effect may be imparted to the atmosphere, in consequence of which they can not always act in a purely dynamic manner.

at the periods of the equinoxes. It is singular that Pliny, at the end of his fanciful theory of earthquakes, names the entire frightful phenom- enon a subterranean storm; not so much in consequence of the rolling sound which frequently accompanies the shock, as because the elastic forces, concussive by their tension, accumulate in the interior of the earth when they are absent in the atmosphere! "Ventos in causa esse non dubium reor. Neque enim unquam intremiscunt terræ, nisi sopito mari, coeloque adeo tranquillo, ut volatus avium non pendeant, subtracto omni spiritu qui vehit; nec unquam nisi post ventos conditos, scilicet in venas et cavernas ejus occulto afflatu. Neque aliud est in terra tremor, quam in nube tonitruum; nec hiatus aliud quam cum fulmen erumpit, incluso spiritu luctante et ad libertatem exire nitente." (Plin., ii., 79.) The germs of almost every thing that has been observed or imagined on the causes of earthquakes, up to the present day, may be found in Seneca, *Nat. Quæst.*, vi., 4–31.

* I have given proof that the course of the horary variations of the barometer is not affected before or after earthquakes, in my *Relat. Hist.*, t. i., p. 311 and 513.

† Humboldt, *Relat. Hist.*, t. i., p. 515–517.

During the long-continued trembling of the ground in the Piedmontese valleys of Pelis and Clusson, the greatest changes in the electric tension of the atmosphere were observed while the sky was cloudless. The intensity of the hollow noise which generally accompanies an earthquake does not increase in the same degree as the force of the oscillations. I have ascertained with certainty that the great shock of the earthquake of Riobamba (4th Feb., 1797)—one of the most fearful phenomena recorded in the physical history of our planet—was not accompanied by any noise whatever. The tremendous noise (*el gran ruido*) which was heard below the soil of the cities of Quito and Ibarra, but not at Tacunga and Hambato, nearer the center of the motion, occurred between eighteen and twenty minutes *after* the actual catastrophe. In the celebrated earthquake of Lima and Callao (28th of October, 1746), a noise resembling a subterranean thunder-clap was heard at Truxillo a quarter of an hour after the shock, and unaccompanied by any trembling of the ground. In like manner, long after the great earthquake in New Granada, on the 16th of November, 1827, described by Boussingault, subterranean detonations were heard in the whole valley of Cauca during twenty or thirty seconds, unattended by motion. The nature of the noise varies also very much, being either rolling, or rustling, or clanking like chains when moved, or like near thunder, as, for instance, in the city of Quito ; or, lastly, clear and ringing, as if obsidian or some other vitrified masses were struck in subterranean cavities. As solid bodies are excellent conductors of sound, which is propagated in burned clay, for instance, ten or twelve times quicker than in the air, the subterranean noise may be heard at a great distance from the place where it has originated. In Caraccas, in the grassy plains of Calabozo, and on the banks of the Rio Apure, which falls into the Orinoco, a tremendously loud noise, resembling thunder, was heard, unaccompanied by an earthquake, over a district of land 9200 square miles in extent, on the 30th of April, 1812, while at a distance of 632 miles to the northeast, the volcano of St. Vincent, in the small Antilles, poured forth a copious stream of lava. With respect to distance, this was as if an eruption of Vesuvius had been heard in the north of France. In the year 1744, on the great eruption of the volcano of Cotopaxi, subterranean noises, resembling the discharge of cannon, were heard in Honda, on the Magdalena River. The crater of Cotopaxi lies not only 18,000 feet higher than Honda, but these two points are separated by the co-

iossal mountain chain of Quito, Pasto, and Popayan, no less
than by numerous valleys and clefts, and they are 436 miles
apart.   The sound was certainly not propagated through the
air, but through the earth, and at a great depth.   During the
violent earthquake of New Granada, in February, 1835, subterranean thunder was heard simultaneously at Popayan, Bogota, Santa Marta, and Caraccas (where it continued for seven
hours without any movement of the ground), in Haiti, Jamai
ca, and on the Lake of Nicaragua.

These phenomena of sound, when unattended by any perceptible shocks, produce a peculiarly deep impression even on
persons who have lived in countries where the earth has been
frequently exposed to shocks.   A striking and unparalleled instance of uninterrupted subterranean noise, unaccompanied by
any trace of an earthquake, is the phenomenon known in the
Mexican elevated plateaux by the name of the "roaring and
the subterranean thunder" (*bramidos y truenos subterraneos*)
of Guanaxuato.*   This celebrated and rich mountain city
lies far removed from any active volcano.   The noise began
about midnight on the 9th of January, 1784, and continued
for a month.   I have been enabled to give a circumstantial

---

* On the *bramidos* of Guanaxuato, see my *Essai Polit. sur la Nouv.
Espagne*, t. i., p. 303.   The subterranean noise, unaccompanied with
any appreciable shock, in the deep mines and on the surface (the town
of Guanaxuato lies 6830 feet above the level of the sea), was not heard
in the neighboring elevated plains, but only in the mountainous parts
of the Sierra, from the Cuesta de los Aguilares, near Marfil, to the north
of Santa Rosa.   There were individual parts of the Sierra 24–28 miles
northwest of Guanaxuato, to the other side of Chichimequillo, near the
boiling spring of San José de Comangillas, to which the waves of sound
did not extend.   Extremely stringent measures were adopted by the
magistrates of the large mountain towns on the 14th of January, 1784,
when the terror produced by these subterranean thunders was at its
height.   " The flight of a wealthy family shall be punished with a fine
of 1000 piasters, and that of a poor family with two months' imprison-
ment.   The militia shall bring back the fugitives."   One of the most
remarkable points about the whole affair is the opinion which the magistrates (el cabildo) cherished of their own superior knowledge.   In
one of their *proclamas*, I find the expression, " The magistrates, in their
wisdom (en su sabiduria), will at once know when there is actual danger, and will give orders for flight; for the present, let processions be
instituted."   The terror excited by the tremor gave rise to a famine,
since it prevented the importation of corn from the table-lands, where
it abounded.   The ancients were also aware that noises sometimes existed without earthquakes.—Aristot., *Meteor.*, ii., p. 802; Plin., ii., **80.**
The singular noise that was heard from March, 1822, to September,
1824, in the Dalmatian island Meleda (sixteen miles from Ragusa), and
on which Partsch has thrown much light, was occasionally accompanied
by shocks.

description of it from the report of many witnesses, and from the documents of the municipality, of which I was allowed to make use. From the 13th to the 16th of January, it seemed to the inhabitants as if heavy clouds lay beneath their feet, from which issued alternate slow rolling sounds and short, quick claps of thunder. The noise abated as gradually as it had begun. It was limited to a small space, and was not heard in a basaltic district at the distance of a few miles. Almost all the inhabitants, in terror, left the city, in which large masses of silver ingots were stored ; but the most courageous, and those more accustomed to subterranean thunder, soon returned, in order to drive off the bands of robbers who had attempted to possess themselves of the treasures of the city. Neither on the surface of the earth, nor in mines 1600 feet in depth, was the slightest shock to be perceived. No similar noise had ever before been heard on the elevated tableland of Mexico, nor has this terrific phenomenon since occurred there. Thus clefts are opened or closed in the interior of the earth, by which waves of sound penetrate to us or are impeded in their propagation.

The activity of an igneous mountain, however terrific and picturesque the spectacle may be which it presents to our contemplation, is always limited to a very small space. It is far otherwise with earthquakes, which, although scarcely perceptible to the eye, nevertheless simultaneously propagate their waves to a distance of many thousand miles. The great earthquake which destroyed the city of Lisbon on the 1st of November, 1755, and whose effects were so admirably investigated by the distinguished philosopher Emmanuel Kant, was felt in the Alps, on the coast of Sweden, in the Antilles, Antigua, Barbadoes, and Martinique ; in the great Canadian Lakes, in Thuringia, in the flat country of Northern Germany, and in the small inland lakes on the shores of the Baltic.* Remote springs were interrupted in their flow, a phenomenon attending earthquakes which had been noticed among the ancients by Demetrius the Callatian. The hot springs of Töplitz dried up, and returned, inundating every thing around, and having their waters colored with iron ocher. In Cadiz

* [It has been computed that the shock of this earthquake pervaded an area of 700,000 miles, or the twelfth part of the circumference of the globe. This dreadful shock lasted only five minutes: it happened about nine o'clock in the morning of the Feast of All Saints, when almost the whole population was within the churches, owing to which circumstance no less than 30,000 persons perished by the fall of these edifices. See Daubeney *On Volcanoes*, p. 514–517.]—*Tr*

the sea rose to an elevation of sixty-four feet, while in the Antilles, where the tide usually rises only from twenty-six to twenty-eight inches, it suddenly rose above twenty feet, the water being of an inky blackness. It has been computed that on the 1st of November, 1755, a portion of the Earth's surface, four times greater than that of Europe, was simultaneously shaken. As yet there is no manifestation of force known to us, including even the murderous inventions of our own race, by which a greater number of people have been killed in the short space of a few minutes : sixty thousand were destroyed in Sicily in 1693, from thirty to forty thousand in the earthquake of Riobamba in 1797, and probably five times as many in Asia Minor and Syria, under Tiberius and Justinian the elder, about the years 19 and 526.

There are instances in which the earth has been shaken for many successive days in the chain of the Andes in South America, but I am only acquainted with the following cases in which shocks that have been felt almost every hour for months together have occurred far from any volcano, as, for instance, on the eastern declivity of the Alpine chain of Mount Cenis, at Fenestrelles and Pignerol, from April, 1808 ; between New Madrid and L ttle Prairie,* north of Cincinnati, in the United States of America, in December, 1811, as well as through the whole winter of 1812 ; and in the Pachalik of Aleppo, in the months of August and September, 1822. As the mass of the people are seldom able to rise to general views, and are consequently always disposed to ascribe great phenomena to local telluric and atmospheric processes, wherever the shaking of the earth is continued for a long time, fears of the eruption of a new volcano are awakened. In some few cases, this apprehension has certainly proved to be well grounded, as, for instance, in the sudden elevation of volcanic islands, and as we see in the elevation of the volcano of Jorullo, a mountain elevated 1684 feet above the ancient level of the neighboring plain, on the 29th of September, 1759, after ninety days of earthquake and subterranean thunder.

If we could obtain information regarding the daily condition of all the earth's surface, we should probably discover that the earth is almost always undergoing shocks at some point of its superficies, and is continually influenced by the reaction

* Drake, *Nat. and Statist. View of Cincinnati*, p. 232–238; Mitchell, in the *Transactions of the Lit. and Philos. Soc. of New York*, vol. i., p. 231–308. In the Piedmontese county of Pignerol, glasses of water, filled to the very brim, exhibited for hours a continuous motion.

of the interior on the exterior. The frequency and general
prevalence of a phenomenon which is probably dependent on
the raised temperature of the deepest molten strata explain
its independence of the nature of the mineral masses in which
it manifests itself. Earthquakes have even been felt in the
loose alluvial strata of Holland, as in the neighborhood of Mid-
dleburg and Vliessingen on the 23d of February, 1828. Granite and mica slate are shaken as well as limestone and sandstone, or as trachyte and amygdaloid. It is not, therefore, the
chemical nature of the constituents, but rather the mechanical
structure of the rocks, which modifies the propagation of the
motion, the wave of commotion. Where this wave proceeds
along a coast, or at the foot and in the direction of a mountain
chain, interruptions at certain points have sometimes been remarked, which manifested themselves during the course of
many centuries. The undulation advances in the depths below, but is never felt at the same points on the surface. The
Peruvians* say of these unmoved upper strata that "they
form a bridge." As the mountain chains appear to be raised
on fissures, the walls of the cavities may perhaps favor the direction of undulations parallel to them; occasionally, however,
the waves of commotion intersect several chains almost per
pendicularly. Thus we see them simultaneously breaking
through the littoral chain of Venezuela and the Sierra Parime.
In Asia, shocks of earthquakes have been propagated from
Lahore and from the foot of the Himalaya (22d of January,
1832) transversely across the chain of the Hindoo Chou to
Badakschan, the upper Oxus, and even to Bokhara.† The
circles of commotion unfortunately expand occasionally in consequence of a single and unusually violent earthquake. It is
only since the destruction of Cumana, on the 14th of December, 1797, that shocks on the southern coast have been felt in
the mica slate rocks of the peninsula of Maniquarez, situated
opposite to the chalk hills of the main land. The advance

* In Spanish they say, *rocas que hacen puente*. With this phenomenon of non-propagation through superior strata is connected the remark
able fact that in the beginning of this century shocks were felt in the
deep silver mines at Marienberg, in the Saxony mining district, while
not the slightest trace was perceptible at the surface. The miners
ascended in a state of alarm. Conversely, the workmen in the mines
of Falun and Persberg felt nothing of the shocks which in November,
1823, spread dismay among the inhabitants above ground.

† Sir Alex. Burnes, *Travels in Bokhara*, vol. i., p. 18; and Wathen,
Mem. on the Usbek State, in the *Journal of the Asiatic Society of Bengal*,
vol. iii., p. 337.

+rom south to north was very striking in the almost uninterrupted undulations of the soil in the alluvial valleys of the Mississippi, the Arkansas, and the Ohio, from 1811 to 1813. It
seemed here as if subterranean obstacles were gradually overcome, and that the way being once opened, the undulatory
movement could be freely propagated.

Although earthquakes appear at first sight to be simply dynamic phenomena of motion, we yet discover, from well-attested facts, that they are not only able to elevate a whole district above its ancient level (as, for instance, the Ulla Bund,
after the earthquake of Cutch, in June, 1819, east of the
Delta of the Indus, or the coast of Chili, in November, 1822),
but we also find that various substances have been ejected during the earthquake, as hot water at Catania in 1818 ; hot
steam at New Madrid, in the Valley of the Mississippi, in
1812 ; irrespirable gases, *Mofettes*, which injured the flocks
grazing in the chain of the Andes ; mud, black smoke, and
even flames, at Messina in 1781, and at Cumana on the 14th
of November, 1797. During the great earthquake of Lisbon,
on the 1st of November, 1755, flames and columns of smoke
were seen to rise from a newly-formed fissure in the rock of
Alvidras, near the city. The smoke in this case became more
dense as the subterranean noise increased in intensity.* At
the destruction of Riobamba, in the year 1797, when the
shocks were not attended by any outbreak of the neighboring
volcano, a singular mass called the *Moya* was uplifted from
the earth in numerous continuous conical elevations, the whole
being composed of carbon, crystals of augite, and the silicious
shields of infusoria. The eruption of carbonic acid gas from
fissures in the Valley of the Magdalene, during the earthquake
of New Granada, on the 16th of November, 1827, suffocated
many snakes, rats, and other animals. Sudden changes of
weather, as the occurrence of the rainy season in the tropics,
at an unusual period of the year, have sometimes succeeded
violent earthquakes in Quito and Peru. Do gaseous fluids rise
from the interior of the earth, and mix with the atmosphere ?
or are these meteorological processes the action of atmospheric
electricity disturbed by the earthquake ? In the tropical regions of America, where sometimes not a drop of rain falls for
ten months together, the natives consider the repeated shocks
of earthquakes, which do not endanger the low reed huts, as
auspicious harbingers of fruitfulness and abundant rain.

* *Philos. Transact.*, vol. xlix. p. 414.

The intimate connection of the phenomena which we have considered is still hidden in obscurity. Elastic fluids are doubt lessly the cause of the slight and perfectly harmless trembling of the earth's surface, which has often continued several days (as in 1816, at Scaccia, in Sicily, before the volcanic elevation of the island of Julia), as well as of the terrific explosions accompanied by loud noise. The focus of this destructive agent, the seat of the moving force, lies far below the earth's surface ; but we know as little of the extent of this depth as we know of the chemical nature of these vapors that are so highly compressed. At the edges of two craters, Vesuvius, and the towering rock which projects beyond the great abyss of Pichincha, near Quito, I have felt periodic and very regular shocks of earthquakes, on each occasion from 20 to 30 seconds before the burning scoriæ or gases were erupted. The intensity of the shocks was increased in proportion to the time intervening between them, and, consequently, to the length of time in which the vapors were accumulating. This simple fact, which has been attested by the evidence of so many travelers, furnishes us with a general solution of the phenomenon, in showing that active volcanoes are to be considered as safety-valves for the immediate neighborhood. The danger of earthquakes increases when the openings of the volcano are closed, and deprived of free communication with the atmosphere ; but the destruction of Lisbon, of Caraccas, of Lima, of Cashmir in 1554,[*] and of so many cities of Calabria, Syria, and Asia Minor, shows us, on the whole, that the force of the shock is not the greatest in the neighborhood of active volcanoes.

As the impeded activity of the volcano acts upon the shocks of the earth's surface, so do the latter react on the volcanic phenomena. Openings of fissures favor the rising of cones of eruption, and the processes which take place in these cones, by forming a free communication with the atmosphere. A column of smoke, which had been observed to rise for months together from the volcano of Pasto, in South America, suddenly disappeared, when, on the 4th of February, 1797, the province of Quito, situated at a distance of 192 miles to the south, suffered from the great earthquake of Riobamba. After the earth had continued to tremble for some time throughout the whole of Syria, in the Cyclades, and in Eubœa, the shocks suddenly ceased on the eruption of a stream of hot mud

---

[*] On the frequency of earthquakes in Cashmir, see Troyer's German translation of the ancient *Radjataringini*, vol. ii., p. 297, and Carl Hügel, *Reisen,* bd. ii., s. 184.

on the Lelantine plains near Chalcis.* The intelligent geographer of Amasea, to whom we are indebted for the notice of
this circumstance, further remarks : " Since the craters of Ætna
have been opened, which yield a passage to the escape of fire,
and since burning masses and water have been ejected, the country near the sea-shore has not been so much shaken as at the
time previous to the separation of Sicily from Lower Italy, when
all communications with the external surface were closed."

We thus recognize in earthquakes the existence of a volcanic force, which, although every where manifested, and as
generally diffused as the internal heat of our planet, attains
but rarely, and then only at separate points, sufficient intensity
to exhibit the phenomenon of eruptions. The formation of
veins, that is to say, the filling up of fissures with crystalline
masses bursting forth from the interior (as basalt, melaphyre,
and greenstone), gradually disturbs the free intercommunication of elastic vapors. This tension acts in three different
ways, either in causing disruptions, or sudden and retroversed
elevations, or, finally, as was first observed in a great part of
Sweden, in producing changes in the relative level of the sea
and land, which, although continuous, are only appreciable at
intervals of long period.

Before we leave the important phenomena which we have
considered, not so much in their individual characteristics as
in their general physical and geognostical relations, I would
advert to the deep and peculiar impression left on the mind by
the first earthquake which we experience, even where it is not
attended by any subterranean noise.† This impression is not,

* Strabo, lib. i., p. 100, Casaub. That the expression πηλοῦ διαπύ-
ρου ποταμόν does not mean erupted mud, but lava, is obvious from a
passage in Strabo, lib. vi., p. 412. Compare Walter, in his *Abnahme der
Vulkanischen Thätigkeit in Historischen Zeiten* (On the Decrease of Volcanic Activity during Historical Times), 1844, s. 25.

† [Dr. Tschudi, in his interesting work, *Travels in Peru*, translated
from the German by Thomasina Ross, p. 170, 1847, describes strikingly the effect of an earthquake upon the native and upon the stranger.
" No familiarity with the phenomenon can blunt this feeling. The inhabitant of Lima, who from childhood has frequently witnessed these
convulsions of nature, is roused from his sleep by the shock, and rushes
from his apartment with the cry of *Misericordia !* The foreigner from
the north of Europe, who knows nothing of earthquakes but by descrip
tion, waits with impatience to feel the movement of the earth, and longs
to hear with his own ear the subterranean sounds which he has hitherto
considered fabulous. With levity he treats the apprehension of a coming convulsion, and laughs at the fears of the natives ; but, as soon as his
wish is gratified, he is terror-stricken, and is involuntarily prompted to
seek safety in flight."]— *Tr.*

in my opinion, the result of a recollection of those fearful pictures of devastation presented to our imaginations by the historical narratives of the past, but is rather due to the sudden revelation of the delusive nature of the inherent faith by which we had clung to a belief in the immobility of the solid parts of the earth.   We are accustomed from early childhood to draw a contrast between the mobility of water and the immobility of the soil on which we tread ; and this feeling is confirmed by the evidence of our senses.   When, therefore, we suddenly feel the ground move beneath us, a mysterious and natural force, with which we are previously unacquainted, is revealed to us as an active disturbance of stability.   A moment destroys the illusion of a whole life ; our deceptive faith in the repose of nature vanishes, and we feel transported, as it were, into a realm of unknown destructive forces.   Every sound—the faintest motion in the air—arrests our attention, and we no longer trust the ground on which we stand.   Animals, especially dogs and swine, participate in the same anxious disquietude ; and even the crocodiles of the Orinoco, which are at other times as dumb as our little lizards, leave the trembling bed of the river, and run with loud cries into the adjacent forests.

To man the earthquake conveys an idea of some universal and unlimited danger.   We may flee from the crater of a volcano in active eruption, or from the dwelling whose destruction is threatened by the approach of the lava stream ; but in an earthquake, direct our flight whithersoever we will, we still feel as if we trod upon the very focus of destruction.   This condition of the mind is not of long duration, although it takes its origin in the deepest recesses of our nature ; and when a series of faint shocks succeed one another, the inhabitants of the country soon lose every trace of fear.   On the coasts of Peru, where rain and hail are unknown, no less than the rolling thunder and the flashing lightning, these luminous explosions of the atmosphere are replaced by the subterranean noises which accompany earthquakes.*   Long habit, and the very

* [" Along the whole coast of Peru the atmosphere is almost uniformly in a state of repose.   It is not illuminated by the lightning's flash, or disturbed by the roar of the thunder; no deluges of rain, no fierce hurricanes, destroy the fruits of the fields, and with them the hopes of the husbandman.   But the mildness of the elements above ground is frightfully counterbalanced by their subterranean fury.   Lima is frequently visited by earthquakes, and several times the city has been reduced to a mass of ruins.   At an average, forty-five shocks may be counted on in the year.   Most of them occur in the latter part of Octo-

prevalent opinion that dangerous shocks are only to be apprehended two or three times in the course of a century, cause
faint oscillations of the soil to be regarded in Lima with scarcely more attention than a hail storm in the temperate zone.

Having thus taken a general view of the activity—the
inner life, as it were—of the Earth, in respect to its internal
heat, its electro-magnetic tension, its emanation of light at the
poles, and its irregularly-recurring phenomena of motion, we
will now proceed to the consideration of the material products,
the chemical changes in the earth's surface, and the composition of the atmosphere, which are all dependent on planetary
vital activity. We see issue from the ground steam and
gaseous carbonic acid, almost always free from the admixture
of nitrogen ;* carbureted hydrogen gas, which has been used
in the Chinese province Sse-tschuan† for several thousand
years, and recently in the village of Fredonia, in the State of
New York, United States, in cooking and for illumination ;
sulphureted hydrogen gas and sulphurous vapors ; and, more
rarely,‡ sulphurous and hydrochloric acids.§ Such effusions

ber, in November, December, January, May, and June. Experience
gives reason to expect the visitation of two desolating earthquakes in a
century. The period between the two is from forty to sixty years. The
most considerable catastrophes experienced in Lima since Europeans
have visited the west coast of South America happened in the years
1586, 1630, 1687, 1713, 1746, 1806. There is reason to fear that in the
course of a few years this city may be the prey of another such visitation."—Tschudi, op. cit.]—*Tr.*

* Bischof's comprehensive work, *Wärmelehre des inneren Erdkörpers.*

† On the Artesian fire-springs (Ho-tsing) in China, and the ancient
use of portable gas (in bamboo canes) in the city of Khiung-tsheu, see
Klaproth, in my *Asie Centrale*, t. iii., p. 519–530.

‡ Boussingault (*Annales de Chimie*, t. lii., p. 181) observed no evolution of hydrochloric acid from the volcanoes of New Granada, while
Monticelli found it in enormous quantity in the eruption of Vesuvius in
1813.

§ [Of the gaseous compounds of sulphur, one, sulphurous acid, appears to predominate chiefly in volcanoes possessing a certain degree
of activity, while the other, sulphureted hydrogen, has been most frequently perceived among those in a dormant condition. The occurrence of abundant exhalations of sulphuric acid, which have been hith
erto noticed chiefly in extinct volcanoes, as, for instance, in a stream
issuing from that of Puracè, between Bogota and Quito, from extinct
volcanoes in Java, is satisfactorily explained in a recent paper by M.
Dumas, *Annales de Chimie*, Dec., 1846. He shows that when sulphureted hydrogen, at a temperature above 100° Fahr., and still better
when near 190°, comes in contact with certain porous bodies, a cata-
lytic action is set up, by which water, sulphuric acid, and sulphur are
produced. Hence probably the vast deposits of sulphur, associated
with sulphates of lime and strontian, which are met with in the
western parts of Sicily.]—*Tr.*

from the fissures of the earth not only occur in the districts of still burning or long-extinguished volcanoes, but they may likewise be observed occasionally in districts where neither trachyte nor any other volcanic rocks are exposed on the earth's surface. In the chain of Quindiu I have seen sulphur deposited in mica slate from warm sulphurous vapor at an elevation of 6832 feet[*] above the level of the sea, while the same species of rock, which was formerly regarded as primitive, contains, in the Cerro Cuello, near Tiscan, south of Quito, an immense deposit of sulphur imbedded in pure quartz.

Exhalations of carbonic acid (*mofettes*) are even in our days to be considered as the most important of all gaseous emanations, with respect to their number and the amount of their effusion. We see in Germany, in the deep valleys of the Eifel, in the neighborhood of the Lake of Laach,[†] in the crater-like valley of the Wehr and in Western Bohemia, exhalations of carbonic acid gas manifest themselves as the last efforts of volcanic activity in or near the foci of an earlier world. In those earlier periods, when a higher terrestrial temperature existed, and when a great number of fissures still remained unfilled, the processes we have described acted more powerfully, and carbonic acid and hot steam were mixed in larger quantities in the atmosphere, from whence it follows, as Adolph Brongniart has ingeniously shown,[‡] that the primitive vegetable world must have exhibited almost every where, and independently of geographical position, the most luxurious abundance and the fullest development of organism. In these constantly warm and damp atmospheric strata, saturated with

[*] Humboldt, *Recueil d'Observ. Astronomiques*, t. i., p. 311 (*Nivellement Barométrique de la Cordillère des Andes*, No. 206).

[†] [The Lake of Laach, in the district of the Eifel, is an expanse of water two miles in circumference. The thickness of the vegetation on the sides of its crater-like basin renders it difficult to discover the nature of the subjacent rock, but it is probably composed of black cellular augitic lava. The sides of the crater present numerous loose masses, which appear to have been ejected, and consist of glassy feldspar, icespar, sodalite, hauyne, spinellane, and leucite. The resemblance between these products and the masses formerly ejected from Vesuvius is most remarkable. (Daubeney *On Volcanoes*, p. 81.) Dr. Hibbert regards the Lake of Laach as formed in the first instance by a crack caused by the cooling of the crust of the earth, which was widened afterward into a circular cavity by the expansive force of elastic vapors. See *History of the Extinct Volcanoes of the Basin of Neuwied*, 1832.] —*Tr.*

[‡] Adolph Brongniart, in the *Annales des Sciences Naturelles*, t. xv., p. 225.

carbonic acid, vegetation must have attained a degree of vital activity, and derived the superabundance of nutrition necessary to furnish materials for the formation of the beds of lignite (coal), constituting the inexhaustible means on which are based the physical power and prosperity of nations. Such masses are distributed in basins over certain parts of Europe, occurring in large quantities in the British Islands, in Belgium, in France, in the provinces of the Lower Rhine, and in Upper Silesia. At the same primitive period of universal volcanic activity, those enormous quantities of carbon must also have escaped from the earth which are contained in limestone rocks, and which, if separated from oxygen and reduced to a solid form, would constitute about the eighth part of the absolute bulk of these mountain masses.* That portion of the carbon which was not taken up by alkaline earths, but remained mixed with the atmosphere, as carbonic acid, was gradually consumed by the vegetation of the earlier stages of the world, so that the atmosphere, after being purified by the processes of vegetable life, only retained the small quantity which it now possesses, and which is not injurious to the present organization of animal life. Abundant eruptions of sulphurous vapor have occasioned the destruction of the species of mollusca and fish which inhabited the inland waters of the earlier world, and have given rise to the formation of the contorted beds of gypsum, which have doubtless been frequently affected by shocks of earthquakes.

Gaseous and liquid fluids, mud, and molten earths, ejected from the craters of volcanoes, which are themselves only a kind of "*intermittent springs*," rise from the earth under precisely analogous physical relations.† All these substances owe their temperature and their chemical character to the place of their origin. The *mean* temperature of aqueous springs is less than that of the air at the point whence they emerge, if the water flow from a height; but their heat increases with the depth of the strata with which they are in contact at their origin. We have already spoken of the numerical law regulating this increase. The blending of waters that have come from the height of a mountain with those that have sprung from the depths of the earth, render it difficult to determine the position of the *isogeothermal lines*‡ (lines of equal internal

* Bischof, op. cit., s. 324, Anm. 2.
† Humboldt, *Asie Centrale*, t. i., p. 43.
‡ On the theory of isogeothermal (chthonisothermal) lines, consult the ingenious labors of Kupffer. in Pogg., *Annalen*, bd xv., s. 184, and bd

terrestrial temperature), when this determination is to be made from the temperature of flowing springs. Such, at any rate, is the result I have arrived at from my own observations and those of my fellow-travelers in Northern Asia. The temperature of springs, which has become the subject of such continuous physical investigation during the last half century, depends, like the elevation of the line of perpetual snow, on very many simultaneous and deeply-involved causes. It is a function of the temperature of the stratum in which they take their rise, of the specific heat of the soil, and of the quantity and temperature of the meteoric water,* which is itself different from the temperature of the lower strata of the atmosphere, according to the different modes of its origin in rain, snow, or hail.†

Cold springs can only indicate the mean atmospheric tem-

---

xxxii., s. 270, in the *Voyage dans l'Oural*, p. 382–398, and in the *Edinburgh Journal of Science*, New Series, vol. iv., p. 355. See, also, Kämtz, *Lehrb. der Meteor.*, bd. ii., s. 217 ; and, on the ascent of the chthonisothermal lines in mountainous districts, Bischof, s. 174–198.

* Leop. v. Buch, in Pogg., *Annalen*, bd. xii., s. 405.

† On the temperature of the drops of rain in Cumana, which fell to 72°, when the temperature of the air shortly before had been 86° and 88°, and during the rain sank to 74°, see my *Relat. Hist.*, t. ii., p. 22. The rain-drops, while falling, change the normal temperature they originally possessed, which depends on the height of the clouds from which they fell, and their heating on their upper surface by the solar rays. The rain-drops, on their first production, have a higher temperature than the surrounding medium in the superior strata of our atmosphere, in consequence of the liberation of their latent heat ; and they continue to rise in temperature, since, in falling through lower and warmer strata, vapor is precipitated on them, and they thus increase in size (Bischof, *Wärmelehre des inneren Erdkörpers*, s. 73) ; but this additional heating is compensated for by evaporation. The cooling of the air by rain (putting out of the question what probably belongs to the electric process in storms) is effected by the drops, which are themselves of lower temperature, in consequence of the cold situation in which they were formed, and bring down with them a portion of the higher colder air, and which finally, by moistening the ground, give rise to evaporation. These are the ordinary relations of the phenomenon. When, as occasionally happens, the rain-drops are warmer than the lower strata of the atmosphere (Humboldt, *Rel. Hist.*, t. iii., p. 513), the cause must probably be sought in higher warmer currents, or in a higher temperature of widely-extended and not very thick clouds, from the action of the sun's rays. How, moreover, the phenomenon of supplementary rainbows, which are explained by the interference of light, is connected with the original and increasing size of the falling drops, and how an optical phenomenon, if we know how to observe it accurately, may enlighten us regarding a meteorological process, according to diversity of zone, has been shown, with much talent and ingenuity, by Arago, in the *Annuaire* for 1836, p. 300.

perature when they are unmixed with the waters rising from great depths, or descending from considerable mountain elevations, and when they have passed through a long course at a depth from the surface of the earth which is equal in our latitudes to 40 or 60 feet, and, according to Boussingault, to about one foot in the equinoctial regions ;[*] these being the depths at which the invariability of the temperature begins in the temperate and torrid zones, that is to say, the depths at which horary, diurnal, and monthly changes of heat in the atmosphere cease to be perceived.

Hot springs issue from the most various kinds of rocks. The hottest permanent springs that have hitherto been observed are, as my own researches confirm, at a distance from all volcanoes. I will here advert to a notice in my journal of the *Aguas Calientes de las Trincheras*, in South America, between Porto Cabello and Nueva Valencia, and the *Aguas de Comangillas*, in the Mexican territory, near Guanaxuato ; the former of these, which issued from granite, had a temperature of 194°·5 ; the latter, issuing from basalt, 205°·5. The depth of the source from whence the water flowed with this temperature, judging from what we know of the law of the increase of heat in the interior of the earth, was probably 7140 feet, or above two miles. If the universally-diffused terrestrial heat be the cause of thermal springs, as of active volcanoes, the rocks can only exert an influence by their different capaci-

---

[*] The profound investigations of Boussingault fully convince me, that in the tropics, the temperature of the ground, at a very slight depth, exactly corresponds with the mean temperature of the air. The following instances are sufficient to illustrate this fact :

Stations within Tropical Zones.	Temperature at 1 French foot [1·066 of the English foot] below the earth's surface.	Mean Temperature of the air.	Height, in English feet, above the level of the sea.
Guayaquil............	78·8	78·1	0
Anserma Nuevo.......	74·6	74·8	3444
Zupia	70·7	70·7	4018
Popayan	64·7	65·6	5929
Quito	59·9	59·9	9559

The doubts about the temperature of the earth within the tropics, of which I am probably, in some degree, the cause, by my observations on the Cave of Caripe (Cueva del Guacharo), *Rel. Hist.*, t. iii., p. 191–196), are resolved by the consideration that I compared the presumed mean temperature of the air of the convent of Caripe, 65°·3, not with the temperature of the air of the cave, 65°·6, but with the temperature of the subterranean stream, 62°·3, although I observed (*Rel. Hist.*, t. iii., p. 146 and 194) that mountain water from a great height might probably be mixed with the water of the cave

ties for heat and by their conducting powers. The hottest of
all permanent springs (between 203° and 209°) are likewise,
in a most remarkable degree, the purest, and such as hold in
solution the smallest quantity of mineral substances. Their
temperature appears, on the whole, to be less constant than
that of springs between 122° and 165°, which in Europe, at
least, have maintained, in a most remarkable manner, their
invariability of heat and mineral contents during the last
fifty or sixty years, a period in which thermometrical measurements and chemical analyses have been applied with increased exactness. Boussingault found in 1823 that the thermal
springs of Las Trincheras had risen 12° during the twenty-
three years that had intervened since my travels in 1800.*
This calmly-flowing spring is therefore now nearly 12° hotter
than the intermittent fountains of the Geyser and the Strokr,
whose temperature has recently been most carefully determined by Krug of Nidda. A very striking proof of the origin
of hot springs by the sinking of cold meteoric water into the
earth, and by its contact with a volcanic focus, is afforded by
the volcano of Jorulla in Mexico, which was unknown before
my American journey. When, in September, 1759, Jorullo
was suddenly elevated into a mountain 1183 feet above the
level of the surrounding plain, two small rivers, the *Rio de
Cuitimba* and *Rio de San Pedro*, disappeared, and some
time afterward burst forth again, during violent shocks of an
earthquake, as hot springs, whose temperature I found in 1803
to be 186°·4.

The springs in Greece still evidently flow at the same places
as in the times of Hellenic antiquity. The spring of Erasinos,
two hours' journey to the south of Argos, on the declivity of
Chaon, is mentioned by Herodotus. At Delphi we still see
Cassotis (now the springs of St. Nicholas) rising south of the
Lesche, and flowing beneath the Temple of Apollo ; Castalia,
at the foot of Phædriadæ ; Pirene, near Acro-Corinth ; and
the hot baths of Ædipsus, in Eubœa, in which Sulla bathed
during the Mithridatic war.† I advert with pleasure to these

---

* Boussingault, in the *Annales de Chimie*, t. lii., p. 181. The spring
of Chaudes Aigues, in Auvergne, is only 176°. It is also to be observed, that while the Aguas Calientes de las Trincheras, south of Porto
Cabello (Venezuela), springing from granite cleft in regular beds, and
far from all volcanoes, have a temperature of fully 206°·6, all the springs
which rise in the vicinity of still active volcanoes (Pasto, Cotopaxi, and
Tunguragua) have a temperature of only 97°–130°.

† Cassotis (the spring of St. Nicholas) and Castalia, at the Phædriadæ,
mentioned in Pausanias, x., 24, 25, and x., 8, 9 ; Pirene (Acro-Corinth),

facts, as they show us that, even in a country subject to frequent and violent shocks of earthquakes, the interior of our planet has retained for upward of 2000 years its ancient configuration in reference to the course of the open fissures that yield a passage to these waters. The *Fontaine jaillissante* of Lillers, in the Department des Pas de Calais, which was bored as early as the year 1126, still rises to the same height and yields the same quantity of water ; and, as another instance, I may mention that the admirable geographer of the Caramanian coast, Captain Beaufort, saw in the district of Phaselis the same flame fed by emissions of inflammable gas which was described by Pliny as the flame of the Lycian Chimera.*

The observation made by Arago in 1821, that the deepest Artesian wells are the warmest,† threw great light on the origin of thermal springs, and on the establishment of the law that terrestrial heat increases with increasing depth. It is a remarkable fact, which has but recently been noticed, that at the close of the third century, St. Patricius,‡ probably Bishop of Pertusa, was led to adopt very correct views regarding the phenomenon of the hot springs at Carthage. On being asked what was the cause of boiling water bursting from the earth, he replied, "Fire is nourished in the clouds and in the interior

in Strabo, p. 379 ; the spring of Erasinos, at Mount Chaon, south of Argos, in Herod., vi., 67, and Pausanias, ii., 24, 7 ; the springs of Ædipsus in Eubœa, some of which have a temperature of 88°, while in others it ranges between 144° and 167°, in Strabo, p. 60 and 447, and Athenæus, ii., 3, 73 ; the hot springs of Thermopylæ, at the foot of Œta, with a temperature of 149°. All from manuscript notes by Professor Curtius, the learned companion of Otfried Müller.

* Pliny, ii., 106 ; Seneca, *Epist.*, 79, § 3, ed. Ruhkopf (Beaufort, *Survey of the Coast of Karamania*, 1820, art. Yanar, near Deliktasch, the ancient Phaselis, p. 24). See, also, Ctesias, *Fragm.*, cap. 10 p. 250, ed. Bähr ; Strabo, lib. xiv., p. 666, Casaub.

["Not far from the Deliktash, on the side of a mountain, is the perpetual fire described by Captain Beaufort. The travelers found it as brilliant as ever, and even somewhat increased ; for, besides the large flame in the corner of the ruins described by Beaufort, there were small jets issuing from crevices in the side of the crater-like cavity five or six feet deep. At the bottom was a shallow pool of sulphureous and turbid water, regarded by the Turks as a sovereign remedy for all skin complaints. The soot deposited from the flames was regarded as efficacious for sore eyelids, and valued as a dye for the eyebrows." See the highly interesting and accurate work, *Travels in Lycia*, by Lieut. Spratt and Professor E. Forbes.]—*Tr.*

† Arago, in the *Annuaire pour* 1835, p. 234.

‡ *Acta S. Patricii*, p. 555, ed. Ruinart, t. ii., p. 385, Mazochi. Dureau de la Malle was the first to draw attention to this remarkable passage in the *Recherches sur la Topographie de Carthage*, 1835, p. 276. (See, also, Seneca, *Nat. Quæst.*, iii., 24.)

of the earth, as Ætna and other mountains near Naples may
teach you.   The subterranean waters rise as if through si-
phons.   The cause of hot springs is this : waters which are
more remote from the subterranean fire are colder, while those
which rise nearer the fire are heated by it, and bring with
them to the surface which we inhabit an insupportable degree
of heat."

As earthquakes are often accompanied by eruptions of water
and vapors, we recognize in the *Salses*,* or small mud volcanoes, a transition from the changing phenomena presented
by these eruptions of vapor and thermal springs to the more
powerful and awful activity of the streams of lava that flow
from volcanic mountains.   If we consider these mountains as
springs of molten earths producing volcanic rocks, we must remember that thermal waters, when impregnated with carbonic
acid and sulphurous gases, are continually forming horizontally ranged strata of limestone (travertine) or conical elevations, as in Northern Africa (in Algeria), and in the Baños
of Caxamarca, on the western declivity of the Peruvian Cordilleras.   The travertine of Van Diemen's Land (near Hobart
Town) contains, according to Charles Darwin, remains of a
vegetation that no longer exists.   Lava and travertine, which
are constantly forming before our eyes, present us with the
two extremes of geognostic relations.

Salses deserve more attention than they have hitherto received from geognosists.   Their grandeur has been overlooked
because of the two conditions to which they are subject ; it is
only the more peaceful state, in which they may continue for
centuries, which has generally been described : their origin is,
however, accompanied by earthquakes, subterranean thunder,
the elevation of a whole district, and lofty emissions of flame
of short duration.   When the mud volcano of Jokmali began
to form on the 27th of November, 1827, in the peninsula of
Abscheron, on the Caspian Sea, east of Baku, the flames
flashed up to an extraordinary height for three hours, while
during the next twenty hours they scarcely rose three feet
above the crater, from which mud was ejected.   Near the
village of Baklichli, west of Baku, the flames rose so high that

---

* [True volcanoes, as we have seen, generate sulphureted hydrogen
and muriatic acid, upheave tracts of land, and emit streams of melted
feldspathic materials ; salses, on the contrary, disengage little else but
carbureted hydrogen, together with bitumen and other products of the
distillation of coal, and pour forth no other torrents except of mud, or
argillaceous materials mixed up with water.   Daubeney, op cit., p.
540.]—*Tr.*

they could be seen at a distance of twenty-four miles. Enormous masses of rock were torn up and scattered around. Similar masses may be seen round the now inactive mud volcano
of Monte Zibio, near Sassuolo, in Northern Italy. The secondary condition of repose has been maintained for upward of
fifteen centuries in the mud volcanoes of Girgenti, the *Maca-
lubi*, in Sicily, which have been described by the ancients.
These salses consist of many contiguous conical hills, from
eight to ten, or even thirty feet in height, subject to variations
of elevation as well as of form. Streams of argillaceous mud,
attended by a periodic development of gas, flow from the small
basins at the summits, which are filled with water; the mud,
although usually cold, is sometimes at a high temperature, as
at Damak, in the province of Samarang, in the island of Java.
The gases that are developed with loud noise differ in their
nature, consisting, for instance, of hydrogen mixed with naph-
tha, or of carbonic acid, or, as Parrot and myself have shown
(in the peninsula of Taman, and in the *Volcancitos de Tur-
baco*, in South America), of almost pure nitrogen.*

Mud volcanoes, after the first violent explosion of fire, which
is not, perhaps, in an equal degree common to all, present to
the spectator an image of the uninterrupted but weak activity
of the interior of our planet. The communication with the
deep strata in which a high temperature prevails is soon closed,
and the coldness of the mud emissions of the salses seems to indicate that the seat of the phenomenon can not be far removed from the surface during their ordinary condition. The
reaction of the interior of the earth on its external surface is
exhibited with totally different force in true volcanoes or igneous mountains, at points of the earth in which a permanent,
or, at least, continually-renewed connection with the volcanic
force is manifested. We must here carefully distinguish between the more or less intensely developed volcanic phenomena, as, for instance, between earthquakes, thermal, aqueous,
and gaseous springs, mud volcanoes, and the appearance of
bell-formed or dome-shaped trachytic rocks without openings;
the opening of these rocks, or of the elevated beds of basalt, as

---

* Humboldt, *Rel. Hist.*, t. iii., p. 562-567; *Asie Centrale*, t. i., p. 43;
t. ii., p. 505-515; *Vues des Cordillères*, pl. xli. Regarding the *Maca-
lubi* (the Arabic *Makhlub*, the *overthrown* or *inverted*, from the word
Khalaba), and on " the Earth ejecting fluid earth," see Solinus, cap. 5:
" idem ager Agrigentinus eructat limosas scaturigenes, et ut venæ fon-
tium sufficiunt rivis subministrandis, ita in hac Siciliæ parte solo nunquam deficiente, æterna rejectatione terram terra evomit."

craters of elevation; and, lastly, the elevation of a permanent volcano in the crater of elevation, or among the *débris* of its earlier formation. At different periods, and in different degrees of activity and force, the permanent volcanoes emit steam, acids, luminous scoriæ, or, when the resistance can be overcome, narrow, band-like streams of molten earths. Elastic vapors sometimes elevate either separate portions of the earth's crust into dome-shaped unopened masses of feldspathic trachyte and dolerite (as in Puy de Dome and Chimborazo), in consequence of some great or local manifestation of force in the interior of our planet, or the upheaved strata are broken through and curved in such a manner as to form a steep rocky ledge on the opposite inner side, which then constitutes the inclosure of a crater of elevation. If this rocky ledge has been uplifted from the bottom of the sea, which is by no means always the case, it determines the whole physiognomy and form of the island. In this manner has arisen the circular form of Palma, which has been described with such admirable accuracy by Leopold von Buch, and that of Nisyros,* in the Ægean Sea. Sometimes half of the annular ledge has been destroyed, and in the bay formed by the encroachment of the sea corallines have built their cellular habitations. Even on continents craters of elevation are often filled with water, and embellish in a peculiar manner the character of the landscape. Their origin is not connected with any determined species of rock: they break out in basalt, trachyte, leucitic porphyry (somma), or in doleritic mixtures of augite and labradorite; and hence arise the different nature and external conformation of these inclosures of craters. No phenomena of eruptions are manifested in such craters, as they open no permanent channel of communication with the interior, and it is but seldom that we meet with traces of volcanic activity either in the neighborhood or in the interior of these craters. The force which was able to produce so important an action must have been long accumulating in the interior before it could overpower the resistance of the mass pressing upon it; it sometimes, for instance, on the origin of new islands, will raise granular rocks and conglomerated masses (strata of tufa filled with marine plants) above the surface of the sea. The compressed vapors escape through the crater of elevation, but a large mass soon falls back and closes the opening, which had been only formed by these manifestations of force. No volcano can, therefore,

* See the interesting little map of the island of Nisyros, in Ross's *Reisen auf den Griechischen Inseln*, bd. ii., 1843, s. 69.

be produced.* A volcano, properly so called, exists only where
a permanent connection is established between the interior of
the earth and the atmosphere, and the reaction of the interior
on the surface then continues during long periods of time. It
may be interrupted for centuries, as in the case of Vesuvius,
Fisove,† and then manifest itself with renewed activity. In
the time of Nero, men were disposed to rank Ætna among
the volcanic mountains which were gradually becoming extinct ;‡ and subsequently Ælian§ even maintained that mariners could no longer see the sinking summit of the mountain
from so great a distance at sea. Where these evidences—
these old scaffoldings of eruption, I might almost say—still
exist, the volcano rises from a crater of elevation, while a high
rocky wall surrounds, like an amphitheater, the isolated conical mount, and forms around it a kind of casing of highly ele-

---

* Leopold von Buch, *Phys. Beschreibung der Canarischen Inseln*, s.
326; and his Memoir *über Erhebungscratere und Vulcane*, in Poggend.,
Annal., bd. xxxvii., s. 169.

In his remarks on the separation of Sicily from Calabria, Strabo gives
an excellent description of the two modes in which islands are formed:
"Some islands," he observes (lib. vi., p. 258, ed. Casaub.), "are frag
ments of the continent, others have arisen from the sea, as even at the
present time is known to happen; for the islands of the great ocean,
lying far from the main land, have probably been raised from its depths,
while, on the other hand, those near promontories appear (according to
reason) to have been separated from the continent."

† Ocre Fisove (Mons Vesuvius) in the Umbrian language. (Lassen,
Deutung der Eugubinischen Tafeln in Rhein. Museum, 1832, s. 387.)
The word *ochre* is very probably genuine Umbrian, and means, according to Festus, *mountain*. Ætna would be a burning and shining mountain, if Voss is correct in stating that Αἴτνη is an Hellenic sound, and is
connected with αἴθω and αἴθινος; but the intelligent writer Parthey
doubts this Hellenic origin on etymological grounds, and also because
Ætna was by no means regarded as a luminous beacon for ships or
wanderers, in the same manner as the ever-travailing Stromboli (Strongyle), to which Homer seems to refer in the Odyssey (xii., 68, 202,
and 219), and its geographical position was not so well determined. I
suspect that Ætna would be found to be a Sicilian word, if we had any
fragmentary materials to refer to. According to Diodorus (v., 6), the
Sicani, or aborigines preceding the Sicilians, were compelled to fly to
the western part of the island, in consequence of successive eruptions
extending over many years. The most ancient eruption of Mount Ætna
on record is that mentioned by Pindar and Æschylus, as occurring under Hiero, in the second year of the 75th Olympiad. It is probable
that Hesiod was aware of the devastating eruptions of Ætna before the
period of Greek immigration. There is, however, some doubt regarding the word Αἴτνη in the text of Hesiod, a subject into which I have
entered at some length in another place. (Humboldt, *Examen Crit
de le Géogr.*, t. i., p. 168.)

‡ Seneca. *Epist.*, 79.   § Ælian, *Var. Hist.*, viii., 11

vated strata. Occasionally not a trace of this inclosure is visible, and the volcano, which is not always conical, rises immediately from the neighboring plateau in an elongated form, as in the case of Pichincha,* at the foot of which lies the city of Quito.

As the nature of rocks, or the mixture (grouping) of simple minerals into granite, gneiss, and mica slate, or into trachyte, basalt, and dolorite, is independent of existing climates, and is the same under the most varied latitudes of the earth, so also we find every where in inorganic nature that the same laws of configuration regulate the reciprocal superposition of the strata of the earth's crust, cause them to penetrate one another in the form of veins, and elevate them by the agency of elastic forces. This constant recurrence of the same phenomena is most strikingly manifested in volcanoes. When the mariner, amid the islands of some distant archipelago, is no longer guided by the light of the same stars with which he had been familiar in his native latitude, and sees himself surrounded by palms and other forms of an exotic vegetation, he still can trace, reflected in the individual characteristics of the landscape, the forms of Vesuvius, of the dome-shaped summits of Auvergne, the craters of elevation in the Canaries and Azores, or the fissures of eruption in Iceland. A glance at the satellite of our planet will impart a wider generalization to this analogy of configuration. By means of the charts that have been drawn in accordance with the observations made with large telescopes, we may recognize in the moon, where water and air are both absent, vast craters of elevation surrounding or supporting conical mountains, thus affording incontrovertible evidence of the effects produced by the reaction of the interior on the surface, favored by the influence of a feebler force of gravitation.

Although volcanoes are justly termed in many languages " fire-emitting mountains," mountains of this kind are not formed by the gradual accumulation of ejected currents of lava, but their origin seems rather to be a general consequence of the sudden elevation of soft masses of trachyte or labradoritic augite. The amount of the elevating force is manifested

* [This mountain contains two funnel-shaped craters, apparently resulting from two sets of eruptions: the western nearly circular, and having in its center a cone of eruption, from the summit and sides of which are no less than seventy vents, some in activity and others extinct. It is probable that the larger number of the vents were produced at periods anterior to history. Daubeney, op. cit., p. 488.]—*Tr*

by the elevation of the volcano, which varies from the inconsiderable height of a hill (as the volcano of Cosima, one of the Japanese Kurile islands) to that of a cone above 19,000 feet in height. It has appeared to me that relations of height have a great influence on the occurrence of eruptions, which are more frequent in low than in elevated volcanoes. I might instance the series presented by the following mountains : Stromboli, 2318 feet ; Guacamayo, in the province of Quixos, from which detonations are heard almost daily (I have myself often heard them at Chillo, near Quito, a distance of eighty-eight miles) ; Vesuvius, 3876 feet ; Ætna, 10,871 feet ; the Peak of Teneriffe, 12,175 feet ; and Cotopaxi, 19,069 feet. If the focus of these volcanoes be at an equal depth below the surface, a greater force must be required where the fused masses have to be raised to an elevation six or eight times greater than that of the lower eminences. While the volcano Stromboli (Strongyle) has been incessantly active since the Homeric ages, and has served as a beacon-light to guide the mariner in the Tyrrhenian Sea, loftier volcanoes have been characterized by long intervals of quiet. Thus we see that a whole century often intervenes between the eruptions of most of the colossi which crown the summits of the Cordilleras of the Andes. Where we meet with exceptions to this law, to which I long since drew attention, they must depend upon the circumstance that the connections between the volcanic foci and the crater of eruption can not be considered as equally permanent in the case of all volcanoes. The channel of communication may be closed for a time in the case of the lower ones, so that they less frequently come to a state of eruption, although they do not, on that account, approach more nearly to their final extinction.

These relations between the absolute height and the frequency of volcanic eruptions, as far as they are externally perceptible, are intimately connected with the consideration of the local conditions under which lava currents are erupted. Eruptions from the crater are very unusual in many mountains, generally occurring from lateral fissures (as was observed in the case of Ætna, in the sixteenth century, by the celebrated historian Bembo, when a youth*), wherever the sides

* Petri Bembi Opuscula (*Ætna Dialogus*), Basil, 1556, p. 63 : " Quicquid in Ætnæ matris utero coalescit, nunquam exit ex cratere superiore, quod vel eo inscondere gravis materia non queat, vel, quia inferius alia spiramenta sunt, non fit opus. Despumant flammis urgentibus ignei rivi pigro fluxu totas delambentes plagas, et in lapidem indurescunt."

of the upheaved mountain were least able, from their configuration and position, to offer any resistance. Cones of eruption
are sometimes uplifted on these fissures ; the larger ones, which
are erroneously termed *new volcanoes*, are ranged together in a
line marking the direction of a fissure, which is soon reclosed,
while the smaller ones are grouped together, covering a whole
district with their dome-like or hive-shaped forms. To the
latter belong the *hornitos de Jorullo*,* the cone of Vesuvius
erupted in October, 1822, that of Awatscha, according to Pos-
tels, and those of the lava-field mentioned by Erman, near the
Baidar Mountains, in the peninsula of Kamtschatka.

When volcanoes are not isolated in a plain, but surrounded,
as in the double chain of the Andes of Quito, by a table-land
having an elevation from nine to thirteen thousand feet, this
circumstance may probably explain the cause why no lava
streams are formed† during the most dreadful eruption of ignited scoriæ accompanied by detonations heard at a distance
of more than a hundred miles. Such are the volcanoes of Popayan, those of the elevated plateau of Los Pastos and of the
Andes of Quito, with the exception, perhaps, in the case of
the latter, of the volcano of Antisana. The height of the cone
of cinders, and the size and form of the crater, are elements
of configuration which yield an especial and individual character to volcanoes, although the cone of cinders and the crater
are both wholly independent of the dimensions of the mountain. Vesuvius is more than three times lower than the Peak
of Teneriffe ; its cone of cinders rises to one third of the height
of the whole mountain, while the cone of cinders of the Peak
is only $\frac{1}{22}$d of its altitude.‡ In a much higher volcano than
that of Teneriffe, the Rucu Pichincha, other relations occur

* See my drawing of the volcano of Jorullo, of its *hornitos*, and of the
uplifted *malpays*, in my *Vues de Cordillères*, pl. xliii., p. 239.
   [Burckhardt states that during the twenty-four years that have intervened since Baron Humboldt's visit to Jorullo, the *hornitos* have either
wholly disappeared or completely changed their forms. See *Aufenthalt
und Reisen in Mexico in* 1825 *und* 1834.]—*Tr.*
  † Humboldt, *Essai sur la Géogr. des Plantes et Tableau Phys. des Régions Equinoxiales*, 1807, p. 130, and *Essai Géogn. sur le Gisement des
Roches*, p. 321. Most of the volcanoes in Java demonstrate that the
cause of the perfect absence of lava streams in volcanoes of incessant
activity is not alone to be sought for in their form, position, and height.
Leop. von Buch, *Descr. Phys. des Iles Canaries*, p. 419 ; Reinwardt and
Hoffmann, in Poggend., *Annalen.*, bd. xii., s. 607.
  ‡ [It may be remarked in general, although the rule is liable to exceptions, that the dimensions of a crater are in an inverse ratio to the
elevation of the mountain. Daubeney, op. cit., p. 444.]—*Tr.*

which approach more nearly to that of Vesuvius. Among all
the volcanoes that I have seen in the two hemispheres, the
conical form of Cotopaxi is the most beautifully regular. A
sudden fusion of the snow at its cone of cinders announces the
proximity of the eruption. Before the smoke is visible in the
rarefied strata of air surrounding the summit and the opening
of the crater, the walls of the cone of cinders are sometimes
in a state of glowing heat, when the whole mountain presents
an appearance of the most fearful and portentous blackness.
The crater, which, with very few exceptions, occupies the
summit of the volcano, forms a deep, caldron-like valley, which
is often accessible, and whose bottom is subject to constant alterations. The great or lesser depth of the crater is in many
volcanoes likewise a sign of the near or distant occurrence of
an eruption. Long, narrow fissures, from which vapors issue
forth, or small rounding hollows filled with molten masses, alternately open and close in the caldron-like valley ; the bottom
rises and sinks, eminences of scoriæ and cones of eruption are
formed, rising sometimes far over the walls of the crater, and
continuing for years together to impart to the volcano a peculiar character, and then suddenly fall together and disappear
during a new eruption. The openings of these cones of eruption, which rise from the bottom of the crater, must not, as is
too often done, be confounded with the crater which incloses
them. If this be inaccessible from extreme depth and from
the perpendicular descent, as in the case of the volcano of
Rucu Pichincha, which is 15,920 feet in height, the traveler
may look from the edge on the summit of the mountains which
rise in the sulphurous atmosphere of the valley at his feet ;
and I have never beheld a grander or more remarkable picture
than that presented by this volcano. In the interval between
two eruptions, a crater may either present no luminous appearance, showing merely open fissures and ascending vapors,
or the scarcely heated soil may be covered by eminences of
scoriæ, that admit of being approached without danger, and
thus present to the geologist the spectacle of the eruption of
burning and fused masses, which fall back on the ledge of the
cone of scoriæ, and whose appearance is regularly announced
by small wholly local earthquakes. Lava sometimes streams
forth from the open fissures and small hollows, without breaking through or escaping beyond the sides of the crater. If,
however, it does break through, the newly-opened terrestrial
stream generally flows in such a quiet and well-defined course,
that the deep valley, which we term the crater, remains acces-

sible even during periods of eruption.   It is impossible, without an exact representation of the configuration—the normal type, as it were, of fire-emitting mountains, to form a just idea of those phenomena which, owing to fantastic descriptions and an undefined phraseology, have long been comprised under the head of *craters, cones of eruption,* and *volcanoes.*   The marginal ledges of craters vary much less than one would be led to suppose.   A comparison of Saussure's measurements with my own yields the remarkable result, for instance, that in the course of forty-nine years (from 1773 to 1822), the elevation of the northwestern margin of Mount Vesuvius (*Rocca del Palo*) may be considered to have remained unchanged.*

Volcanoes which, like the chain of the Andes, lift their summits high above the boundaries of the region of perpetual snow, present peculiar phenomena.   The masses of snow, by their sudden fusion during eruptions, occasion not only the most fearful inundations and torrents of water, in which smoking scoriæ are borne along on thick masses of ice, but they likewise exercise a constant action, while the volcano is in a state of perfect repose, by infiltration into the fissures of the trachytic rock. Cavities which are either on the declivity or at the foot of the mountain are gradually converted into subterranean reservoirs of water, which communicate by numerous narrow openings with mountain streams, as we see exemplified in the highlands of Quito.   The fishes of these rivulets multiply, especially in the obscurity of the hollows ; and when the shocks of earthquakes, which precede all eruptions in the Andes, have violently shaken the whole mass of the volcano, these subterranean caverns are suddenly opened, and water, fishes, and tufaceous mud are all ejected together.   It is through this singular phenomenon† that the inhabitants of the highlands of Quito became acquainted with the existence of the little cyclopic fishes, termed by them the preñadilla.   On the night between the 19th and 20th of June, 1698, when the summit of Carguairazo, a mountain 19,720 feet in height, fell in, leaving only two huge masses of rock remaining of the ledge of the crater, a space of nearly thirty-two square miles was overflowed and devastated by streams of liquid tufa and argillaceous mud (*lodazales*), containing large quantities of dead fish.

* See the ground-work of my measurements compared with those of Saussure and Lord Minto, in the *Abhandlungen der Akademie der Wiss. zu Berlin* for the years 1822 and 1823.

† Pimelodes cyclopum. See Humboldt, *Recueil d'Observations de Zoologie et d'Anatomie Comparée,* t. i., p. 21–25.

In like manner, the putrid fever, which raged seven years previously in the mountain town of Ibarra, north of Quito, was ascribed to the ejection of fish from the volcano of Imbaburu.*

Water and mud, which flow not from the crater itself, but from the hollows in the trachytic mass of the mountain, can not, strictly speaking, be classed among volcanic phenomena. They are only indirectly connected with the volcanic activity of the mountain, resembling, in that respect, the singular meteorological process which I have designated in my earlier writings by the term of *volcanic storm*. The hot stream which rises from the crater during the eruption, and spreads itself in the atmosphere, condenses into a cloud, and surrounds the column of fire and cinders which rises to an altitude of many thousand feet. The sudden condensation of the vapors, and, as Gay-Lussac has shown, the formation of a cloud of enormous extent, increase the electric tension. Forked lightning flashes from the column of cinders, and it is then easy to distinguish (as at the close of the eruption of Mount Vesuvius, in the latter end of October, 1822) the rolling thunder of the volcanic storm from the detonations in the interior of the mountain. The flashes of lightning that darted from the volcanic cloud of steam, as we learn from Olafsen's report, killed eleven horses and two men, on the eruption of the volcano of Katlagia, in Iceland, on the 17th of October, 1755.

Having thus delineated the structure and dynamic activity of volcanoes, it now remains for us to throw a glance at the differences existing in their material products. The subterranean forces sever old combinations of matter in order to produce new ones, and they also continue to act upon matter as long as it is in a state of liquefaction from heat, and capable of being displaced. The greater or less pressure under which merely softened or wholly liquid fluids are solidified, appears to constitute the main difference in the formation of Plutonic and volcanic rocks. The mineral mass which flows in narrow, elongated streams from a volcanic opening (an earth-spring), is called lava. Where many such currents meet and are arrested in their course, they expand in width, filling large basins, in which they become solidified in superimposed strata. These few sentences describe the general character of the products of volcanic activity.

* [It would appear, as there is no doubt that these fishes proceed from the mountain itself, that there must be large lakes in the interior, which in ordinary seasons are out of the immediate influence of the volcanic action See Daubeney, op. cit., p. 488, 497.]—*Tr*.

Rocks which are merely broken through by the volcanic action are often inclosed in the igneous products. Thus I have found angular fragments of feldspathic syenite imbedded in the black augitic lava of the volcano of Jorullo, in Mexico ; but the masses of dolomite and granular limestone, which contain magnificent clusters of crystalline fossils (vesuvian and garnets, covered with mejonite, nepheline, and sodalite), are not the ejected products of Vesuvius, these belonging rather to very generally distributed formations, viz., strata of tufa, which are more ancient than the elevation of the Somma and of Vesuvius, and are probably the products of a deep-seated and concealed submarine volcanic action.[*] We find five metals among the products of existing volcanoes, iron, copper, lead, arsenic, and selenium, discovered by Stromeyer in the crater of Volcano.[†] The vapors that rise from the *fumarolles* cause the sublimation of the chlorids of iron, copper, lead, and ammonium ; iron glance[‡] and chlorid of sodium (the latter often in large quantities) fill the cavities of recent lava streams and the fissures of the margin of the crater.

The mineral composition of lava differs according to the nature of the crystalline rock of which the volcano is formed, the height of the point where the eruption occurs, whether at the foot of the mountain or in the neighborhood of the crater, and the condition of temperature of the interior. Vitreous volcanic formations, obsidian, pearl-stone, and pumice, are entirely wanting in some volcanoes, while in the case of others they only proceed from the crater, or, at any rate, from very considerable heights. These important and involved relations can only be explained by very accurate crystallographic and chemical investigations. My fellow-traveler in Siberia, Gustav Rose, and subsequently Hermann Abich, have already been able, by their fortunate and ingenious researches, to throw much light on the structural relations of the various kinds of volcanic rocks.

* Leop. von Buch, in Poggend., *Annalen*, bd. xxxvii., s. 179.

† [The little island of Volcano is separated from Lipari by a narrow channel. It appears to have exhibited strong signs of volcanic activity long before the Christian era, and still emits gaseous exhalations. Stromeyer detected the presence of selenium in a mixture of sal ammoniac and sulphur. Another product, supposed to be peculiar to this volcano, is boracic acid, which lines the sides of the cavities in beautiful white silky crystals. Daubeney, op. cit., p. 257.]—*Tr.*

‡ Regarding the chemical origin of iron glance in volcanic masses, see Mitscherlich, in Poggend., *Annalen*, bd. xv., s. 630 ; and on the liberation of hydrochloric acid in the crater, see Gay-Lussac, in the *Annales de Chimique et de Physique*, t. xxii., p. 423.

The greater part of the ascending vapor is mere steam. When condensed, this forms springs, as in Pantellaria,* where they are used by the goatherds of the island.   On the morning of the 26th of October, 1822, a current was seen to flow from a lateral fissure of the crater of Vesuvius, and was long supposed to have been boiling water; it was, however, shown, by Monticelli's accurate investigations, to consist of dry ashes, which fell like sand, and of lava pulverized by friction.   The ashes, which sometimes darken the air for hours and days together, and produce great injury to the vineyards and olive groves by adhering to the leaves, indicate by their columnar ascent, impelled by vapors, the termination of every great earthquake.   This is the magnificent phenomenon which Pliny the younger, in his celebrated letter to Cornelius Tacitus, compares, in the case of Vesuvius, to the form of a lofty and thickly-branched and foliaceous pine.   That which is described as flames in the eruption of scoriæ, and the radiance of the glowing red clouds that hover over the crater, can not be ascribed to the effect of hydrogen gas in a state of combustion.   They are rather reflections of light which issue from molten masses, projected high in the air, and also reflections from the burning depths, whence the glowing vapors ascend. We will not, however, attempt to decide the nature of the flames, which are occasionally seen now, as in the time of Strabo, to rise from the deep sea during the activity of littoral volcanoes, or shortly before the elevation of a volcanic island.

When the questions are asked, what is it that burns in the volcano ? what excites the heat, fuses together earths and metals, and imparts to lava currents of thick layers a degree of heat that lasts for many years ?† it is necessarily implied that volcanoes must be connected with the existence of substances capable of maintaining combustion, like the beds of coal in subterranean fires.   According to the different phases of chemical science, bitumen, pyrites, the moist admixture of finely-pulverized sulphur and iron, pyrophoric substances, and the metals of the alkalies and earths, have in turn been designated as the cause of intensely active volcanic phenomena. The great chemist, Sir Humphrey Davy, to whom we are indebted for the knowledge of the most combustible metallic

* [Steam issues from many parts of this insular mountain, and several hot springs gush forth from it, which form together a lake 6000 feet in circumference.   Daubeney, op. cit.]—*Tr.*

† See the beautiful experiments on the cooling of masses of rock, in Bischof's *Wärmelehre*, s. 384, 443, 500–512.

substances, has himself renounced his bold chemical hypothesis in his last work (*Consolation in Travel, and last Days of a Philosopher*)—a work which can not fail to excite in the reader a feeling of the deepest melancholy. The great mean density of the earth (5·44), when compared with the specific weight of potassium (0·865), of sodium (0·972), or of the metals of the earths (1·2), and the absence of hydrogen gas in the gaseous emanations from the fissures of craters, and from still warm streams of lava, besides many chemical considerations, stand in opposition with the earlier conjectures of Davy and Ampère.* If hydrogen were evolved from erupted lava, how great must be the quantity of the gas disengaged, when, the seat of the volcanic activity being very low, as in the case of the remarkable eruption at the foot of the Skaptar Jokul in Iceland (from the 11th of June to the 3d of August, 1783, described by Mackenzie and Soemund Magnussen), a space of many square miles was covered by streams of lava, accumulated to the thickness of several hundred feet! Similar difficulties are opposed to the assumption of the penetration of the atmospheric air into the crater, or, as it is figuratively expressed, the *inhalation of the earth*, when we have regard to the small quantity of nitrogen emitted. So general, deep-seated, and far-propagated an activity as that of volcanoes, can not assuredly have its source in chemical affinity, or in the mere contact of individual or merely locally distributed substances. Modern geognosy† rather seeks the cause of this activity in the increased temperature with the increase of depth at all degrees of latitude, in that powerful internal heat which our planet owes to its first solidification, its formation in the regions of space, and to the spherical contraction of

* See Berzelius and Wöhler, in Poggend., *Annalen*, bd. i., s. 221, and bd. xi., s. 146; Gay-Lussac, in the *Annales de Chimie*, t. x., xii., p. 422; and Bischof's *Reasons against the Chemical Theory of Volcanoes*, in the English edition of his *Wärmelehre*, p. 297–309.

† [On the various theories that have been advanced in explanation of volcanic action, see Daubeney *On Volcanoes*, a work to which we have made continual reference during the preceding pages, as it constitutes the most recent and perfect compendium of all the important facts relating to this subject, and is peculiarly adapted to serve as a source of reference to the *Cosmos*, since the learned author in many instances enters into a full exposition of the views advanced by Baron Humboldt. The appendix contains several valuable notes with reference to the most recent works that have appeared on the Continent, on subjects relating to volcanoes; among others, an interesting notice of Professor Bischof's views "on the origin of the carbonic acid discharged from volcanoes," as enounced in his recently published work, *Lehrbuch der Chemischen und Physikalischen Geologie.*]— *Tr.*

matter revolving elliptically in a gaseous condition. We have thus mere conjecture and supposition side by side with certain knowledge. A philosophical study of nature strives ever to elevate itself above the narrow requirements of mere natural description, and does not consist, as we have already remarked, in the mere accumulation of isolated facts. The inquiring and active spirit of man must be suffered to pass from the present to the past, to conjecture all that can not yet be known with certainty, and still to dwell with pleasure on the ancient myths of geognosy which are presented to us under so many various forms. If we consider volcanoes as irregular intermittent springs, emitting a fluid mixture of oxydized metals, alkalies, and earths, flowing gently and calmy wherever they find a passage, or being upheaved by the powerful expansive force of vapors, we are involuntarily led to remember the geognostic visions of Plato, according to which hot springs, as well as all volcanic igneous streams, were eruptions that might be traced back to one generally distributed subterranean cause, *Pyriphlegethon.**

* According to Plato's geognostic views, as developed in the *Phædo*, Pyriphlegethon plays much the same part in relation to the activity of volcanoes that we now ascribe to the augmentation of heat as we descend from the earth's surface, and to the fused condition of its internal strata. (*Phædo*, ed. Ast, p. 603 and 607; Annot., p. 808 and 817.) "Within the earth, and all around it, are larger and smaller caverns. Water flows there in abundance; also much fire and large streams of fire, and streams of moist mud (some purer and others more filthy), like those in Sicily, consisting of mud and fire, preceding the great eruption. These streams fill all places that fall in the way of their course. Pyriphlegethon flows forth into an extensive district burning with a fierce fire, where it forms a lake larger than our sea, boiling with water and mud. From thence it moves in circles round the earth, turbid and muddy." This stream of molten earth and mud is so much the general cause of volcanic phenomena, that Plato expressly adds, "thus is Pyriphlegethon constituted, from which also the streams of fire (οἱ ῥύακες), wherever they reach the earth (ὅπη ἂν τυχωσι τῆς γῆς), inflate such parts (detached fragments)." Volcanic scoriæ and lava streams are therefore portions of Pyriphlegethon itself, portions of the subterranean molten and ever-undulating mass. That οἱ ῥύακες are lava streams, and not, as Schueider, Passow, and Schleiermacher will have it, "fire-vomiting mountains," is clear enough from many passages, some of which have been collected by Ukert (*Geogr. der Griechen und Römer*, th. ii., s. 200); ῥύαξ is the volcanic phenomenon in reference to its most striking characteristic, the lava stream. Hence the expression, the ῥύακες of Ætna. Aristot., *Mirab. Ausc.*, t. ii., p. 833; sect. 38, Bekker; Thucyd., iii., 116; Theophrast., *De Lap.*, 22, p. 427, Schneider; Diod., v., 6, and xiv., 59, where are the remarkable words, "Many places near the sea, in the neighborhood of Ætna, were leveled to the ground, ὑπὸ τοῦ καλουμένου ῥύακος;" Strabo, vi., p. 269; xiii., p. 268, and

The different volcanoes over the earth's surface, when they are considered independently of all climatic differences, are acutely and characteristically classified as central and linear volcanoes. Under the first name are comprised those which constitute the central point of many active mouths of eruption, distributed almost regularly in all directions ; under the second, those lying at some little distance from one another, forming, as it were, chimneys or vents along an extended fissure. Linear volcanoes again admit of further subdivision, namely, those which rise like separate conical islands from the bottom of the sea, being generally parallel with a chain of primitive mountains, whose foot they appear to indicate, and those volcanic chains which are elevated on the highest ridges of these mountain chains, of which they form the summits.* The Peak of Teneriffe, for instance, is a central volcano, being the central point of the volcanic group to which the eruption of Palma and Lancerote may be referred. The long, rampart-like chain of the Andes, which is sometimes single, and sometimes divided into two or three parallel branches, connected by various transverse ridges, presents, from the south of Chili to the northwest coast of America, one of the grandest instances of a continental volcanic chain. The proximity of

where there is a notice of the celebrated burning mud of the Lelantine plains, in Eubœa, i., p. 58, Casaub. ; and Appian, *De Bello Civili*, v., 114. The blame which Aristotle throws on the geognostical fantasies of the Phædo (*Meteor.*, ii., 2, 19) is especially applied to the sources of the rivers flowing over the earth's surface. The distinct statement of Plato, that "in Sicily eruptions of wet mud precede the glowing (lava) stream," is very remarkable. Observations on Ætna could not have led to such a statement, unless pumice and ashes, formed into a mud-like mass by admixture with melted snow and water, during the volcano-electric storm in the crater of eruption, were mistaken for ejected mud. It is more probable that Plato's streams of moist mud (ὑγροῦ πηλοῦ ποταμοι) originated in a faint recollection of the salses (mud volcanoes) of Agrigentum, which, as I have already mentioned, eject argillaceous mud with a loud noise. It is much to be regretted, in reference to this subject, that the work of Theophrastus περι ρυακος του εν Σικελια, *On the Volcanic Stream in Sicily*, to which Diog. Laert., v., 49, refers, has not come down to us.

* Leopold von Buch, *Physikal. Beschreib. der Canarischen Inseln*, s. 326–407. I doubt if we can agree with the ingenious Charles Darwin (*Geological Observations on Volcanic Islands*, 1844, p. 127) in regarding central volcanoes in general as volcanic chains of small extent on parallel fissures. Friedrich Hoffman believes that in the group of the Lipari Islands, which he has so admirably described, and in which two eruption fissures intersect near Panaria, he has found an intermediate link between the two principal modes in which volcanoes appear, namely, the central volcanoes and volcanic chains of Von Buch (Poggendorf, *Annalen der Physik*, bd. xxvi., s. 81–88).

active volcanoes is always manifested in the chain of the Andes by the appearance of certain rocks (as dolerite, melaphyre,
trachyte, andesite, and dioritic porphyry), which divide the so-
called primitive rocks, the transition slates and sandstones, and
the stratified formations. The constant recurrence of this
phenomenon convinced me long since that these sporadic rocks
were the seat of volcanic phenomena, and were connected with
volcanic eruptions. At the foot of the grand Tunguragua,
near Penipe, on the banks of the Rio Puela, I first distinctly
observed mica slate resting on granite, broken through by a
volcanic rock.

In the volcanic chain of the New Continent, the separate
volcanoes are occasionally, when near together, in mutual dependence upon one another ; and it is even seen that the volcanic activity for centuries together has moved on in one and
the same direction, as, for instance, from north to south in the
province of Quito.* The focus of the volcanic action lies below the whole of the highlands of this province ; the only
channels of communication with the atmosphere are, howev
er, those mountains which we designate by special names, as
the mountains of Pichincha, Cotopaxi, and Tunguragua, and
which, from their grouping, elevation, and form, constitute the
grandest and most picturesque spectacle to be found in any
volcanic district of an equally limited extent. Experience
shows us, in many instances, that the extremities of such
groups of volcanic chains are connected together by subterranean communications ; and this fact reminds us of the ancient
and true expression made use of by Seneca,† that the igneous
mountain is only the issue of the more deeply-seated volcanic
forces. In the Mexican highlands a mutual dependence is

* Humboldt, *Geognost. Beobach, über die Vulkane des Hochlandes von
Quito*, in Poggend., *Annal. der Physik*, bd. xliv., s. 194.

† Seneca, while he speaks very clearly regarding the problematical
sinking of Ætna, says in his 79th letter, " Though this might happen,
not because the mountain's height is lowered, but because the fires are
weakened, and do not blaze out with their former vehemence ; and for
which reason it is that such vast clouds of smoke are not seen in the
day-time. Yet neither of these seem incredible, for the mountain may
possibly be consumed by being daily devoured, and the fire not be so
large as formerly, since it is not self-generated here, but is kindled in
the distant bowels of the earth, and there rages, being fed with continual fuel, not with that of the mountain, through which it only makes
its passage." The subterranean communication, " by galleries," between the volcanoes of Sicily, Lipari, Pithecusa (Ischia), and Vesuvius,
" of the last of which we may conjecture that it formerly burned and
presented a fiery circle," seems fully understood by Strabo (lib. i., p.
247 and 248). He terms the whole district " sub-igneous."

also observed to exist among the volcanic mountains Orizaba, Popocatepetl, Jorullo, and Colima ; and I have shown*
that they all lie in one direction between 18° 59' and 19° 12'
north latitude, and are situated in a transverse fissure running
from sea to sea.   The volcano of Jorullo broke forth on the
29th of September, 1759, exactly in this direction, and over
the same transverse fissure, being elevated to a height of 1604
feet above the level of the surrounding plain.   The mountain
only once emitted an eruption of lava, in the same manner as
is recorded of Mount Epomeo in Ischia, in the year 1302
But although Jorullo, which is eighty miles from any active
volcano, is in the strict sense of the word a new mountain, it
must not be compared with Monte Nuovo, near Puzzuolo,
which first appeared on the 19th of September, 1538, and is
rather to be classed among craters of elevation.   I believe
that I have furnished a more natural explanation of the eruption of the Mexican volcano, in comparing its appearance to
the elevation of the Hill of Methone, now Methana, in the
peninsula of Trœzene.   The description given by Strabo and
Pausanias of this elevation, led one of the Roman poets, most
celebrated for his richness of fancy, to develop views which
agree in a remarkable manner with the theory of modern
geognosy.   " Near Trœzene is a tumulus, steep and devoid of
trees, once a plain, now a mountain.   The vapors inclosed in
dark caverns in vain seek a passage by which they may escape.
The heaving earth, inflated by the force of the compressed
vapors, expands like a bladder filled with air, or like a goat-
skin.   The ground has remained thus inflated, and the high
projecting eminence has been solidified by time into a naked
rock."   Thus picturesquely, and, as analogous phenomena
justify us in believing, thus truly has Ovid described that
great natural phenomenon which occurred 282 years before
our era, and, consequently, 45 years before the volcanic separation of Thera (Santorino) and Therasia, between Trœzene
and Epidaurus, on the same spot where Russegger has found
veins of trachyte.†

      * Humboldt, *Essai Politique sur la Nouv. Espagne*, t. ii., p. 173–175.
      † Ovid's description of the eruption of Methone (*Metam.*, xv., p. 296
306) :
                    " Near Trœzene stands a hill, exposed in air
                    To winter winds, of leafy shadows bare :
                    This once was level ground ; but (strange to tell)
                    Th' included vapors, that in caverns dwell,
                    Laboring with colic pangs, and close confined,
                    In vain sought issue for the rumbling wind :
                    Yet still they heaved for vent, and heaving still,
                    Enlarged the concave and shot up the hill,

Santorino is the most important of all the *islands of erup tion* belonging to volcanic chains.* " It combines within it

> As breath extends a bladder, or the skins
> Of goats are blown t' inclose the hoarded wines ;
> The mountain yet retains a mountain's face,
> And gathered rubbish heads the hollow space."
>
> *Dryden's Translation.*

This description of a dome-shaped elevation on the continent is of great importance in a geognostical point of view, and coincides to a remarkable degree with Aristotle's account (*Meteor.*, ii., 8, 17–19) of the upheaval of islands of eruption: " The heaving of the earth does not cease till the wind (ἄνεμος) which occasions the shocks has made its escape into the crust of the earth. It is not long ago since this actually happened at Heraclea in Pontus, and a similar event formerly occurred at Hiera, one of the Æolian Islands. A portion of the earth swelled up, and with loud noise rose into the form of a hill, till the mighty urging blast (πνεῦμα) found an outlet, and ejected sparks and ashes which covered the neighborhood of Lipari, and even extended to several Italian cities." In this description, the vesicular distension of the earth's crust (a stage at which many trachytic mountains have remained) is very well distinguished from the eruption itself. Strabo, lib. i., p. 59 (Casaubon), likewise describes the phenomenon as it occurred at Methone: near the town, in the Bay of Hermione, there arose a flaming eruption; a fiery mountain, seven (?) stadia in height, was then thrown up, which during the day was inaccessible from its heat and sulphureous stench, but at night evolved an agreeable odor (?), and was so hot that the sea boiled for a distance of five stadia, and was turbid for full twenty stadia, and also was filled with detached masses of rock. Regarding the present mineralogical character of the peninsula of Methana, see Fiedler, *Reise durch Griechenland*, th. i., s. 257–263.

* [I am indebted to the kindness of Professor E. Forbes for the following interesting account of the island of Santorino, and the adjacent islands of Neokaimeni and Microkaimeni. " The aspect of the bay is that of a great crater filled with water, Thera and Therasia forming its walls, and the other islands being after-productions in its center. We sounded with 250 fathoms of line in the middle of the bay, between Therasia and the main islands, but got no bottom. Both these islands appear to be similarly formed of successive strata of volcanic ashes, which, being of the most vivid and variegated colors, present a striking contrast to the black and cindery aspect of the central isles. Neokaimeni, the last-formed island, is a great heap of obsidian and scoriæ. So, also, is the greater mass, Microkaimeni, which rises up in a conical form, and has a cavity or crater. On one side of this island, however, a section is exposed, and cliffs of fine pumiceous ash appear stratified in the greater islands. In the main island, the volcanic strata abut against the limestone mass of Mount St. Elias in such a way as to lead to the inference that they were deposited in a sea bottom in which the present mountain rose as a submarine mass of rock. The people at Santorino assured us that subterranean noises are not unfrequently heard, especially during calms and south winds, when they say the water of parts of the bay becomes the color of sulphur. My own impression is, that this group of islands constitutes a crater of elevation, of which the outer ones are the remains of the walls, while the central group are of later origin, and consist partly of upheaved sea bottoms

self the history of all islands of elevation. For upward of 2000 years, as far as history and tradition certify, it would appear as if nature were striving to form a volcano in the midst of the crater of elevation."* Similar insular eleva- tions, and almost always at regular intervals of 80 or 90 years,† have been manifested in the island of St. Michael, in the Azores ; but in this case the bottom of the sea has not been elevated at exactly the same parts.‡ The island which Captain Tillard named *Sabrina,* appeared unfortunately at a time (the 30th of January, 1811) when the political rela- tions of the maritime nations of Western Europe prevented that attention being bestowed upon the subject by scientific institutions which was afterward directed to the sudden ap- pearance (the 2d of July, 1831), and the speedy destruction of the igneous island of Ferdinandea in the Sicilian Sea, between the limestone shores of Sciacca and the purely volcanic island of Pantellaria.§

and partly of erupted matter—erupted, however, beneath the surface of the water."]—*Tr.*

* Leop. von Buch, *Physik. Beschr. der Canar. Inseln,* s. 356–358, and particularly the French translation of this excellent work, p. 402 ; and his memoir in Poggendorf's *Annalen,* bd. xxxviii., s. 183. A sub- marine island has quite recently made its appearance within the crater of Santorino. In 1810 it was still fifteen fathoms below the surface of the sea, but in 1830 it had risen to within three or four. It rises steeply, like a great cone, from the bottom of the sea, and the continuous ac tivity of the submarine crater is obvious from the circumstance that sul phurous acid vapors are mixed with the sea water, in the eastern bay of Neokaimeni, in the same manner as at Vromolimni, near Methana. Coppered ships lie at anchor in the bay in order to get their bottoms cleaned and polished by this natural (volcanic) process. (Virlet, in the *Bulletin de la Société Géologique de France,* t. iii., p. 109, and Fiedler, *Reise durch Griechenland,* th. ii., s. 469 and 584.)

† Appearance of a new island near St. Miguel, one of the Azores, 11th of June, 1638, 31st of December, 1719, 13th of June, 1811.

‡ [My esteemed friend, Dr. Webster, professor of Chemistry and Mineralogy at Harvard College, Cambridge, Massachusetts, U. S., in his *Description of the Island of St. Michael, &c.,* Boston, 1822, gives an interesting account of the sudden appearance of the island named Sa- brina, which was about a mile in circumference, and two or three hundred feet above the level of the ocean. After continuing for some weeks, it sank into the sea. Dr. Webster describes the whole of the island of St. Michael as volcanic, and containing a number of conical hills of trachyte, several of which have craters, and appear at some former time to have been the openings of volcanoes. The hot springs which abound in the island are impregnated with sulphureted hydro- gen and carbonic acid gases, appearing to attest the existence of vol- canic action.]—*Tr.*

§ Prévost, in the *Bulletin de la Société Géologique,* t. iii., p. 34; Fried- rich Hoffman, *Hinterlassene Werke,* bd. ii., s. 451–456.

The geographical distribution of the volcanoes which have been in a state of activity during historical times, the great number of insular and littoral volcanic mountains, and the occasional, although ephemeral, eruptions in the bottom of the sea, early led to the belief that volcanic activity was connected with the neighborhood of the sea, and was dependent upon it for its continuance. " For many hundred years," says Justinian, or rather Trogus Pompeius, whom he follows,* " Ætna and the Æolian Islands have been burning, and how could this have continued so long if the fire had not been fed by the

* " Accedunt vicini et perpetui Ætnæ montis ignes et insularum Æolidum, veluti ipsis undis alatur incendium; neque enim aliter durare tot seculis tantus ignis potuisset, nisi humoris nutrimentis aleretur." (Justin, *Hist. Philipp.*, iv., i.) The volcanic theory with which the physical description of Sicily here begins is extremely intricate. Deep strata of sulphur and resin; a very thin soil full of cavities and easily fissured; violent motion of the waves of the sea, which, as they strike together, draw down the air (the wind) for the maintenance of the fire: such are the elements of the theory of Trogus. Since he seems from Pliny (xi., 52) to have been a physiognomist, we may presume that his numerous lost works were not confined to history alone. The opinion that air is forced into the interior of the earth, there to act on the volcanic furnaces, was connected by the ancients with the supposed influence of winds from different quarters on the intensity of the fires burning in Ætna, Hiera, and Stromboli. (See the remarkable passage in Strabo, lib. vi., p. 275 and 276.) The mountain island of Stromboli (Strongyle) was regarded, therefore, as the dwelling-place of Æolus, " the regulator of the winds," in consequence of the sailors foretelling the weather from the activity of the volcanic eruptions of this island. The connection between the eruption of a small volcano with the state of the barometer and the direction of the wind is still generally recognized (Leop. von Buch, *Descr. Phys. des Iles Canaries*, p. 334; Hoffmann, in Poggend., *Annalen*, bd. xxvi., s. viii.), although our present knowledge of volcanic phenomena, and the slight changes of atmospheric pressure accompanying our winds, do not enable us to offer any satisfactory explanation of the fact. Bembo, who during his youth was brought up in Sicily by Greek refugees, gave an agreeable narrative of his wanderings, and in his *Ætna Dialogus* (written in the middle of the sixteenth century) advances the theory of the penetration of sea water to the very center of the volcanic action, and of the necessity of the proximity of the sea to active volcanoes. In ascending Ætna the following question was proposed: " Explana potius nobis quæ petimus, ea incendia unde oriantur et orta quomodo perdurent. In omni tellure nuspiam majores fistulæ aut meatus ampliores sunt quam in locis, quæ vel mari vicina sunt, vel a mari protinus alluuntur: mare erodit illa facillime pergitque in viscera terræ. Itaque cum in aliena regna sibi viam faciat, ventis etiam facit; ex quo fit, ut loca quæque maritima maxime terræ motibus subjecta sint, parum mediterranea. Habes quum in sulfuris venas venti furentes inciderint, unde incendia oriantur Ætnæ tuæ. Vides, quæ mare in radicibus habeat, quæ sulfurea sit. quæ cavernosa, quæ a mari aliquando perforata ventos admiserit æstuantes, per quos idonea flammæ materies incenderetur."

neighboring sea ?"*   In order to explain the necessity of the vicinity of the sea, recourse has been had, even in modern times, to the hypothesis of the penetration of sea water into the foci of volcanic agency, that is to say, into deep-seated terrestrial strata.   When I collect together all the facts that may be derived from my own observation and the laborious researches of others, it appears to me that every thing in this involved investigation depends upon the questions whether the great quantity of aqueous vapors, which are unquestionably exhaled from volcanoes even when in a state of rest, be derived from sea water impregnated with salt, or rather, perhaps, with fresh meteoric water; or whether the expansive force of the vapors (which, at a depth of nearly 94,000 feet, is equal to 2800 atmospheres) would be able at different depths to counterbalance the hydrostatic pressure of the sea, and thus afford them, under certain conditions, a free access to the focus;† or whether the formation of metallic chlorids, the presence of chlorid of sodium in the fissures of the crater, and the frequent mixture of hydrochloric acid with the aqueous vapors, necessarily imply access of sea water ; or, finally, whether the repose of volcanoes (either when temporary, or permanent and complete) depends upon the closure of the channels by which the sea or meteoric water was conveyed, or whether the absence of flames and of exhalations of hydrogen (and sulphureted hydrogen gas seems more characteristic of solfataras than of active volcanoes) is not directly at variance

* [Although extinct volcanoes seem by no means confined to the neighborhood of the present seas, being often scattered over the most inland portions of our existing continents, yet it will appear that, at the time at which they were in an active state, the greater part were in the neighborhood either of the sea, or of the extensive salt or fresh water lakes, which existed at that period over much of what is now dry land. This may be seen either by referring to Dr. Boué's map of Europe, or to that published by Mr. Lyell in the recent edition of his *Principles of Geology* (1847), from both of which it will become apparent that, at a comparatively recent epoch, those parts of France, of Germany, of Hungary, and of Italy, which afford evidences of volcanic action now extinct, were covered by the ocean.   Daubeney *On Volcanoes*, p. 605.] —*Tr.*

† Compare Gay-Lussac, *Sur les Volcans*, in the *Annales de Chimie*, t. xxii., p. 427, and Bischof, *Wärmelehre*, s. 272.   The eruptions of smoke and steam which have at different periods been seen in Lance rote, Iceland, and the Kurile Islands, during the eruption of the neigh boring volcanoes, afford indications of the reaction of volcanic foci through tense columns of water; that is to say, these phenomena oc cur when the expansive force of the vapor exceeds the hydrostatic pressure.

with the hypothesis of the decomposition of great masses of water ?*

The discussion of these important physical questions does not come within the scope of a work of this nature ; but, while we are considering these phenomena, we would enter somewhat more into the question of the geographical distribution of still active volcanoes. We find, for instance, that in the New World, three, viz., Jorullo, Popocatepetl, and the volcano of De la Fragua, are situated at the respective distances of 80, 132, and 196 miles from the sea-coast, while in Central Asia, as Abel Rémusat† first made known to geognosists, the Thian-schan (Celestial Mountains), in which are situated the lava-emitting mountain of Pe-schan, the solfatara of Urumtsi, and the still active igneous mountain (Ho-tscheu) of Turfan, lie at an almost equal distance (1480 to 1528 miles) from the shores of the Polar Sea and those of the Indian Ocean.   Pe-schan is also fully 1360 miles distant from the Caspian Sea,‡ and 172 and 218 miles from the seas of Issikul and Balkasch.   It is a fact worthy of notice, that among the four great parallel mountain chains which traverse the Asiatic continent from east to west, the Altai, the Thianschan, the Kuen-lun, and the Himalaya, it is not the latter chain, which is nearest to the ocean, but the two inner ranges, the Thianschan and the Kuen-lun, at the distance of 1600 and 720 miles from the sea, which have fire-emitting mountains like Ætna and Vesuvius, and generate ammonia like the volcano of Guatimala.   Chinese writers undoubtedly speak of lava streams when they describe the emissions of smoke and flame, which, issuing from Pe-schan, devastated a space measuring ten li§ in the first and seventh centuries of our era.   Burning masses of stone flowed, according to their description, "like thin melted fat." The facts that have been enumerated, and to which sufficient attention has not been bestowed, render it probable that the vicinity of the sea, and the penetration of sea water to the foci of volcanoes, are not absolutely necessary to the eruption of

---

* [See Daubeney *On Volcanoes*, Part iii., ch. xxxvi., xxxviii., xxxix.] —*Tr.*

† Abel Rémusat, *Lettre à M. Cordier*, in the *Annales de Chimie*, t. v., p. 137.

‡ Humboldt, *Asie Centrale*, t. ii., p. 30–33, 38–52, 70–80, and 426–428. The existence of active volcanoes in Kordofan, 540 miles from the Red Sea, has been recently contradicted by Rüppell, *Reisen in Nubien*, 1829, s. 151.

§ [A *li* is a Chinese measurement, equal to about one thirtieth of a mile.]—*Tr.*

subterranean fire, and that littoral situations only favor the eruption by forming the margin of a deep sea basin, which, covered by strata of water, and lying many thousand feet lower than the interior continent, can offer but an inconsiderable degree of resistance.

The present active volcanoes, which communicate by permanent craters simultaneously with the interior of the earth and with the atmosphere, must have been formed at a subsequent period, when the upper chalk strata and all the tertiary formations were already present : this is shown to be the fact by the trachytic and basaltic eruptions which frequently form the walls of the crater of elevation. Melaphyres extend to the middle tertiary formations, but are found already in the Jura limestone, where they break through the variegated sandstone.* We must not confound the earlier outpourings of granite, quartzose porphyry, and euphotide from temporary fissures in the old transition rocks with the present active volcanic craters.

The extinction of volcanic activity is either only partial— in which case the subterranean fire seeks another passage of escape in the same mountain chain—or it is total, as in Auvergne. More recent examples are recorded in historical times, of the total extinction of the volcano of Mosychlos,† on the island sacred to Hephæstos (Vulcan), whose "high whirling flames" were known to Sophocles ; and of the volcano of Medina, which, according to Burckhardt, still continued to pour out a stream of lava on the 2d of November, 1276. Every stage of volcanic activity, from its first origin to its extinction, is characterized by peculiar products ; first by ignited scoriæ, streams of lava consisting of trachyte, pyroxene, and obsidian, and by rapilli and tufaceous ashes, accompanied by the devel-

* Dufrénoy et Elie de Beaumont, *Explication de la Carte Géologique de la France*, t. i., p. 89.

† Sophocl., *Philoct.*, v. 971 and 972. On the supposed epoch of the extinction of the Lemnian fire in the time of Alexander, compare Buttmann, in the *Museum der Alterthumswissenschaft*, bd. i., 1807, s. 295 ; Dureau de la Malle, in Malte-Brun, *Annales des Voyages*, t. ix., 1809, p. 5 ; Ukert, in Bertuch, *Geogr. Ephemeriden*, bd. xxxix., 1812, s. 361 ; Rhode, *Res Lemnicæ*, 1829, p. 8 ; and Walter, *Ueber Abnahme der Vulkan. Thätigkeit in Historischen Zeiten*, 1844, s. 24. The chart of Lemnos, constructed by Choiseul, makes it extremely probable that the extinct crater of Mosychlos, and the island of Chryse, the desert habitation of Philoctetes (Otfried Müller, *Minyer*, s. 300), have been long swallowed up by the sea. Reefs and shoals, to the northeast of Lemnos, still indicate the spot where the Ægean Sea once possessed an active volcano like Ætna, Vesuvius, Stromboli, and Volcano (in the Lipari Isles).

opment of large quantities of pure aqueous vapor ; subsequently, when the volcano becomes a solfatara, by aqueous vapors
mixed with sulphureted hydrogen and carbonic acid gases ;
and, finally, when it is completely cooled, by exhalations of
carbonic acid alone. There is a remarkable class of igneous
mountains which do not eject lava, but merely devastating
streams of hot water,* impregnated with burning sulphur and
rocks reduced to a state of dust (as, for instance, the Galun-
gung in Java) ; but whether these mountains present a normal
condition, or only a certain transitory modification of the volcanic process, must remain undecided until they are visited by
geologists possessed of a knowledge of chemistry in its present
condition.

I have endeavored in the above remarks to furnish a general description of volcanoes—comprising one of the most important sections of the history of terrestrial activity—and I
have based my statements partly on my own observations, but
more in their general bearing on the results yielded by the labors of my old friend, Leopold von Buch, the greatest geognosist of our own age, and the first who recognized the intimate
connection of volcanic phenomena, and their mutual dependence upon one another, considered with reference to their relations in space.

Volcanic action, or the reaction of the interior of a planet on
its external crust and surface, was long regarded only as an
isolated phenomenon, and was considered solely with respect
to the disturbing action of the subterranean force ; and it is
only in recent times that—greatly to the advantage of geognostical views based on physical analogies—volcanic forces
have been regarded as *forming new rocks, and transforming
those that already existed.* We here arrive at the point to
which I have already alluded, at which a well-grounded study
of the activity of volcanoes, whether igneous or merely such
as emit gaseous exhalations, leads us, on the one hand, to the
mineralogical branch of geognosy (the science of the texture
and the succession of terrestrial strata), and, on the other, to
the science of geographical forms and outlines—the configuration of continents and insular groups elevated above the level

* Compare Reinwardt and Hoffmann, in Poggendorf's *Annalen*, bd.
xii., s. 607 ; Leop. von Buch, *Descr. des Iles Canaries*, p. 424–426. The
eruptions of argillaceous mud at Carguairazo, when that volcano was
destroyed in 1698, the Lodazales of Igualata, and the Moya of Pelileo
—all on the table-land of Quito—are volcanic phenomena of a similar
nature.

of the sea. This extended insight into the co ction of natural phenomena is the result of the philosop ical direction
which has been so generally assumed by the more earnest
study of geognosy. Increased cultivation of science and enlargement of political views alike tend to unite elements that
had long been divided.

If, instead of classifying rocks according to their varieties of
form and superposition into stratified and unstratified, schistose
and compact, normal and abnormal, we investigate those phenomena of formation and transformation which are still going
on before our eyes, we shall find that rocks admit of being arranged according to four modes of origin.

Rocks of eruption, which have issued from the interior of
the earth either in a state of fusion from volcanic action, or
in a more or less soft, viscous condition, from Plutonic action.

Sedimentary rocks, which have been precipitated and deposited on the earth's surface from a fluid, in which the most
minute particles were either dissolved or held in suspension
constituting the greater part of the secondary (or flötz) and
tertiary groups.

*Transformed or metamorphic rocks,** in which the internal
texture and the mode of stratification have been changed, ei-

* [As the doctrine of mineral metamorphism is now exciting very
general attention, we subjoin a few explanatory observations by the
celebrated Swiss philosopher, Professor Studer, taken from the *Edinb.
New Philos. Journ.*, Jan., 1848: " In its widest sense, mineral metamorphism means every change of aggregation, structure, or chemical
condition which rocks have undergone subsequently to their deposition
and stratification, or the effects which have been produced by other
forces than gravity and cohesion. There fall under this definition, the
discoloration of the surface of black limestone by the loss of carbon;
the formation of brownish-red crusts on rocks of limestone, sandstone,
many slate stones, serpentine, granite, &c., by the decomposition of iron
pyrites, or magnetic iron, finely disseminated in the mass of the rock;
the conversion of anhydrite into gypsum, in consequence of the absorption of water; the crumbling of many granites and porphyries into
gravel, occasioned by the decomposition of the mica and feldspar. In
its more limited sense, the term metamorphic is confined to those
changes of the rock which are produced, not by the effect of the atmosphere or of water on the exposed surfaces, but which are produced,
directly or indirectly, by agencies seated in the interior of the earth.
In many cases the mode of change may be explained by our physical
or chemical theories, and may be viewed as the effect of temperature
or of electro-chemical actions. Adjoining rocks, or connecting communications with the interior of the earth, also distinctly point out the
seat from which the change proceeds. In many other cases the metamorphic process itself remains a mystery, and from the nature of the
products alone do we conclude that such a metamorphic action has
taken place.]—*Tr.*

ther by contact or proximity with a Plutonic or volcanic endogenous rock of eruption,* or, what is more frequently the
case, by a gaseous sublimation of substances† which accompany certain masses erupted in a hot, fluid condition.

Conglomerates; coarse or finely granular sandstones, or
breccias composed of mechanically-divided masses of the three
previous species.

These four modes of formation—by the emission of volcanic
masses, as narrow lava streams; by the action of these masses
on rocks previously hardened; by mechanical separation or
chemical precipitation from liquids impregnated with carbonic
acid; and, finally, by the cementation of disintegrated rocks
of heterogeneous nature—are phenomena and formative processes which must merely be regarded as a faint reflection of
that more energetic activity which must have characterized
the chaotic condition of the earlier world under wholly different conditions of pressure and at a higher temperature, not
only in the whole crust of the earth, but likewise in the more

---

* In a plan of the neighborhood of Tezcuco, Totonilco, and Moran
(*Atlas Géographique et Physique,* pl. vii.), which I originally (1803)
intended for a work which I never published, entitled *Pasigrafia Geog-
nostica destinada al uso de los Jovenes del Colegio de Mineria de Mexico,* I named (in 1832) the Plutonic and volcanic eruptive rocks *endogenous* (generated in the interior), and the sedimentary and flötz rocks
exogenous (or generated externally on the surface of the earth). Pasi-
graphically, the former were designated by an arrow directed upward †, and the latter by the same symbol directed downward ↓.
These signs have at least some advantage over the ascending lines,
which in the older systems represent arbitrarily and ungracefully the
horizontally ranged sedimentary strata, and their penetration through
masses of basalt, porphyry, and syenite. The names proposed in the
pasigraphico-geognostic plan were borrowed from De Candolle's nomen
clature, in which *endogenous* is synonymous with monocotyledonous,
and *exogenous* with dicotyledonous plants. Mohl's more accurate examination of vegetable tissues has, however, shown that the growth of
monocotyledons from within, and dicotyledons from without, is not
strictly and generally true for vegetable organisms (Link, *Elementa
Philosophiæ Botanicæ,* t. i., 1837, p. 287; Endlicher and Unger, *Grundzüge der Botanik,* 1843, s. 89; and Jussieu, *Traité de Botanique,* t. i.,
p. 85). The rocks which I have termed endogenous are characteristically distinguished by Lyell, in his *Principles of Geology,* 1833, vol. iii.,
p. 374, as "nether-formed" or "hypogene rocks."

† Compare Leop. von Buch, *Ueber Dolomit als Gebirgsart,* 1823, s.
36; and his remarks on the degree of fluidity to be ascribed to Plutonic
rocks at the period of their eruption, as well as on the formation of
gneiss from schist, through the action of granite and of the substances
upheaved with it, to be found in the *Abhandl. der Akad. der Wissensch. zu Berlin* for the year 1842, s. 58 und 63, and in the *Jahrbuch für
Wissenschaftliche Kritik,* 1840, s. 195.

extended atmosphere, overloaded with vapors. The vast fissures which were formerly open in the solid crust of the earth
have since been filled up or closed by the protrusion of elevated mountain chains, or by the penetration of veins of rocks of
eruption (granite, porphyry, basalt, and melaphyre); and while,
on a superficial area equal to that of Europe, there are now
scarcely more than four volcanoes remaining through which
fire and stones are erupted, the thinner, more fissured, and unstable crust of the earth was anciently almost every where
covered by channels of communication between the fused interior and the external atmosphere. Gaseous emanations, rising from very unequal depths, and therefore conveying substances differing in their chemical nature, imparted greater
activity to the Plutonic processes of formation and transformation. The sedimentary formations, the deposits of liquid fluids
from cold and hot springs, which we daily see producing the
travertine strata near Rome, and near Hobart Town in Van
Diemen's Land, afford but a faint idea of the flötz formation.
In our seas, small banks of limestone, almost equal in hardness
at some parts to Carrara marble,* are in the course of formation, by gradual precipitation, accumulation, and cementation
—processes whose mode of action has not been sufficiently
well investigated. The Sicilian coast, the island of Ascension,
and King George's Sound in Australia, are instances of this
mode of formation. On the coasts of the Antilles, these
formations of the present ocean contain articles of pottery,
and other objects of human industry, and in Guadaloupe even
human skeletons of the Carib tribes.† The negroes of the
French colonies designate these formations by the name of
Maconne-bon-Dieu.‡ A small oolitic bed, formed in Lancerote, one of the Canary Islands, and which, notwithstand-

---

* Darwin, *Volcanic Islands*, 1844, p. 49 and 154.

† [In most instances the bones are dispersed; but a large slab of rock,
in which a considerable portion of the skeleton of a female is imbedded,
is preserved in the British Museum. The presence of these bones has
been explained by the circumstance of a battle, and the massacre of a
tribe of Gallibis by the Caribs, which took place near the spot in which
they are found, about 120 years ago; for, as the bodies of the slain
were interred on the sea-shore, their skeletons may have been subsequently covered by sand-drift, which has since consolidated into limestone. Dr. Moultrie, of the Medical College, Charleston, South Caro-
lina, U. S., is, however, of opinion that these bones did not belong to
individuals of the Carib tribe, but of the Peruvian race, or of a tribe
possessing a similar craniological development.]—*Tr.*

‡ Moreau de Jonnès, *Hist. Phys. des Antilles*, t. i., p. 136, 138, and
543; Humboldt, *Relation Historique*, t. iii., p. 367.

ing its recent formation, bears a resemblance to Jura limestone, has been recognized as a product of the sea and of tem-
pests.*

Composite rocks are definite associations of certain oryctognostic, simple minerals, as feldspar, mica, solid silex, augite,
and nepheline. Rocks very similar to these, consisting of the
same elements, but grouped differently, are still formed by
volcanic processes, as in the earlier periods of the world. The
character of rocks, as we have already remarked, is so independent of geographical relations of space,† that the geologist
recognizes with surprise, alike to the north or the south of
the equator, in the remotest and most dissimilar zones, the
familiar aspect, and the repetition of even the most minute
characteristics in the periodic stratification of the silurian
strata, and in the effects of contact with augitic masses of
eruption.

We will now enter more fully into the consideration of the
four modes in which rocks are formed—the four phases of
their formative processes manifested in the stratified and unstratified portions of the earth's surface ; thus, in the *endogenous* or *erupted rocks*, designated by modern geognosists as
compact and abnormal rocks, we may enumerate the following principal groups as immediate products of terrestrial activity :

1. *Granite and syenite* of very different respective ages ;
the granite is frequently the more recent,‡ traversing the syenite in veins, and being, in that case, the active upheaving
agent. "Where the granite occurs in large, insulated masses
of a faintly-arched, ellipsoidal form, it is covered by a crust or
shell cleft into blocks, instances of which are met with alike
in the Hartz district, in Mysore, and in Lower Peru. This
sea of rocks probably owes its origin to a contraction of the
surface of the granite, owing to the great expansion that accompanied its first upheaval."§

Both in Northern Asia,‖ on the charming and romantic
shores of the Lake of Kolivan, on the northwest declivity of

---

* Near Teguiza. Leop. von Buch, *Canarische Inseln*, s. 301.
† Leop. von Buch, op. cit., p. 9.
‡ Bernhard Cotta, *Geognosie*, 1839, s. 273.
§ Leop. von Buch, *Ueber Granit und Gneiss*, in the *Abhandl. der Berl.
Akad.* for the year 1842, s. 60.
‖ In the projecting mural masses of granite of Lake Kolivan, divided
into narrow parallel beds, there are numerous crystals of feldspar and
albite, and a few of titanium (Humboldt, *Asie Centrale*, t. i., p. 295,
Gustav Rose, *Reise nach dem Ural*, bd. i., s. 524).

the Altai Mountains, and at Las Trincheras, on the slope of
the littoral chain of Caraccas,* I have seen granite divided
into ledges, owing probably to a similar contraction, although
the divisions appeared to penetrate far into the interior. Further to the south of Lake Kolivan, toward the boundaries of
the Chinese province Ili (between Buchtarminsk and the
River Narym), the formation of the erupted rock, in which
there is no gneiss, is more remarkable than I ever observed in
any other part of the earth. The granite, which is always
covered with scales and characterized by tabular divisions,
rises in the steppes, either in small hemispherical eminences,
scarcely six or eight feet in height, or like basalt, in mounds,
terminating on either side of their bases in narrow streams.†
At the cataracts of the Orinoco, as well as in the district
of the Fichtelgebirge (Seissen), in Galicia, and between the
Pacific and the highlands of Mexico (on the Papagallo), I
have seen granite in large, flattened spherical masses, which
could be divided, like basalt, into concentric layers. In the
valley of Irtysch, between Buchtarminsk and Ustkamenogorsk,
granite covers transition slate for a space of four miles,‡ penetrating into it from above in narrow, variously ramified,
wedge-like veins. I have only instanced these peculiarities
in order to designate the individual character of one of the
most generally diffused erupted-rocks. As granite is superposed on slate in Siberia and in the Département de Finisterre
(Isle de Mihau), so it covers the Jura limestone in the mountains of Oisons (Fermonts), and syenite, and indirectly also
chalk, in Saxony, near Weinböhla.§ Near Mursinsk, in the
Uralian district, granite is of a drusous character, and here
the pores, like the fissures and cavities of recent volcanic products, inclose many kinds of magnificent crystals, especially
beryls and topazes.

2. *Quartzose porphyry* is often found in the relation of
veins to other rocks. The base is generally a finely granular
mixture of the same elements which occur in the larger im-

---

* Humboldt, *Relation Historique*, t. ii., p. 99.

† See the sketch of Biri-tau, which I took from the south side, where
the Kirghis tents stood, and which is given in Rose's *Reise*, bd. i., s. 584.
On spheres of granite scaling off concentrically, see my *Relat. Hist.*, t.
ii., p. 497, and *Essai Géogn. sur les Gisement des Roches*, p. 78.

‡ Humboldt, *Asie Centrale*, t. i., p. 299–311, and the drawings in
Rose's *Reise*, bd. i., s. 611, in which we see the curvature in the layers
of granite which Leop. von Buch has pointed out as characteristic.

§ This remarkable superposition was first described by Weiss in
Karsten's *Archiv für Bergbau und Hüttenwesen*, bd. xvi., 1827, s. 5.

bedded crystals. In granitic porphyry that is very poor in quartz, the feldspathic base is almost granular and laminated.*

3. *Greenstones, Diorite,* are granular mixtures of white albite and blackish-green hornblende, forming dioritic porphyry when the crystals are deposited in a base of denser tissue. The greenstones, either pure, or inclosing laminæ of diallage (as in the Fichtelgebirge), and passing into serpentine, have sometimes penetrated, in the form of strata, into the old stratified fissures of green argillaceous slate, but they more frequently traverse the rocks in veins, or appear as globular masses of greenstone, similar to domes of basalt and porphyry.†

Hypersthene rock is a granular mixture of labradorite and hypersthene.

Euphotide and serpentine, containing sometimes crystals of augite and uralite instead of diallage, are thus nearly allied to another more frequent, and, I might almost say, more *en ergetic* eruptive rock—augitic porphyry.‡

Melaphyre, augitic, uralitic, and oligoklastic porphyries To the last-named species belongs the genuine *verd-antique,* so celebrated in the arts.

Basalt, containing olivine and constituents which gelatinize in acids ; phonolithe (porphyritic slate), trachyte, and dolerite ; the first of these rocks is only partially, and the second always, divided into thin laminæ, which give them an appearance of stratification when extended over a large space. Mesotype and nepheline constitute, according to Girard, an important part in the composition and internal texture of basalt. The nepheline contained in basalt reminds the geognosist both of the miascite of the Ilmen Mountains in the Ural,§ which has been confounded with granite, and sometimes contains zirconium, and of the pyroxenic nepheline discovered by Gumprecht near Lobau and Chemnitz.

To the second or sedimentary rocks belong the greater part of the formations which have been comprised under the old

---

* Dufrenoy et Elie de Beaumont, *Géologie de la France,* t. i., p. 130.

† These intercalated beds of diorite play an important part in the mountain district of Nailau, near Steben, where I was engaged in mining operations in the last century, and with which the happiest associations of my early life are connected. Compare Hoffmann, in Poggendorf's *Annalen,* bd. xvi., s. 558.

‡ In the southern and Bashkirian portion of the Ural. Rose, *Reise,* bd. ii., s. 171.

§ G. Rose, *Reise nach dem Ural,* bd. ii., s. 47–52. Respecting the identity of eleolite and nepheline (the latter containing rather the more lime), see Scheerer, in Poggend., *Annalen,* bd. xlix., s. 359–381.

systematic, but not very correct designation of *transition, flötz* or *secondary*, and *tertiary formations*. If the erupted rocks had not exercised an elevating, and, owing to the simultane· ous shock of the earth, a disturbing influence on these sedimentary formations, the surface of our planet would have consisted of strata arranged in a uniformly horizontal direction above one another. Deprived of mountain chains, on whose declivities the gradations of vegetable forms and the scale of the diminishing heat of the atmosphere appear to be picturesquely reflected—furrowed only here and there by valleys of erosion, formed by the force of fresh water moving on in gentle undulations, or by the accumulation of detritus, resulting from the action of currents of water—continents would have presented no other appearance from pole to pole than the dreary uniformity of the llanos of South America or the steppes of Northern Asia. The vault of heaven would every where have appeared to rest on vast plains, and the stars to rise as if they emerged from the depths of ocean. Such a condition of things could not, however, have generally prevailed for any length of time in the earlier periods of the world, since subterranean forces must have striven in all epochs to exert a counteracting influence.

Sedimentary strata have been either precipitated or deposited from liquids, according as the materials entering into their composition are supposed, whether as limestone or argillaceous slate, to be either chemically dissolved or suspended and commingled. But earths, when dissolved in fluids impregnated with carbonic acid, must be regarded as undergoing a mechanical process while they are being precipitated, deposited, and accumulated into strata. This view is of some importance with respect to the envelopment of organic bodies in petrifying calcareous beds. The most ancient sediments of the transition and secondary formations have probably been formed from water at a more or less high temperature, and at a time when the heat of the upper surface of the earth was still very considerable. Considered in this point of view, a Plutonic action seems to a certain extent also to have taken place in the sedimentary strata, especially the more ancient ; but these strata appear to have been hardened into a schistose structure, and under great pressure, and not to have been solidified by cooling, like the rocks that have issued from the interior, as, for instance, granite, porphyry, and basalt. By degrees, as the waters lost their temperature, and were able to absorb a copious supply of the carbonic acid gas with which the atmosphere was overcharged, they became fitted to hold in solution a larger quantity of lime.

The sedimentary strata, setting aside all other exogenous, purely mechanical deposits of sand or detritus, are as follows :

Schist, of the lower and upper transition rock, composing the silurian and devonian formations ; from the lower silurian strata, which were once termed cambrian, to the upper strata of the old red sandstone or devonian formation, immediately in contact with the mountain limestone.

Carboniferous deposits :

Limestones imbedded in the transition and carboniferous formations ; zechstein, muschelkalk, Jura formation and chalk, also that portion of the tertiary formation which is not included in sandstone and conglomerate.

Travertine, fresh-water limestone, and silicious concretions of hot springs, formations which have not been produced under the pressure of a large body of sea water, but almost in immediate contact with the atmosphere, as in shallow marshes and streams.

Infusorial deposits : geognostical phenomena, whose great importance in proving the influence of organic activity in the formation of the solid part of the earth's crust was first discovered at a recent period by my highly-gifted friend and fellow-traveler, Ehrenberg.

If, in this short and superficial view of the mineral constituents of the earth's crust, I do not place immediately after the simple sedimentary rocks the conglomerates and sandstone formations which have also been deposited as sedimentary strata from liquids, and which have been imbedded alternately with schist and limestone, it is only because they contain, together with the detritus of eruptive and sedimentary rocks, also the detritus of gneiss, mica slate, and other metamorphic masses.   The obscure process of this metamorphism, and the action it produces, must therefore compose the third class of the fundamental forms of rock.

Endogenous or erupted rocks (granite, porphyry, and melaphyre) produce, as I have already frequently remarked, not only dynamical, shaking, upheaving actions, either vertically or laterally displacing the strata, but they also occasion changes in their chemical composition as well as in the nature of their internal structure ; new rocks being thus formed, as gneiss, mica slate, and granular limestone (Carrara and Parian marble).   The old silurian or devonian transition schists, the belemnitic limestone of Tarantaise, and the dull gray calcareous sandstone (*Macigno*), which contains algæ found in the northern Apennines, often assume a new and more brilliant appearance after their metamorphosis, which renders it difficult to recognize them. The theory of metamorphism was not established until the individual phases of the change were followed step by step, and direct chemical experiments on the difference in the fusion point, in the pressure and time of cooling, were brought in aid of mere inductive conclusions. Where the study of chemical combinations is regulated by leading ideas,* it may be the means of throwing a clear light on the wide field of geognosy, and over the vast laboratory of nature in which rocks are continually being formed and modified by the agency of subterranean forces. The philosophical inquirer will escape the deception of apparent analogies, and the danger of being led astray by a narrow view of natural phenomena, if he constantly bear in view the complicated conditions which may, by the intensity of their force, have modified the counteracting effect of those individual substances whose nature is better known to us. Simple bodies have, no doubt, at all periods, obeyed the same laws of attraction, and, wherever apparent contradictions present themselves, I am confident that chemistry will in most cases be able to trace the cause to some corresponding error in the experiment.

Observations made with extreme accuracy over large tracts of land, show that erupted rocks have not been produced in an irregular and unsystematic manner. In parts of the globe most remote from one another, we often find that granite, basalt, and diorite have exercised a regular and uniform metamorphic action, even in the minutest details, on the strata of argillaceous slate, dense limestone, and the grains of quartz in sandstones. As the same endogenous rock manifests almost every where the same degree of activity, so, on the contrary, different rocks belonging to the same class, whether to the endogenous or the erupted, exhibit great differences in their character. Intense heat has undoubtedly influenced all these phenomena, but the degree of fluidity (the more or less perfect mobility of the particles—their more viscous composition) has varied very considerably from the granite to the basalt, while at different geo-

* See the admirable researches of Mitscherlich, in the *Abhandl. der Berl. Akad.* for the years 1822 and 1823, s. 25–41; and in Poggend., *Annalen*, bd. x., s. 137–152; bd. xi., s. 323–332; bd. xli., s. 213–216 (Gustav Rose, *Ueber Bildung des Kalkspaths und Aragonits*, in Poggend., *Annalen*, bd. xli., s, 353–366; Haidinger, in the *Transactions of the Royal Society of Edinburgh*, 1827, p. 148.)

.ogical periods (or metamorphic phases of the earth's crust)
other substances dissolved in vapors have issued from the interior of the earth simultaneously with the eruption of granite,
basalt, greenstone porphyry, and serpentine. This seems a
fitting place again to draw attention to the fact that, according to the admirable views of modern geognosy, the metamorphism of rocks is not a mere phenomenon of contact, limited to the effect produced by the apposition of two rocks, since
it comprehends all the generic phenomena that have accompanied the appearance of a particular erupted mass. Even
where there is no immediate contact, the proximity of such a
mass gives rise to modifications of solidification, cohesion, granulation, and crystallization.

All eruptive rocks penetrate, as ramifying veins, either into
the sedimentary strata, or into other equally endogenous masses ; but there is a special importance to be attached to the
difference manifested between *Plutonic* rocks* (granite, porphyry, and serpentine) and those termed *volcanic* in the strict
sense of the word (as trachyte, basalt, and lava). The rocks
produced by the activity of our present volcanoes appear as
band-like streams, but by the confluence of several of them
they may form an extended basin. Wherever it has been
possible to trace basaltic eruptions, they have generally been
found to terminate in slender threads. Examples of these
narrow openings may be found in three places in Germany :
in the " *Pflaster-kaute*," at Marksuhl, eight miles from Ei-
senach ; in the blue " *Kuppe*," near Eschwege, on the banks
of the Werra ; and in the Druidical stone on the Hollert road
(Siegen), where the basalt has broken through the variegated
sandstone and graywacke slate, and has spread itself into cup-
like fungoid enlargements, which are either grouped together
like rows of columns, or are sometimes stratified in thin laminæ. The case is otherwise with granite, syenite, quartzose
porphyry, serpentine, and the whole series of unstratified compact rocks, to which, from a predilection for a mythological
nomenclature, the term Plutonic has been applied. These,
with the exception of occasional veins, were probably not
erupted in a state of fusion, but merely in a softened condition ; not from narrow fissures, but from long and widely-ex-
tending gorges. They have been protruded, but have not
flowed forth, and are found, not in streams like lava, but in
extended masses.† Some groups of dolerite and trachyte in-

* [Lyell, *Principles of Geology*, vol. i.i., p. 353 and 359.]—*Tr*
† The description here given of the relations of position under which

dicate a certain degree of basaltic fluidity ; others, which have
been expanded into vast craterless domes, appear to have been
only in a softened condition at the time of their elevation.
Other trachytes, like those of the Andes, in which I have frequently perceived a striking analogy with the greenstones and
syenitic porphyries (which are argentiferous, and without
quartz), are deposited in the same manner as granite and
quartzose porphyry.

Experiments on the changes which the texture and chemical constitution of rocks experience from the action of heat,
have shown that volcanic masses* (diorite, augitic porphyry,
basalt, and the lava of Ætna) yield different products, according to the difference of the pressure under which they have
been fused, and the length of time occupied during their cooling ; thus, where the cooling was rapid, they form a black
glass, having a homogeneous fracture, and where the cooling
was slow, a stony mass of granular crystalline structure.   In
the latter case, the crystals are formed partly in cavities and
partly inclosed in the matrix.   The same materials yield the
most dissimilar products, a fact that is of the greatest import
ance in reference to the study of the nature of erupted rocks, and
of the metamorphic action which they occasion.   Carbonate of
lime, when fused under great pressure, does not lose its carbonic
acid, but becomes, when cooled, granular limestone ; when
the crystallization has been effected by the dry method, saccharoidal marble ; while by the humid method, calcareous
spar and aragonite are produced, the former under a lesser degree of temperature than the latter.†   Differences of temper-

granite occurs, expresses the general or leading character of the whole
formation.   But its aspect at some places leads to the belief that it was
occasionally more fluid at the period of its eruption.   The description
given by Rose, in his *Reise nach dem Ural*, bd. i., s. 599, of part of the
Narym chain, near the frontiers of the Chinese territories, as well as the
evidence afforded by trachyte, as described by Dufrénoy and Elie de
Beaumont, in their *Description Géologique de la France*, t. i., p. 70.
Having already spoken in the text of the narrow apertures through
which the basalts have sometimes been effused, I will here notice the
large fissures, which have acted as conducting passages for melaphyres,
which must not be confounded with basalts.   See Murchison's interesting account (*The Silurian System*, p. 126) of a fissure 480 feet wide,
through which melaphyre has been ejected, at the coal-mine at Corn-
brook, Hoar Edge.
    * Sir James Hall, in the *Edin. Trans.*, vol. v., p. 43, and vol. vi., p.
71; Gregory Watt, in the *Phil. Trans. of the Roy. Soc. of London for*
1804, Part ii., p. 279 ; Dartigues and Fleurieu de Bellevue, in the *Journal de Physique*, t. lx., p. 456; Bischof, *Wärmelehre*, s. 313 und 443.
    † Gustav Rose, in Poggend., *Annalen*, bd. xlii., s 364.

arure likewise modify the direction in which the different particles arrange themselves in the act of crystallization, and also
affect the form of the crystal.* Even when a body is not in
a fluid condition, the smallest particles may undergo certain
relations in their various modes of arrangement, which are
manifested by the different action on light.† The phenomena presented by devitrification, and by the formation of steel
by cementation and casting—the transition of the fibrous into
the granular tissue of the iron, from the action of heat,‡ and
probably, also, by regular and long-continued concussions—
likewise throw a considerable degree of light on the geological
process of metamorphism. Heat may even simultaneously induce opposite actions in crystalline bodies ; for the admirable
experiments of Mitscherlich have established the fact§ that
calcareous spar, without altering its condition of aggregation,
expands in the direction of one of its axes and contracts in
the other.

If we pass from these general considerations to individual
examples, we find that schist is converted, by the vicinity of
Plutonic erupted rocks, into a bluish-black, glistening roofing
slate. Here the planes of stratification are intersected by another system of divisional stratification, almost at right angles
with the former,‖ and thus indicating an action subsequent to
the alteration. The penetration of silica causes the argillaceous schist to be traversed by quartz, transforming it, in part,
into whetstone and silicious schist ; the latter sometimes containing carbon, and being then capable of producing galvanic
effects on the nerves. The highest degree of silicification of
schist is that observed in ribbon jasper, a material highly valuable in the arts,¶ and which is produced in the Oural Mount-

* On the dimorphism of sulphur, see Mitscherlich, *Lehrbuch der
Chemie,* § 55–63.

† On gypsum as a uniaxal crystal, and on the sulphate of magnesia,
and the oxyds of zinc and nickel, see Mitscherlich, in Poggend., *Annalen,* bd. xi., s. 328.

‡ Coste, *Versuche am Creusot über das brüchig werden des Stabeisens.*
Elie de Beaumont, *Mém. Géol.,* t. ii., p. 411.

§ Mitscherlich, *Ueber die Ausdehnung der Krystallisirten Körper durch
die Wärmelehre,* in Poggend., *Annalen,* bd. x., s. 151.

‖ On the double system of divisional planes, see Elie de Beaumont,
Géologie de la France, p. 41 ; Credner, *Geognosie Thüringens und des
Harzes,* s. 40 ; and Römer, *Das Rheinische Uebergangsgebirge,* 1844.
s. 5 und 9.

¶ The silica is not merely colored by peroxyd of iron, but is accompanied by clay, lime, and potash. Rose, *Reise,* bd. ii.. s. 187. On the
formation of jasper by the action of dioritic porphyry, augite, and by

ains by the contact and eruption of augitic porphyry (at Orsk), of dioritic porphyry (at Aufschkul), or of a mass of hyper-sthenic rock conglomerated into spherical masses (at Bogoslowsk). At Monte Serrato, in the island of Elba, according to Frederic Hoffman, and in Tuscany, according to Alexander Brongniart, it is formed by contact with euphotide and serpentine.

The contact and Plutonic action of granite have sometimes made argillaceous schist granular, as was observed by Gustav Rose and myself in the Altai Mountains (within the fortress of Buchtarminsk),* and have transformed it into a mass resembling granite, consisting of a mixture of feldspar and mica, in which larger laminæ of the latter were again imbedded.† Most geognosists adhere, with Leopold von Buch, to the well-known hypothesis " that all the gneiss in the silurian strata of the transition formation, between the Icy Sea and the Gulf of Finland, has been produced by the metamorphic action of granite.‡ In the Alps, at St. Gothard, calcareous marl is likewise changed from granite into mica slate, and then transformed into gneiss." Similar phenomena of the formation of gneiss and mica slate through granite present themselves in the oolitic group of the Tarantaise,§ in which belemnites are

persthene rock, see Rose, bd. ii., s. 169, 187, und 192. See, also, bd. i., s. 427, where there is a drawing of the porphyry spheres between which jasper occurs, in the calcareous graywacke of Bogoslowsk, being produced by the Plutonic influence of the augitic rock; bd. ii., s. 545; and likewise Humboldt, *Asie Centrale*, t. i., p. 486.

* Rose, *Reise nach dem Ural*, bd. i., s. 586–588.

† In respect to the volcanic origin of mica, it is important to notice that crystals of mica are found in the basalt of the Bohemian Mittelge-birge, in the lava that in 1822 was ejected from Vesuvius (Monticelli, *Storia del Vesuvio negli Anni* 1821 *e* 1822, § 99), and in fragments of argillaceous slate imbedded in scoriaceous basalt at Hohenfels, not far from Gerolstein, in the Eifel (see Mitscherlich, in Leonhard, *Basalt-Gebilde*, s. 244). On the formation of feldspar in argillaceous schist, through contact with porphyry, occurring between Urval and Poïet (Forez), see Dufrénoy, in *Géol. de la France*, t. i., p. 137. It is probably to a similar contact that certain schists near l'aimpol, in Brittany, with whose appearance I was much struck, while making a geological pedestrian tour through that interesting country with Professor Kunth, owe their amygdaloid and cellular character, t. i., p. 234.

‡ Leopold von Buch, in the *Abhandlungen der Akad. der Wissen-schaft zu Berlin, aus dem Jahr* 1842, s. 63, and in the *Jahrbüchern für Wissenschaftliche Kritik Jahrg.* 1840, s. 196.

§ Elie de Beaumont, in the *Annales des Sciences Naturelles*, t. xv., p. 362–372. "In approaching the primitive masses of Mont Rosa, and the mountains situated to the west of Coni, we perceive that the secondary strata gradually lose the characters inherent in their mode of deposition. Frequently assuming a character apparently arising from a perfectly

found in rocks, which have some claim to be considered as mica slate, and in the schistose group in the western part of the island of Elba, near the promontory of Calamita, and the Fichtelgebirge in Baireuth, between Lomitz and Markleiten.[*]

Jasper, which,[†] as I have already remarked, is a production formed by the volcanic action of augitic porphyry, could only be obtained in small quantities by the ancients, while another material, very generally and efficiently used by them in the arts, was granular or saccharoidal marble, which is likewise to be regarded solely as a sedimentary stratum altered by terrestrial heat and by proximity with erupted rocks. This opinion is corroborated by the accurate observations on the phenomena of contact, by the remarkable experiments on fusion

distinct cause, but not losing their stratification, they somewhat resemble in their physical structure a brand of half-consumed wood, in which we can follow the traces of the ligneous fibers beyond the spots which continue to present the natural characters of wood." (See, also, the *Annales des Sciences Naturelles*, t. xiv., p. 118–122, and von Dechen, *Geognosie*, s. 553.) Among the most striking proofs of the transformation of rocks by Plutonic action, we must place the belemnites in the schists of Nuffenen (in the Alpine valley of Eginen and in the Griesglaciers), and the belemnites found by M. Charpentier in the so-called primitive limestone on the western descent of the Col de la Seigne, between the Enclove de Monjovet and the *châlet* of La Lanchette, and which he showed to me at Bex in the autumn of 1822 (*Annales de Chimie*, t. xxiii., p. 262).

* Hoffmann, in Poggend., *Annalen*, bd. xvi., s. 552, "Strata of transition argillaceous schist in the Fichtelgebirge, which can be traced for a length of 16 miles, are transformed into gneiss only at the two extremities, where they come in contact with granite. We can there follow the gradual formation of the gneiss, and the development of the mica and of the feldspathic amygdaloids, in the interior of the argillaceous schist, which indeed contains in itself almost all the elements of these substances."

† Among the works of art which have come down to us from the ancient Greeks and Romans, we observe that none of any size—as columns or large vases—are formed from jasper; and even at the present day, this substance, in large masses, is only obtained from the Ural Mountains. The material worked as jasper from the Rhubarb Mountain (Raveniaga Sopka), in Altai, is a beautiful ribboned porphyry. The word *jasper* is derived from the Semitic languages; and from the confused descriptions of Theophrastus (*De Lapidibus*, 23 and 27) and Pliny (xxxvii., 8 and 9), who rank jasper among the "opaque gems," the name appears to have been given to fragments of *jaspachat*, and to a substance which the ancients termed *jasponyx*, which we now know as *opal-jasper*. Pliny considers a piece of jasper eleven inches in length so rare as to require his mentioning that he had actually seen such a specimen: "Magnitudinem jaspidis undecim unciarum vidimus, formatamque inde effigiem Neronis thoracatam." According to Theophrastus, the stone which he calls emerald, and from which large obelisks were cut, must have been an imperfect jasper.

made by Sir James Hall more than half a century ago, and
by the attentive study of granitic veins, which has contributed
so largely to the establishment of modern geognosy.   Sometimes the erupted rock has not transformed the compact into
granular limestone to any great depth from the point of contact.   Thus, for instance, we meet with a slight transformation—a penumbra—as at Belfast, in Ireland, where the basaltic veins traverse the chalk, and, as in the compact calcareous beds, which have been partially inflected by the contact of syenitic granite, at the Bridge of Boscampo and the
Cascade of Conzocoli, in the Tyrol (rendered celebrated by
the mention made of it by Count Mazari Peucati).*   Another
mode of transformation occurs where all the strata of the compact limestone have been changed into granular limestone by
the action of granite, and syenitic or dioritic porphyry.†

I would here wish to make special mention of Parian and
Carrara marbles, which have acquired such celebrity from the
noble works of art into which they have been converted, and
which have too long been considered in our geognostic collections as the main types of primitive limestone.   The action
of granite has been manifested sometimes by immediate contact, as in the Pyrenees,‡ and sometimes, as in the main land
of Greece, and in the insular groups in the Ægean Sea, through
the intermediate layers of gneiss or mica slate.   Both cases
presuppose a simultaneous but heterogeneous process of trans

---

* Humboldt, *Lettre à M. Brochant de Villiers*, in the *Annales de
Chimie et de Physique*, t. xxiii., p. 261; Leop. von Buch, *Geog. Briefe
über das südliche Tyrol*, s. 101, 105, und 273.

† On the transformation of compact into granular limestone by the
action of granite, in the Pyrenees at the *Montagnes de Rancie*, see
Dufrénoy, in the *Mémoires Géologiques*, t. ii., p. 440; and on similar
changes in the *Montagnes de l'Oisans*, see Elie de Beaumont, in the
Mém. Géolog., t. ii., p. 379–415; on a similar effect produced by the
action of dioritic and pyroxenic porphyry (the *ophite* described by Elie
de Beaumont, in the *Géologie de la France*, t. i., p. 72), between Tolosa
and St. Sebastian, see Dufrénoy, in the *Mém. Géolog.*, t. ii., p. 130; and
by syenite in the Isle of Skye, where the fossils in the altered limestone
may still be distinguished, see Von Dechen, in his *Géognosie*, p. 573.
In the transformation of chalk by contact with basalt, the transposition
of the most minute particles in the processes of crystallization and
granulation is the more remarkable, because the excellent microscopic
investigations of Ehrenberg have shown that the particles of chalk previously existed in the form of closed rings.   See Poggend., *Annalen der
Physik*, bd. xxxix., s. 105; and on the rings of aragonite deposited
from solution, see Gustav Rose in vol. xlii., p. 354, of the same journal.

‡ Beds of granular limestone in the granite at Port d'Oo and in the
Mont de Labourd.   See Charpentier, *Constitution Géologique des Pyré-
nées*, p. 144, 146.

formation. In Attica, in the island of Euboea, and in the Peloponnesus, it has been remarked, "that the limestone, when superposed on mica slate, is beautiful and crystalline in proportion to the purity of the latter substance and to the smallness of its argillaceous contents ; and, as is well known, this rock, together with beds of gneiss, appears at many points, at a considerable depth below the surface, in the islands of Paros and Antiparos."* We may here infer the existence of an imperfectly metamorphosed flötz formation, if faith can be yielded to the testimony of Origen, according to whom, the ancient Eleatic, Xenophanes of Colophon† (who supposed the whole earth's crust to have been once covered by the sea), de-clared that marine fossils had been found in the quarries of Syracuse, and the impression of a fish (a sardine) in the deepest rocks of Paros. The Carrara or Luna marble quarries, which constituted the principal source from which statuary marble was derived even prior to the time of Augustus, and which will probably continue to do so until the quarries of Paros shall be reopened, are beds of calcareous sandstone—macigno —altered by Plutonic action, and occurring in the insulated mountain of Apuana, between gneiss-like mica and talcose schist.‡ Whether at some points granular limestone may not have been formed in the interior of the earth, and been raised by gneiss and syenite to the surface, where it forms vein-like fissures,§ is a question on which I can not hazard an opinion, owing to my own want of personal knowledge of the subject.

* Leop. von Buch, *Descr. des Canaries*, p. 394 ; Fiedler, *Reise durch das Königreich Griechenland*, th. ii., s., 181, 190, und 516.

† I have previously alluded to the remarkable passage in Origen's *Philosophumena*, cap. 14 (*Opera*, ed. Delarue, t. i., p. 893). From the whole context, it seems very improbable that Xenophanes meant an impression of a laurel (τυπον δάφνες) instead of an impression of a fish (τύπον ἀφύης). Delarue is wrong in blaming the correction of Jacob Gronovius in changing the laurel into a sardel. The petrifaction of a fish is also much more probable than the natural picture of Silenus, which, according to Pliny (lib. xxxvi., 5), the quarry-men are stated to have met with in Parian marble from Mount Marpessos. *Servius ad Virg., Æn.*, vi., 471.

‡ On the geognostic relations of Carrara (*The City of the Moon*, Strabo, lib. v., p. 222), see Savi, *Osservazioni sui terreni antichi Toscani*, in the *Nuovo Giornale de' Letterati di Pisa*, and Hoffmann, in Karsten's *Archiv für Mineralogie*, bd. vi., s. 258–263, as well as in his *Geogn Reise durch Italien*, s. 244–265.

§ According to the assumption of an excellent and very experienced observer, Karl von Leonhard. See his *Jahrbuch für Mineralogie*, 1834 s. 329, and Bernhard Cotta, *Geognosie*, s. 310.

According to the admirable observations of Leopold von Buch, the masses of dolomite found in Southern Tyrol, and on the Italian side of the Alps, present the most remarkable instance of metamorphism produced by massive eruptive rocks on compact calcareous beds. This formation of the limestone seems to have proceeded from the fissures which traverse it in all directions. The cavities are every where covered with rhomboidal crystals of magnesian bitter spar, and the whole formation, without any trace of stratification, or of the fossil remains which it once contained, consists only of a granular aggregation of crystals of dolomite. Talc laminæ lie scattered here and there in the newly-formed rock, traversed by masses of serpentine. In the valley of the Fassa, dolomite rises perpendicularly in smooth walls of dazzling whiteness to a height of many thousand feet. It forms sharply-pointed conical mountains, clustered together in large numbers, but yet not in contact with each other. The contour of their forms recalls to mind the beautiful landscape with which the rich imagination of Leonardi da Vinci has embellished the back-ground of the portrait of Mona Lisa.

The geognostic phenomena which we are now describing, and which excite the imagination as well as the powers of the intellect, are the result of the action of augitic porphyry manifested in its elevating, destroying, and transforming force.[*] The process by which limestone is converted into dolomite is not regarded by the illustrious investigator who first drew attention to the phenomenon as the consequence of the talc being derived from the black porphyry, but rather as a transformation simultaneous with the appearance of this erupted stone through wide fissures filled with vapors. It remains for future inquirers to determine how transformation can have been effected without contact with the endogenous stone, where strata of dolomite are found to be interspersed in limestone. Where, in this case, are we to seek the concealed channels by which the Plutonic action is conveyed? Even here it may not, however, be necessary, in conformity with the old Roman adage, to believe " that much that is alike in nature may have been formed in wholly different ways." When we find, over widely extended parts of the earth, that two phenomena are always associated together, as, for instance, the occurrence of mela-

---

[*] Leop. von Buch, *Geognostische Briefe an Alex. von Humboldt,* 1824, s. 86 and 82; also in the *Annalen de Chemie,* t. xxiii., p. 276, and in the *Abhandl. der Berliner Akad. aus der Jahren* 1822 *und* 1823, s. 83–136; Von Dechen, *Geognosie.* s. 574–576.

phyre and the transformation of compact limestone into a crystalline mass differing in its chemical character, we are, to a certain degree, justified in believing, where the second phenomenon is manifested unattended by the appearance of the first, that this apparent contradiction is owing to the absence, in certain cases, of some of the conditions attendant upon the exciting causes. Who would call in question the volcanic nature and igneous fluidity of basalt merely because there are some rare instances in which basaltic veins, traversing beds of coal or strata of sandstone and chalk, have not materially deprived the coal of its carbon, nor broken and slacked the sandstone, nor converted the chalk into granular marble? Wherever we have obtained even a faint light to guide us in the obscure domain of mineral formation, we ought not ungratefully to disregard it, because there may be much that is still unexplained in the history of the relations of the transitions, or in the isolated interposition of beds of unaltered strata.

After having spoken of the alteration of compact carbonate of lime into granular limestone and dolomite, it still remains for us to mention a third mode of transformation of the same mineral, which is ascribed to the emission, in the ancient periods of the world, of the vapors of sulphuric acid. This transformation of limestone into gypsum is analogous to the penetration of rock salt and sulphur, the latter being deposited from sulphureted aqueous vapor. In the lofty Cordilleras of Quindiu, far from all volcanoes, I have observed deposits of sulphur in fissures in gneiss, while in Sicily (at Cattolica, near Girgenti), sulphur, gypsum, and rock salt belong to the most recent secondary strata, the chalk formations.* I have also seen, on the edge of the crater of Vesuvius, fissures filled with rock salt, which occurred in such considerable masses as occasionally to lead to its being disposed of by contraband trade. On both declivities of the Pyrenees, the connection of diorite and pyroxene, and dolomite, gypsum, and rock salt, can not be questioned;† and here, as in the other phenomena which we have been considering, every thing bears evidence of the action of subterranean forces on the sedimentary strata of the ancient sea.

There is much difficulty in explaining the origin of the beds of pure quartz, which occur in such large quantities in South America, and impart so peculiar a character to the chain of

---

* Hoffman, *Geogn. Reise*, edited by Von Dechen, s. 113–119, and 380–386; Poggend., *Annalen der Physik*, bd. xxvi., s. 41.

† Dufrénoy, in the *Mémoires Géologiques*, t. ii., p. 145 and 179.

the Andes.* In descending toward the South Sea, from Cax-amarca toward Guangamarca, I have observed vast masses of quartz, from 7000 to 8000 feet in height, superposed some-times on porphyry devoid of quartz, and sometimes on diorite. Can these beds have been transformed from sandstone, as Elie de Beaumont conjectures in the case of the quartz strata on the Col de la Poissonnière, east of Briançon ?† In the Brazils, in the diamond district of Minas Geraes and St. Paul, which has recently been so accurately investigated by Clausen, Plutonic action has developed in dioritic veins sometimes ordi-nary mica, and sometimes specular iron in quartzose itacol-umite. The diamonds of Grammagoa are imbedded in strata of solid silica, and are occasionally enveloped in laminæ of mica, like the garnets found in mica slate. The diamonds that occur furthest to the north, as those discovered in 1829 at 58° lat., on the European slope of the Uralian Mountains, bear a geognostic relation to the black carboniferous dolomite of Adolffskoi‡ and to augitic porphyry, although more accu-rate observations are required in order fully to elucidate this subject.

Among the most remarkable phenomena of contact, we must, finally, enumerate the formation of garnets in argilla-ceous schist in contact with basalt and dolerite (as in Northum-berland and the island of Anglesea), and the occurrence of a vast number of beautiful and most various crystals, as garnets, vesuvian, augite, and ceylanite, on the surfaces of contact be-tween the erupted and sedimentary rock, as, for instance, on the junction of the syenite of Monzon with dolomite and com-pact limestone.§ In the island of Elba, masses of serpentine, which perhaps nowhere more clearly indicate the character of erupted rocks, have occasioned the sublimation of iron glance and red oxyd of iron in fissures of calcareous sandstone.‖ We still daily find the same iron glance formed by sublimation from the vapors and the walls of the fissures of open veins on the margin of the crater, and in the fresh lava currents of the volcanoes of Stromboli, Vesuvius, and Ætna.¶ The veins that

* Humboldt, *Essai Geogn. sur le Gisement des Roches*, p. 93 ; *Asie Centrale*, t. iii., p. 532.

† Elie de Beaumont, in the *Annales des Sciences Naturelles*, t. xv., p 362 ; Murchison, *Silurian System*, p. 286.

‡ Rose, *Reise nach dem Ural*, bd. i., s. 364 und 367.

§ Leop. von Buch, *Briefe*, s. 109–129. See, also, Elie de Beaumont, *On the Contact of Granite with the Beds of the Jura*, in the *Mém. Géol.* t. ii., p. 408.   ‖ Hoffman, *Reise*, s. 30 und 37.

¶ On the chemical process in the formation of specular iron, see Gay

are thus formed beneath our eyes by volcanic forces, where the contiguous rock has already attained a certain degree of solidification, show us how, in a similar manner, mineral and metallic veins may have been every where formed in the more ancient periods of the world, where the solid but thinner crust of our planet, shaken by earthquakes, and rent and fissured by the change of volume to which it was subjected in cooling, may have presented many communications with the interior, and many passages for the escape of vapors impregnated with earthy and metallic substances. The arrangement of the particles in layers parallel with the margins of the veins, the regular recurrence of analogous layers on the opposite sides of the veins (on their different walls), and, finally, the elongated cellular cavities in the middle, frequently afford direct evidence of the Plutonic process of sublimation in metalliferous veins. As the traversing rocks must be of more recent origin than the traversed, we learn from the relations of stratification existing between the porphyry and the argentiferous ores in the Saxon mines (the richest and most important in Germany), that these formations are at any rate more recent than the vegetable remains found in carboniferous strata and in the red sandstone.*

All the facts connected with our geological hypotheses on the formation of the earth's crust and the metamorphism of rocks have been unexpectedly elucidated by the ingenious idea which led to a comparison of the slags or scoriæ of our smelting furnaces with natural minerals, and to the attempt of reproducing the latter from their elements.† In all these operations, the same affinities manifest themselves which determine chemical combinations both in our laboratories and in the interior of the earth. The most considerable part of

Lussac, in the *Annales de Chimie*, t. xxii., p. 415, and Mitscherlich, in Poggend., *Annalen*, bd. xv., s. 630. Moreover, crystals of olivine have been formed (probably by sublimation) in the cavities of the obsidian of Cerro del Jacal, which I brought from Mexico (Gustav Rose, in Poggend., *Annalen*, bd. x., s. 323). Hence olivine occurs in basalt, lava, obsidian, artificial scoriæ, in meteoric stones, in the syenite of Elfdale, and (as hyalosiderite) in the wacke of the Kaiserstuhl.

* Constantin von Beust, *Ueber die Porphyrgebilde*, 1835, s. 89–96; also his *Beleuchtung der Werner'schen Gangtheorie*, 1840, s. 6; and C. von Wissenbach, *Abbildungen merkwürdiger Gangverhältnisse*, 1836, fig. 12. The ribbon-like structure of the veins is, however, no more to be regarded of general occurrence than the periodic order of the different members of these masses.

† Mitscherlich, *Ueber die künstliche Darstellung der Mineralien*, in the *Abhandl. der Akademie der Wiss. zu Berlin*, 1822–3, s. 25–41

the simple minerals which characterize the more generally diffused Plutonic and erupted rocks, as well as those on which they have exercised a metamorphic action, have been produced in a crystalline state, and with perfect identity, in artificial mineral products. We must, however, distinguish here between the scoriæ accidentally formed, and those which have been designedly produced by chemists. To the former belong feldspar, mica, augite, olivine, hornblende, crystallized oxyd of iron, magnetic iron in octahedral crystals, and metallic titanium ;* to the latter, garnets, idocrase, rubies (equal in hardness to those found in the East), olivine, and augite.† These minerals constitute the main constituents of granite, gneiss, and mica schist, of basalt, dolerite, and many porphyries. The artificial production of feldspar and mica is of most especial geognostic importance with reference to the theory of the formation of gneiss by the metamorphic agency of argillaceous schist, which contains all the constituents of granite,

* In scoriæ, crystals of feldspar have been discovered by Heine in the refuse of a furnace for copper fusing, near Sangerhausen, and analyzed by Kersten (Poggend., *Annalen*, bd. xxxiii., s. 337); crystals of augite in scoriæ, at Sahle (Mitscherlich, in the *Abhandl. der Akad. zu Berlin*, 1822–23, s. 40); of olivine by Seifström (Leonhard, *Basalt-Gebilde*, bd. ii., s. 495); of mica in old scoriæ of Schloss Garpenberg (Mitscherlich, in Leonhard, op. cit., s. 506); of magnetic iron in the scoriæ of Chatillon sur Seine (Leonhard, s. 441); and of micaceous iron in potter's clay (Mitscherlich, in Leonhard, op. cit., s. 234).
[See Ebelmer's papers in *Ann. de Chimie et de Physique*, 1847 ; also *Report on the Crystalline Slags*, by John Percy, M.D., F.R.S., and William Hallows Miller, M.A., 1847. Dr. Percy, in a communication with which he has kindly favored me, says that the minerals which he has found artificially produced and proved by analysis are Humboldtilite, gehlenite, olivine, and magnetic oxyd of iron, in octahedral crystals. He suggests that the circumstance of the production of gehlenite at a high temperature in an iron furnace may possibly be made available by geologists in explaining the formation of the rocks in which the natural mineral occurs, as in Fassathal in the Tyrol.]—*Tr.*
† Of minerals purposely produced, we may mention idocrase and garnet (Mitscherlich, in Poggend., *Annalen der Physik*, bd. xxxii., s. 340); ruby (Gaudin, in the *Comptes Rendus de l'Académie de Science*, t. iv., Part i., p. 999); olivine and augite (Mitscherlich and Berthier, in the *Annales de Chimie et de Physique*, t. xxiv., p. 376). Notwithstanding the greatest possible similarity in crystalline form, and perfect identity in chemical composition, existing, according to Gustav Rose, between augite and hornblende, hornblende has never been found accompanying augite in scoriæ, nor have chemists ever succeeded in artificially producing either hornblende or feldspar (Mitscherlich in Poggend., *Annalen*, bd. xxxiii., s. 340, and Rose, *Reise nach dem Ural*, bd. ii., s. 358 und 363). See, also, Beudant, in the *Mem. de l'Acad. des Sciences*, t. viii., p. 221, and Becquerel's ingenious experiments in his *Trait{ de l'Electricité*, t. i., p. 334 ; t. iii., p. 218; and t. v., p. 148 and 185

potash not excepted.* It would not be very surprising, therefore, as is well observed by the distinguished geognosist, Von
Dechen, if we were to meet with a fragment of gneiss formed
on the walls of a smelting furnace which was built of argillaceous slate and graywacke.

After having taken this general view of the three classes
of erupted, sedimentary, and metamorphic rocks of the earth's
crust, it still remains for us to consider the fourth class, comprising *conglomerates*, or *rocks of detritus*. The very term
recalls the destruction which the earth's crust has suffered,
and likewise, perhaps reminds us of the process of cementation,
which has connected together, by means of oxyd of iron, or of
some argillaceous and calcareous substances, the sometimes
rounded and sometimes angular portions of fragments. Conglomerates and rocks of detritus, when considered in the widest
sense of the term, manifest characters of a double origin. The
substances which enter into their mechanical composition have
not been alone accumulated by the action of the waves of the
sea or currents of fresh water, for there are some of these rocks
the formation of which can not be attributed to the action of
water. " When basaltic islands and trachytic rocks rise on
fissures, friction of the elevated rock against the walls of the
fissures causes the elevated rock to be inclosed by conglomerates composed of its own matter. The granules composing
the sandstones of many formations have been separated rather
by friction against the erupted volcanic or Plutonic rock than
destroyed by the erosive force of a neighboring sea. The existence of these friction *conglomerates*, which are met with in
enormous masses in both hemispheres, testifies the intensity
of the force with which the erupted rocks have been propelled
from the interior through the earth's crust. This detritus
has subsequently been taken up by the waters, which have
then deposited it in the strata which it still covers."† Sandstone formations are found imbedded in all strata, from the
lower silurian transition stone to the beds of the tertiary formations, superposed on the chalk. They are found on the
margin of the boundless plains of the New Continent, both
within and without the tropics, extending like breast-works
along the ancient shore, against which the sea once broke in
foaming waves.

* D'Aubuisson, in the *Journal de Physique*, t. lxviii., p. 128.
† Leop. von Buch, *Geognost. Briefe*, s. 75–82, where it is also shown
why the new red sandstone (the *Todtliegende* of the Thuringian flötz
formation) and the coal measures must be regarded as produced by
erupted porphyry.

If we cast a glance on the geographical distribution of rocks, and their relations in space, in that portion of the earth's crust which is accessible to us, we shall find that the most universally distributed chemical substance is *silicic acid*, generally in a variously-colored and opaque form. Next to solid silicic acid we must reckon carbonate of lime, and then the combinations of silicic acid with alumina, potash, and soda, with lime, magnesia, and oxyd of iron.

The substances which we designate as *rocks* are determinate associations of a small number of minerals, in which some combine parasitically, as it were, with others, but only under definite relations ; thus, for instance, although quartz (silica), feldspar, and mica are the principal constituents of granite, these minerals also occur, either individually or collectively, in many other formations. By way of illustrating how the quantitative relations of one feldspathic rock differ from another, richer in mica than the former, I would mention that, according to Mitscherlich, three times more alumina and one third more silica than that possessed by feldspar, give the constituents that enter into the composition of mica. Potash is contained in both—a substance whose existence in many kinds of rocks is probably antecedent to the dawn of vegetation on the earth's surface.

The order of succession, and the relative age of the different formations, may be recognized by the superposition of the sedimentary, metamorphic, and conglomerate strata ; by the nature of the formations traversed by the erupted masses, and —with the greatest certainty—by the presence of organic remains and the differences of their structure. The application of botanical and zoological evidence to determine the relative age of rocks—this chronometry of the earth's surface, which was already present to the lofty mind of Hooke—indicates one of the most glorious epochs of modern geognosy, which has finally, on the Continent at least, been emancipated from the sway of Semitic doctrines. Palæontological investigations have imparted a vivifying breath of grace and diversity to the science of the solid structure of the earth.

The fossiliferous strata contain, entombed within them, the floras and faunas of by-gone ages. We ascend the stream of time, as in our study of the relations of superposition we descend deeper and deeper through the different strata, in which lies revealed before us a past world of animal and vegetable life. Far-extending disturbances, the elevation of great mountain chains, whose relative ages we are able to define, attest the

destruction of ancient and the manifestation of recent organisms.   A few of these older structures have remained in the
midst of more recent species.   Owing to the limited nature of
our knowledge of existence, and from the figurative terms by
which we seek to hide our ignorance, we apply the appellation
recent structure to the historical phenomena of transition manifested in the organisms as well as in the forms of primitive
seas and of elevated lands.   In some cases these organized
structures have been preserved perfect in the minutest details
of tissues, integument, and articulated parts, while in others,
the animal, passing over soft argillaceous mud, has left nothing but the traces of its course,* or the remains of its undi-
gested food, as in the coprolites.†   In the lower Jura formations (the lias of Lyme Regis), the ink bag of the sepia has
been so wonderfully preserved, that the material, which myr-

---

* [In certain localities of the new red sandstone, in the Valley of the
Connecticut, numerous tridactyl markings have been occasionally observed on the surface of the slabs of stone when split asunder, in like
manner as the ripple-marks appear on the successive layers of sandstone
in Tilgate Forest.   Some remarkably distinct impressions of this kind,
at Turner's Falls (Massachusetts), happening to attract the attention of
Dr. James Deane, of Greenfield, that sagacious observer was struck
with their resemblance to the foot-marks left on the mud-banks of the
adjacent river by the aquatic birds which had recently frequented the
spot.   The specimens collected were submitted to Professor G. Hitch-
cock, who followed up the inquiry with a zeal and success that have
led to the most interesting results.   No reasonable doubt now exists
that the imprints in question have been produced by the tracks of bi-
peds impressed on the stone when in a soft state.   The announcement
of this extraordinary phenomenon was first made by Professor Hitch-
cock, in the *American Journal of Science* (January, 1836), and that
eminent geologist has since published full descriptions of the different
species of imprints which he has detected, in his splendid work on the
geology of Massachusetts.—Mantell's *Medals of Creation*, vol. ii., p. 810.
In the work of Dr. Mantell above referred to, there is, in vol. ii., p. 815,
an admirable diagram of a slab from Turner's Falls, covered with numerous foot-marks of birds, indicating the track of ten or twelve individuals of different sizes.]—*Tr.*
    † [From the examination of the fossils spoken of by geologists under
the name of *Coprolites*, it is easy to determine the nature of the food of
the animals, and some other points; and when, as happened occasionally, the animal was killed while the process of digestion was going on,
the stomach and intestines being partly filled with half-digested food,
and exhibiting the coprolites actually *in situ*, we can make out with
certainty not only the true nature of the food, but the proportionate size
of the stomach, and the length and nature of the intestinal canal.   Within the cavity of the rib of an extinct animal, the palæontologist thus
finds recorded, in indelible characters, some of those hieroglyphics upon
which he founds his history.—*The Ancient World*, by D. T. Ansted,
1847, p. 173.]—*Tr.*

iads of years ago might have served the animal to conceal itself from its enemies, still yields the color with which its image
may be drawn.* In other strata, again, nothing remains but
the faint impression of a muscle shell ; but even this, if it belong to a main division of mollusca,† may serve to show the
traveler, in some distant land, the nature of the rock in which
it is found, and the organic remains with which it is associated. Its discovery gives the history of the country in which it
occurs.

The analytic study of primitive animal and vegetable life
has taken a double direction : the one is purely morpholog-
ical, and embraces, especially, the natural history and physiology of organisms, filling up the chasms in the series of still
living species by the fossil structures of the primitive world.
The second is more specially geognostic, considering fossil remains in their relations to the superposition and relative age
of the sedimentary formations. The former has long predominated over the latter, and an imperfect and superficial comparison of fossil remains with existing species has led to errors,
which may still be traced in the extraordinary names applied
to certain natural bodies. It was sought to identify all fossil
species with those still extant in the same manner as, in the
sixteenth century, men were led by false analogies to compare the animals of the New Continent with those of the Old.
Peter Camper, Sömmering, and Blumenbach had the merit
of being the first, by the scientific application of a more ac-

---

* A discovery made by Miss Mary Anning, who was likewise the
discoverer of the coprolites of fish. These coprolites, and the excrements of the Ichthyosauri, have been found in such abundance in England (as, for instance, near Lyme Regis), that, according to Buckland's
expression, they lie like potatoes scattered in the ground. See Buckland, *Geology considered with reference to Natural Theology*, vol. i., p.
188–202 and 305. With respect to the hope expressed by Hooke "to
raise a chronology" from the mere study of broken and fossilized shells
"and to state the interval of time wherein such or such catastrophes
and mutations have happened," see his *Posthumous Works, Lecture,*
Feb. 29, 1688.

[Still more wonderful is the preservation of the substance of the animal of certain Cephalopodes in the Oxford clay. In some specimens
recently obtained, and described by Professor Owen, not only the ink
bag, but the muscular mantle, the head, and its crown of arms, are all
preserved in connection with the belemnite shell, while one specimen
exhibits the large eyes and the funnel of the animal, and the remains of
two fins, in addition to the shell and the ink bag. See Ansted's *Ancient
World*, p. 147.]— *Tr.*

† Leop. von Buch, in the *Abhandlungen der Akad. der Wiss. zu Berlin in dem Jahr* 1837, s. 64.

curate comparative anatomy, to throw light on the osteological branch of palæontology—the archæology of organic life ; but the actual geognostic views of the doctrine of fossil remains, the felicitous combination of the zoological character with the order of succession, and the relative ages of strata, are due to the labors of George Cuvier and Alexander Brongniart.

The ancient sedimentary formations and those of transition rocks exhibit, in the organic remains contained within them, a mixture of structures very variously situated on the scale of progressively-developed organisms. These strata contain but few plants, as, for instance, some species of Fuci, Lycopodiaceæ which were probably arborescent, Equisetaceæ, and tropical ferns ; they present, however, a singular association of animal forms, consisting of Crustacea (trilobites with reticulated eyes, and Calymene), Brachiopoda (*Spirifer, Orthis*), elegant Sphæronites, nearly allied to the Crinoidea,[*] Orthoceratites, of the family of the Cephalopoda, corals, and, blended with these low organisms, fishes of the most singular forms, imbedded in the upper silurian formations. The family of the Cephalaspides, whose fragments of the species *Pterichtys* were long held to be trilobites, belongs exclusively to the devonian period (the old red), manifesting, according to Agassiz, as peculiar a type among fishes as do the Ichthyosauri and Plesiosauri among reptiles.[†] The Goniatites, of the tribe of Ammonites,[‡] are manifested in the transition chalk, in the graywacke of the devonian periods, and even in the latest silurian formations.

The dependence of physiological gradation upon the age of the formations, which has not hitherto been shown with perfect certainty in the case of invertebrata,[§] is most regularly manifested in vertebrated animals. The most ancient of these, as we have already seen, are fishes ; next in the order of succession of formation, passing from the lower to the upper, come reptiles and mammalia. The first reptile (a Saurian, the Monitor of Cuvier), which excited the attention of Leibnitz,[||] is found in cuperiferous schist of the Zechstein of

* Leop. von Buch, *Gebirgsformationen von Russland*, 1840, s. 24–40.

† Agassiz, *Monographie des Poissons Fossiles du vieux Grès Rouge*, p. vi. and 4.

‡ Leop. von Buch, in the *Abhandl. der Berl. Akad.*, 1838, s. 149–168 ; Beyrich, *Beitr. zur Kenntniss des Rheinischen Uebergangsgebirges*, 1837, s. 45.

§ Agassiz, *Recherches sur les Poissons Fossiles*, t. i., *Introd.*, p. xviii. ; Davy, *Consolation in Travel*, dial. iii.

|| A Protosaurus, according to Hermann von Meyer. The rib of a

Thuringia ; the Palæosaurus and Thecodontosaurus of Bristol are, according to Murchison, of the same age.   The Saurians are found in large numbers in the muschelkalk,* in the
keuper, and in the oolitic formations, where they are the most
numerous.   At the period of these formations there existed
Plesiosauri, having long, swan-like necks consisting of thirty
vertebræ ; Megalosauri, monsters resembling the crocodile,
forty-five feet in length, and having feet whose bones were
like those of terrestrial mammalia, eight species of large-eyed
Ichthyosauri, the Geosaurus or *Lacerta gigantea* of Sömmering, and, finally, seven remarkable species of Pterodac-
tyles,† or Saurians furnished with membranous wings.   In
the chalk the number of the crocodilial Saurians diminishes,
although this epoch is characterized by the so-called crocodile
of Maestricht (the Mososaurus of Conybeare), and the colossal, probably graminivorous Iguanodon.   Cuvier has found
animals belonging to the existing families of the crocodile in
the tertiary formation, and Scheuchzer's *antediluvian man*
(*homo diluvii testis*), a large salamander allied to the Ax-
olotl, which I brought with me from the large Mexican lakes,
belongs to the most recent fresh-water formations of Œnin-
gen.‡
     The determination of the relative ages of organisms by the
superposition of the strata has led to important results regarding the relations which have been discovered between extinct
families and species (the latter being but few in number) and
those which still exist.   Ancient and modern observations
concur in showing that the fossil floras and faunas differ more
from the present vegetable and animal forms in proportion as
they belong to lower, that is, more ancient sedimentary formations.   The numerical relations first deduced by Cuvier

Saurian asserted to have been found in the mountain limestone (carbonate of lime) of Northumberland (Herm. von Meyer, *Palæologica*, s.
299), is regarded by Lyell (*Geology*, 1832, vol. i., p. 148) as very doubtful.   The discoverer himself referred it to the alluvial strata which
cover the mountain limestone.
     * F. von Alberti, *Monographie des Bunten Sandsteins, Muschelkalks
und Keupers*, 1834, s. 119 und 314.
     † See Hermann von Meyer's ingenious considerations regarding the
organization of the flying Saurians, in his *Palæologica*, s. 228–252.   In
the fossil specimen of the Pterodactylus crassirostris, which, as well as
the longer known P. longirostris (Ornithocephalus of Sömmering), was
found at Solenhofen, in the lithographic slate of the upper Jura formation, Professor Goldfuss has even discovered traces of the membranous
wing, " with the impressions of curling tufts of hair, in some places a
full inch in length."   ‡ [Ansted's *Ancient World*, p. 56.]— *Tr.*

from the great phenomena of the metamorphism of organic life,* have led, through the admirable labors of Deshayes and Lyell, to the most marked results, especially with reference to the different groups of the tertiary formations, which contain a considerable number of accurately investigated structures. Agassiz, who has examined 1700 species of fossil fishes, and who estimates the number of living species which have either been described or are preserved in museums at 8000, expressly says, in his masterly work, that, "with the exception of a few small fossil fishes peculiar to the argillaceous geodes of Greenland, he has not found any animal of this class in all the tran sition, secondary or tertiary formations, which is specifically identical with any still extant fish." He subjoins the important observation "that in the lower tertiary formations, for instance, in the coarse granular calcareous beds, and in the London clay,† one third of the fossil fishes belong to wholly extinct families. Not a single species of a still extant family is to be found under the chalk, while the remarkable family of the *Sauroidi* (fishes with enameled scales), almost allied to reptiles, and which are found from the coal beds—in which the larger species lie—to the chalk, where they occur individually, bear the same relation to the two families (the Lepidosteus and Polypterus) which inhabit the American rivers and the Nile, as our present elephants and tapirs do to the Mastodon and Anaplotheriun of the primitive world."‡

The beds of chalk which contain two of these sauroid fishes and gigantic reptiles, and a whole extinct world of corals and muscles, have been proved by Ehrenberg's beautiful discoveries to consist of microscopic Polythalamia, many of which still exist in our seas, and in the middle latitudes of the North Sea and Baltic. The first group of tertiary formations above the chalk, which has been designated as belonging to the *Eocene Period*, does not, therefore, merit that designation, since "the *dawn of the world* in which we live extends much further back in the history of the past than we have hitherto supposed."§

As we have already seen, fishes, which are the most ancient of all vertebrata, are found in the silurian transition strata,

* Cuvier, *Recherches sur les Ossemens Fossiles*, t. i., p. 52–57. See, also, the geological scale of epochs in Phillips's *Geology*, 1837, p. 166–185.　　　† [See *Wonders of Geology*, vol. i., p. 230.]—*Tr*

‡ Agassiz, *Poissons Fossiles*, t. i., p. 30, and t. iii., p. 1–52; Buckland, *Geology*, vol. i., p. 273–277.

§ Ehrenberg, *Ueber noch jetzt lebende Thierarten der Kreidebildung*, in the *Abhandl. der Berliner Akad.*, 1839, s. 164.

and then uninterruptedly on through all formations to the strata of the tertiary period, while Saurians begin with the zechstone.   In like manner, we find the first mammalia (*Thylacotherium Prevostii*, and *T. Bucklandii*, which are nearly allied, according to Valenciennes,* with marsupial animals) in the oolitic formations (Stonesfield schist), and the first birds in the most ancient cretaceous strata.†   Such are, according to the present state of our knowledge, the lowest‡ limits of fishes, Saurians, mammalia, and birds.

Although corals and Serpulidæ occur in the most ancient formations simultaneously with highly-developed Cephalopodes and Crustaceans, thus exhibiting the most various orders grouped together, we yet discover very determinate laws in the case of many individual groups of one and the same orders.   A single species of fossil, as Goniatites, Trilobites, or Nummulites, sometimes constitutes whole mountains.   Where different families are blended together, a determinate succession of organisms has not only been observed with reference to the superposition of the formations, but the association of certain families and species has also been noticed in the lower strata of the same formation.   By his acute discovery of the arrangement of the lobes of their chamber-sutures, Leopold von Buch has been enabled to divide the innumerable quantity of Ammonites into well-characterized families, and to show that Ceratites appertain to the muschelkalk, Arietes to the lias, and Goniatites to transition limestone and graywacke.§ The lower limits of Belemnites are, in the keuper, covered by Jura limestone, and their upper limits in the chalk formations.‖   It appears, from what we now know of this subject, that the waters must have been inhabited at the same epoch, and in the most widely-remote districts of the world, by shell-fish, which were, at any rate, in part, identical with the fossil remains found in England.   Leopold von Buch has discovered exogyra and trigonia in the southern hemisphere (volcano of

* Valenciennes, in the *Comptes Rendus de l'Académie des Sciences*, t. vii., 1838, Part ii., p. 580.

† In the Weald clay; Beudant, *Géologie*, p. 173.   The ornitholites increase in number in the gypsum of the tertiary formations.   Cuvier, *Ossemens Fossiles*, t. ii., p. 302–328.

‡ [Recent collections from the southern hemisphere show that this distribution was not so universal during the earlier epochs as has generally been supposed.   See papers by Darwin, Sharpe, Morris, and M‘Coy, in the *Geological Journal*.]—*Tr.*

§ Leop. von Buch, in the *Abhandl. der Berl. Akad.*, 1830, s. 135–187

‖ Quenstedt, *Flötzgebirge Würtembergs*, 1843, s. 135.

Maypo in Chili), and D'Orbigny has described Ammonites and Gryphites from the Himalaya and the Indian plains of Cutch, these remains being identical with those found in the old Jurassic sea of Germany and France.

The strata which are distinguished by definite kinds of petrifactions, or by the fragments contained within them, form a geognostic horizon, by which the inquirer may guide his steps, and arrive at certain conclusions regarding the identity or relative age of the formations, the periodic recurrence of certain strata, their parallelism, or their total suppression. If we classify the type of the sedimentary structures in the simplest mode of generalization, we arrive at the following series in proceeding from below upward :

1. The so-called *transition rocks*, in the two divisions of upper and lower graywacke (silurian and devonian systems), the latter being formerly designated as old red sandstone.

2. The *lower trias*,* comprising mountain limestone, coalmeasures, together with the lower new red sandstone (Todtliegende and Zechstein).†

3. The *upper trias*, including variegated sandstone,† muschelkalk, and keuper.

4. *Jura limestone* (lias and oolite).

5. *Green sandstone*, the quader sanstein, upper and lower chalk, terminating the secondary formations, which begin with limestone.

6. *Tertiary formations* in three divisions, distinguished as granular limestone, the lignites, and the sub-Apennine gravel of Italy.

Then follow, in the alluvial beds, the colossal bones of the mammalia of the primitive world, as the mastodon, dinothe-

* Quenstedt, *Flötzgebirge Würtembergs*, 1843, s. 13.

† Murchison makes two divisions of the *bunter sandstone*, the upper being the same as the *trias* of Alberti, while of the lower division, to which the *Vosges sandstone* of Elie de Beaumont belongs—the *zechstein* and the *todtliegende*—he forms his *Permian* system. He makes the secondary formations commence with the *upper trias*, that is to say, with the upper division of our (German) bunter sandstone, while the Permian system, the carboniferous or mountain limestone, and the devonian and silurian strata, constitute his *palæozoic formations*. According to these views, the chalk and Jura constitute the upper, and the keuper, the muschelkalk, and the bunter sandstone the lower secondary formations, while the Permian system and the carboniferous limestone are the upper, and the devonian and silurian strata are the lower palæozoic formation. The fundamental principles of this general classification are developed in the great work in which this indefatigable British geologist purposes to describe the geology of a large part of Eastern Europe.

rium, missurium, and the megatherides, among which is
Owen's sloth-like mylodon, eleven feet in length.* Besides
these extinct families, we find the fossil remains of still extant
animals, as the elephant, rhinoceros, ox, horse, and stag. The
field near Bogota, called the *Campo de Gigantes*, which is
filled with the bones of mastodons, and in which I caused excavations to be made, lies 8740 feet above the level of the
sea, while the osseous remains, found in the elevated plateaux
of Mexico, belong to true elephants of extinct species.† The
projecting spurs of the Himalaya, the Sewalik Hills, which
have been so zealously investigated by Captain Cautley‡ and
Dr. Falconer, and the Cordilleras, whose elevations are, probably, of very different epochs, contain, besides numerous mastodons, the sivatherium, and the gigantic land tortoise of the
primitive world (*Colossochelys*), which is twelve feet in length
and six in height, and several extant families, as elephants,
rhinoceroses, and giraffes ; and it is a remarkable fact, that
these remains are found in a zone which still enjoys the same
tropical climate which must be supposed to have prevailed at
the period of the mastodons.§

Having thus passed in review both the inorganic formations
of the earth's crust and the animal remains which are contained within it, another branch of the history of organic life
still remains for our consideration, viz., the epoch of vegeta
tion, and the successive floras that have occurred simultaneously with the increasing extent of the dry land and the
modifications of the atmosphere. The oldest transition strata,
as we have already observed, contain merely cellular marine
plants, and it is only in the devonian system that a few cryptogamic forms of vascular plants (Calamites and Lycopodiaceæ) have been observed.‖ Nothing appears to corroborate

---

* [See Mantell's *Wonders of Geology*, vol. i., p. 168.]—*Tr.*

† Cuvier, *Ossemens Fossiles*, 1821, t. i., p. 157, 261, and 264. See,
also, Humboldt, *Ueber die Hochebene von Bogota*, in the *Deutschen
Vierteljahrs-schrift*, 1839, bd. i., s. 117.

‡ [The fossil fauna of the Sewalik range of hills, skirting the southern base of the Himalaya, has proved more abundant in genera and
species of mammalia than that of any other region yet explored. As
a general expression of the leading features, it may be stated, that it
appears to have been composed of representative forms of all ages,
from the *oldest of the tertiary period down to the modern*, and of *all the
geographical* divisions of the Old Continent grouped together into one
comprehensive fauna. *Fauna Antiqua Sivaliensis*, by Hugh Falconer,
M.D., and Major P. T. Cautley.]—*Tr.*

§ *Journal of the Asiatic Society*, 1844, No. 15, p. 109.

‖ Beyrich, in Karsten's *Archiv für Mineralogie*, 1844, bd. xviii., s. 218

the theoretical views that have been started regarding the simplicity of primitive forms of organic life, or that vegetable preceded animal life, and that the former was necessarily dependent upon the latter. The existence of races of men inhabiting the icy regions of the North Polar lands, and whose nutriment is solely derived from fish and cetaceans, shows the possibility of maintaining life independently of vegetable substances. After the devonian system and the mountain limestone, we come to a formation, the botanical analysis of which has made such brilliant advances in modern times.* The coal measures contain not only fern-like cryptogamic plants and phanerogamic monocotyledons (grasses, yucca-like Liliareæ, and palms), but also gymnospermic dicotyledons (Coniferæ and Cycadeæ), amounting in all to nearly 400 species, as characteristic of the coal formations. Of these we will only enumerate arborescent Calamites and Lycopodiaceæ, scaly Lepidodendra, Sigillariæ, which attain a height of sixty feet, and are sometimes found standing upright, being distinguished by a double system of vascular bundles, cactus-like Stigmariæ, a great number of ferns, in some cases the stems, and in others the fronds alone being found, indicating by their abundance the insular form of the dry land,† Cycadeæ,‡ especially palms, although fewer in number,§ Asterophyllites, having whorl-like leaves, and allied to the Naiades, with araucaria-like Coniferæ,‖ which exhibit faint traces of annual rings. This difference of character from our present vegetation, manifested in the vegetative forms which were so luxuriously developed on the drier

* By the important labors of Count Sternberg, Adolphe Brongniart, Göppert, and Lindley.

† See Robert Brown's *Botany of Congo*, p. 42, and the Memoir of the unfortunate D'Urville, *De la Distribution des Fougères sur la Surface du Globe Terrestre*.

‡ Such are the Cycadeæ discovered by Count Sternberg in the old carboniferous formation at Radnitz, in Bohemia, and described by Corda (two species of Cycatides and Zamites Cordai. See Göppert, *Fossile Cycadeen in den Arbeiten der Schles. Gesellschaft, für vaterl. Cultur im Jahr* 1843, s. 33, 37, 40, and 50). A Cycadea (Pterophyllum gonorrhachis, Göpp.) has also been found in the carboniferous formations in Upper Silesia, at Königshütte.

§ Lindley, *Fossil Flora*, No. xv., p. 163.

‖ *Fossil Coniferæ*, in Buckland's *Geology*, p. 483–490. Witham has the great merit of having first recognized the existence of Coniferæ in the early vegetation of the old carboniferous formation. Almost all the trunks of trees found in this formation were previously regarded as palms. The species of the genus *Araucaria* are, however, not peculiar to the coal formations of the British Islands; they likewise occur in Upper Silesia.

and more elevated portions of the old red sandstone, was maintained through all the subsequent epochs to the most recent
chalk formations ; amid the peculiar characteristics exhibited
in the vegetable forms contained in the coal measures, there
is, however, a strikingly-marked prevalence of the same families, if not of the same species,* in all parts of the earth as it
then existed, as in New Holland, Canada, Greenland, and
Melville Island.

The vegetation of the primitive period exhibits forms which,
from their simultaneous affinity with several families of the
present world, testify that many intermediate links must have
become extinct in the scale of organic development. Thus,
for example, to mention only two instances, we would notice
the Lepidodendra, which, according to Lindley, occupy a place
between the Coniferæ and the Lycopodiaceæ,† and the Arau-
cariæ and pines, which exhibit some peculiarities in the union
of their vascular bundles. Even if we limit our consideration
to the present world alone, we must regard as highly important the discovery of Cycadeæ and Coniferæ side by side with
Sagenariæ and Lepidodendra in the ancient coal measures.
The Coniferæ are not only allied to Cupuliferæ and Betulinæ,
with which we find them associated in lignite formations, but
also with Lycopodiaceæ. The family of the sago-like Cycadeæ approaches most nearly to palms in its external appearance, while these plants are specially allied to Coniferæ in respect to the structure of their blossoms and seed.‡ Where
many beds of coal are superposed over one another, the families and species are not always blended, being most frequently
grouped together in separate genera ; Lycopodiaceæ and certain ferns being alone found in one bed, and Stigmariæ and
Sigillariæ in another. In order to give some idea of the luxuriance of the vegetation of the primitive world, and of the
immense masses of vegetable matter which was doubtlessly
accumulated in currents and converted in a moist condition
into coal,§ I would instance the Saarbrücker coal measures,

* Adolphe Brongniart, *Prodrome d'une Hist. des Végétaux Fossiles*, p.
179 ; Buckland, *Geology*, p. 479 ; Endlicher and Unger, *Grundzüge der
Botanik*, 1843, s. 455.

† " By means of Lepidodendron, a better passage is established from
flowering to flowerless plants than by either Equisetum or Cycas, or
any other known genus."—Lindley and Hutton, *Fossil Flora*, vol. ii.,
p. 53.

‡ Kunth, *Anordnung der Pflanzenfamilien*, in his *Handb. der Botanik*,
s. 307 und 314.

§ That coal has not been formed from vegetable fibers charred by

where 120 beds are superposed on one another, exclusive of a great many which are less than a foot in thickness ; the coal beds at Johnstone, in Scotland, and those in the Creuzot, in Burgundy, are some of them, respectively, thirty and fifty feet in thickness,* while in the forests of our temperate zones, the carbon contained in the trees growing over a certain area would hardly suffice, in the space of a hundred years, to cover it with more than a stratum of seven French lines in thickness.† Near the mouth of the Mississippi, and in the "wood hills" of the Siberian Polar Sea, described by Admiral Wrangel, the vast number of trunks of trees accumulated by river and sea water currents affords a striking instance of the enormous quantities of drift-wood which must have favored the formation of carboniferous depositions in the inland waters and insular bays. There can be no doubt that these beds owe a considerable portion of the substances of which they consist to grasses, small branching shrubs, and cryptogamic plants.

The association of palms and Coniferæ, which we have indicated as being characteristic of the coal formations, is discoverable throughout almost all formations to the tertiary period. In the present condition of the world, these genera

fire, but that it has more probably been produced in the moist way by the action of sulphuric acid, is strikingly demonstrated by the excellent observation made by Göppert (Karsten, *Archiv für Mineralogie,* bd. xviii., s. 530), on the conversion of a fragment of amber-tree into black coal. The coal and the unaltered amber lay side by side. Regarding the part which the lower forms of vegetation may have had in the formation of coal beds, see Link, in the *Abhandl. der Berliner Akademie der Wissenschaften,* 1838, s. 38.

* [The actual total thickness of the different beds in England varies considerably in different districts, but appears to amount in the Lancashire coal field to as much as 150 feet.—Ansted's *Ancient World,* p. 78. For an enumeration of the thickness of coal measures in America and the Old Continent, see Mantell's *Wonders of Geology,* vol. ii., p. 69.]—*Tr.*

† See the accurate labors of Chevandier, in the *Comptes Rendus de l'Académie des Sciences,* 1844, t. xviii., Part i., p. 285. In comparing this bed of carbon, seven lines in thickness, with beds of coal, we must not omit to consider the enormous pressure to which the latter have been subjected from superimposed rock, and which manifests itself in the flattened form of the stems of the trees found in these subterranean regions. "The so-called *wood-hills* discovered in 1806 by Sirowatskoi, on the south coast of the island of New Siberia, consist, according to Hedenström, of horizontal strata of sandstone, alternating with bituminous trunks of trees, forming a mound thirty fathoms in height; at the summit the stems were in a vertical position. The bed of drift-wood is visible at five wersts' distance."—See Wrangel, *Reise längs der Nordküste von Sibirien, in den Jahren* 1820–24, th. i., s. 102.

appear to exhibit no tendency whatever to occur associated
together. We have so accustomed ourselves, although erroneously, to regard Coniferæ as a northern form, that I experienced a feeling of surprise when, in ascending from the shores
of the South Pacific toward Chilpansingo and the elevated
valleys of Mexico, between the *Venta de la Moxonera* and the
Alto de los Caxones, 4000 feet above the level of the sea, I
rode a whole day through a dense wood of Pinus occidentalis,
where I observed that these trees, which are so similar to the
Weymouth pine, were associated with fan palms* (*Corypha
dulcis*), swarming with brightly-colored parrots. South America has oaks, but not a single species of pine ; and the first
time that I again saw the familiar form of a fir-tree, it was
thus associated with the strange appearance of the fan palm.†
Christopher Columbus, in his first voyage of discovery, saw
Coniferæ and palms growing together on the northeastern extremity of the island of Cuba, likewise within the tropics, and
scarcely above the level of the sea. This acute observer,
whom nothing escaped, mentions the fact in his journal as a
remarkable circumstance, and his friend Anghiera, the secretary of Ferdinand the Catholic, remarks with astonishment
" that *palmeta* and *pineta* are found associated together in
the newly-discovered land." It is a matter of much importance to geology to compare the present distribution of plants
over the earth's surface with that exhibited in the fossil floras
of the primitive world. The temperate zone of the southern
hemisphere, which is so rich in seas and islands, and where

* This corypha is the *soyate* (in Aztec, *zoyatl*), or the *Palma dulce* of
the natives. See Humboldt and Bonpland, *Synopsis Plant. Æquinoct.
Orbis Novi*, t. i., p. 302. Professor Buschmann, who is profoundly acquainted with the American languages, remarks, that the *Palma soyate*
is so named in Yepe's *Vocabulario de la Lengua Othomi*, and that the
Aztec word zoyatl (Molina, *Vocabulario en Lengua Mexicana y Castel-
lana*, p. 25) recurs in names of places, such as Zoyatitlan and Zoya-
panco, near Chiapa.

† Near Baracoa and Cayos de Moya. See the Admiral's journal of
the 25th and 27th of November, 1492, and Humboldt, *Examen Critique
de l'Hist. de la Géographie du Nouveau Continent*, t. ii., p. 252, and t.
iii., p. 23. Columbus, who invariably paid the most remarkable attention to all natural objects, was the first to observe the difference between *Podocarpus* and *Pinus*. " I find," said he, " en la tierra aspera
del Cibao pinos que no llevan pinas (fir cones), pero portal orden com-
puestos por naturaleza, que (los frutos) parecen azeytunas del Axarafe
de Sevilla." The great botanist, Richard, when he published his excellent Memoir on Cycadeæ and Coniferæ, little imagined that before
the time of L'Héritier, and even before the end of the fifteenth century, a navigator had separated *Podocarpus* from the Abietineæ.

tropical forms blend so remarkably with those of colder parts of the earth, presents, according to Darwin's beautiful and animated descriptions,[*] the most instructive materials for the study of the present and the past geography of plants. The history of the primordial ages is, in the strict sense of the word, a part of the history of plants.

Cycadeæ, which, from the number of their fossil species, must have occupied a far more important part in the extinct than in the present vegetable world, are associated with the nearly allied Coniferæ from the coal formations upward. They are almost wholly absent in the epoch of the variegated sandstone which contains Coniferæ of rare and luxuriant structure (*Voltizia, Haidingera, Albertia*); the Cycadeæ, however, occur most frequently in the keuper and lias strata, in which more than twenty different forms appear. In the chalk, marine plants and naiades predominate. The forests of Cycadeæ of the Jura formations had, therefore, long disappeared, and even in the more ancient tertiary formations they are quite subordinate to the Coniferæ and palms.[†]

The lignites, or beds of brown coal[‡] which are present in all divisions of the tertiary period, present, among the most ancient cryptogami land plants, some few palms, many Coniferæ having distinct annual rings, and foliaceous shrubs of a more or less tropical character. In the middle tertiary period we again find palms and Cycadeæ fully established, and finally a great similarity with our existing flora, manifested in the sudden and abundant occurrence of our pines and firs, Cupuliferæ, maples, and poplars. The dicotyledonous stems found in lignite are occasionally distinguished by colossal size and great age. In the trunk of a tree found at Bonn, Nöggerath counted 792 annual rings.[§] In the north of France, at Yseux, near Abbeville, oaks have been discovered in the turf moors of the Somme which measured fourteen feet in diameter, a thickness which is very remarkable in the Old Continent and without the tropics. According to Göppert's excellent investigations, which, it is hoped, may soon be illustrated by plates, it would appear that "all the amber of the Baltic comes from

[*] Charles Darwin, *Journal of the Voyages of the Adventure and Beagle*, 1839, p. 271.

[†] Göppert describes three other Cycadeæ (species of Cycadites and Pterophyllum), found in the brown carboniferous schistose clay of Altsattel and Commotau, in Bohemia. They very probably belong to the Eocene Period. Göppert, *Fossile Cycadeen*, s. 61.

[‡] [*Medals of Creation*, vol. i., ch. v., &c. *Wonders of Geology*, vol. i., p. 278, 392.]—*Tr.*   [§] Buckland, *Geology*, p. 509.

a coniferous tree, which, to judge by the still extant remains
of the wood and the bark at different ages, approaches very
nearly to our white and red pines, although forming a distinct
species. The amber-tree of the ancient world (*Pinites succifer*) abounded in resin to a degree far surpassing that manifested by any extant coniferous tree ; for not only were large
masses of amber deposited in and upon the bark, but also in
the wood itself, following the course of the medullary rays,
which, together with ligneous cells, are still discernible under
the microscope, and peripherally between the rings, being some
times both yellow and white."

" Among the vegetable forms inclosed in amber are male
and female blossoms of our native needle-wood trees and Cupuliferæ, while fragments which are recognized as belonging tc
thuia, cupressus, ephedera, and castania vesca, blended with
those of junipers and firs, indicate a vegetation different from
that of the coasts and plains of the Baltic."*

We have now passed through the whole series of formations
comprised in the geological portion of the present work, proceeding from the oldest erupted rock and the most ancient sedimentary formations to the alluvial land on which are scattered those large masses of rock, the causes of whose general
distribution have been so long and variously discussed, and
which are, in my opinion, to be ascribed rather to the penetration and violent outpouring of pent-up waters by the elevation of mountain chains than to the motion of floating blocks
of ice.† The most ancient structures of the transition forma-

---

* [The forests of amber-pines, *Pinites succifer*, were in the southeast-
ern part of what is now the bed of the Baltic, in about 55° N. lat.,
and 37° E. long. The different colors of amber are derived from local
chemical admixture. The amber contains fragments of vegetable matter, and from these it has been ascertained that the amber-pine forests
contained four other species of pine (besides the *Pinites succifer*), several cypresses, yews, and junipers, with oaks, poplars, beeches, &c.—
altogether forty-eight species of trees and shrubs, constituting a flora
of North American character. There are also some ferns, mosses, fungi,
and liverworts. See Professor Göppert, *Geol. Trans.*, 1845. Insects, spi-
ders, small crustaceans, leaves, and fragments of vegetable tissue, are
imbedded in some of the masses. Upward of 800 species of insects
have been observed; most of them belong to species, and even genera,
that appear to be distinct from any now known, but others are nearly
related to indigenous species, and some are identical with existing forms,
that inhabit more southern climes.— *Wonders of Geology*, vol. i., p. 242,
&c.]—*Tr*.

† Leopold von Buch, in the *Abhandl. der Akad. der Wissensch. zu
Berlin*, 1814–15, s. 161 ; and in Poggend., *Annalen*, bd. ix., s. 575 ; Elie
de Beaumont, in the *Annales des Sciences Naturelles*, t. xix., p. 60.

tion with which we are acquainted are slate and graywacke, which contain some remains of sea weeds from the silurian or cambrian sea. On what did these so-called *most ancient* formations rest, if gneiss and mica schist must be regarded as changed sedimentary strata ?   Dare we hazard a conjecture on that which can not be an object of actual geognostic observation ?   According to an ancient Indian myth, the earth is borne up by an elephant, who in his turn is supported by a gigantic tortoise, in order that he may not fall ; but it is not permitted to the credulous Brahmins to inquire on what the tortoise rests.   We venture here upon a somewhat similar problem, and are prepared to meet with opposition in our endeavors to arrive at its solution.   In the first formation of the planets, as we stated in the astronomical portion of this work, it is probable that nebulous rings revolving round the sun were agglomerated into spheroids, and consolidated by a gradual condensation proceeding from the exterior toward the center. What we term the ancient silurian strata are thus only the upper portions of the solid crust of the earth.   The erupted rocks which have broken through and upheaved these strata have been elevated from depths that are wholly inaccessible to our research ; they must, therefore, have existed under the silurian strata, and been composed of the same association of minerals which we term granite, augite, and quartzose porphyry, when they are made known to us by eruption through the surface. Basing our inquiries on analogy, we may assume that the substances which fill up deep fissures and traverse the sedimentary strata are merely the ramifications of a lower deposit.   The foci of active volcanoes are situated at enormous depths, and, judging from the remarkable fragments which I have found in various parts of the earth incrusted in lava currents, I should deem it more than probable that a primordial granite rock forms the substratum of the whole stratified edifice of fossil remains.*   Basalt containing olivine first shows itself in the period of the chalk, trachyte still later, while erup- tions of granite belong, as we learn from the products of their metamorphic action, to the epoch of the oldest sedimentary strata of the transition formation. Where knowledge can not be attained from immediate perceptive evidence, we may be allowed from induction, no less than from a careful comparison of facts, to hazard a conjecture by which granite would be re-

* See Elie de Beaumont, *Descr. Géol. de la France,* t. i., p. 65 ; Beu- dant, *Géologie,* 1844. p. 200.

stored to a portion of its contested right and title to be considered as a *primordial* rock.

The recent progress of geognosy, that is to say, the more
extended knowledge of the geognostic epochs characterized by
difference of mineral formations, by the peculiarities and succession of the organisms contained within them, and by the
position of the strata, whether uplifted or inclined horizontally,
leads us, by means of the causal connection existing among all
natural phenomena, to the distribution of solids and fluids into
the continents and seas which constitute the upper crust of our
planet. We here touch upon a point of contact between geological and geographical geognosy which would constitute the
complete history of the form and extent of continents. The
limitation of the solid by the fluid parts of the earth's surface
and their mutual relations of area, have varied very considerably in the long series of geognostic epochs. They were very
different, for instance, when carboniferous strata were horizontally deposited on the inclined beds of the mountain limestone
and old red sandstone ; when lias and oolite lay on a substratum of keuper and muschelkalk, and the chalk rested on the
slopes of green sandstone and Jura limestone. If, with Elie
de Beaumont, we term the waters in which the Jura limestone
and chalk formed a soft deposit the *Jurassic or oolitic*, and the
cretaceous seas, the outlines of these formations will indicate,
for the two corresponding epochs, the boundaries between the
already dried land and the ocean in which these rocks were
forming. An ingenious attempt has been made to draw maps
of this physical portion of primitive geography, and we may
consider such diagrams as more correct than those of the wanderings of Io or the Homeric geography, since the latter are
merely graphic representations of mythical images, while the
former are based upon positive facts deduced from the science
of geology.

The results of the investigations made regarding the areal
relations of the solid portions of our planet are as follows : in
the most ancient times, during the silurian and devonian transition epochs, and in the secondary formations, including the
trias, the continental portions of the earth were limited to insular groups covered with vegetation ; these islands at a subsequent period became united, giving rise to numerous lakes
and deeply-indented bays ; and, finally, when the chains of
the Pyrenees, Apennines, and Carpathian Mountains were
elevated about the period of the more ancient tertiary formations, large continents appeared, having almost their present

size.* In the silurian epoch, as well as in that in which the Cycadeæ flourished in such abundance, and gigantic saurians were living, the dry land, from pole to pole, was probably less than it now is in the South Pacific and the Indian Ocean. We shall see, in a subsequent part of this work, how this preponderating quantity of water, combined with other causes, must have contributed to raise the temperature and induce a greater uniformity of climate. Here we would only remark, in considering the gradual extension of the dry land, that, shortly before the *disturbances* which at longer or shorter intervals caused the sudden destruction of so great a number of colossal vertebrata in the *diluvial period*, some parts of the present continental masses must have been completely separated from one another. There is a great similarity in South America and Australia between still living and extinct species of animals. In New Holland fossil remains of the kangaroo have been found, and in New Zealand the semi-fossilized bones of an enormous bird, resembling the ostrich, the dinornis of Owen,† which is nearly allied to the present apteryx, and but little so to the recently extinct dronte (dodo) of the island of Rodriguez.

The form of the continental portions of the earth may, perhaps, in a great measure, owe their elevation above the surrounding level of the water to the eruption of quartzose porphyry, which overthrew with violence the first great vegetation from which the material of our present coal measures was formed. The portions of the earth's surface which we term plains are nothing more than the broad summits of hills and mountains whose bases rest on the bottom of the ocean. Every plain is, therefore, when considered according to its submarine relations, an *elevated plateau*, whose inequalities have been covered over by horizontal deposition of new sedimentary formations and by the accumulation of alluvium.

---

* [These movements, described in so few words, were doubtless going on for many thousands and tens of thousands of revolutions of our planet. They were accompanied, also, by vast but slow changes of other kinds. The expansive force employed in lifting up, by mighty movements, the northern portion of the continent of Asia, found partial vent; and from partial subaqueous fissures there were poured out the tabular masses of basalt occurring in Central India, while an extensive area of depression in the Indian Ocean, marked by the coral islands of the Laccadives, the Maldives, the great Chagos Bank, and some others, were in the course of depression by a counteracting movement.—Ansted's *Ancient World*, p. 346, &c.]—*Tr.*

† [See *American Journal of Science*, vol. xlv., p. 187 ; and *Medals of Creation*, vol. ii., p. 817 ; *Trans. Zoolog. Society of London*, vol. ii.; *Wonders of Geology*, vol. i., p. 129.]—*Tr.*

Among the general subjects of contemplation appertaining to a work of this nature, a prominent place must be given, first, to the consideration of the *quantity* of the land raised above the level of the sea, and, next, to the individual configuration of each part, either in relation to horizontal extension (relations of form) or to vertical elevation (hypsometrical relations of mountain-chains). Our planet has two envelopes, of which one, which is general—the atmosphere—is composed of an elastic fluid, and the other—the sea—is only locally distributed, surrounding, and therefore modifying, the form of the land. These two envelopes of air and sea constitute a natural whole, on which depend the difference of climate on the earth's surface, according to the relative extension of the aqueous and solid parts, the form and aspect of the land, and the direction and elevation of mountain chains. A knowledge of the reciprocal action of air, sea, and land teaches us that great meteorological phenomena can not be comprehended when considered independently of geognostic relations. Meteorology, as well as the geography of plants and animals, has only begun to make actual progress since the mutual dependence of the phenomena to be investigated has been fully recognized. The word climate has certainly special reference to the character of the atmosphere, but this character is itself dependent on the perpetually concurrent influences of the ocean, which is universally and deeply agitated by currents having a totally opposite temperature, and of radiation from the dry land, which varies greatly in form, elevation, color, and fertility, whether we consider its bare, rocky portions, or those that are covered with arborescent or herbaceous vegetation.

In the present condition of the surface of our planet, the area of the solid is to that of the fluid parts as $1 : 2\frac{4}{5}$ths (according to Rigaud, as $100 : 270$).[*] The islands form scarcely $\frac{1}{22}$d of the continental masses, which are so unequally divided that they consist of three times more land in the northern than in the southern hemisphere; the latter being, therefore, pre-eminently oceanic. From $40°$ south latitude to the Antarctic pole the earth is almost entirely covered with water. The fluid element predominates in like manner between the eastern shores of the Old and the western shores of the New Continent, being only interspersed with some few insular groups. The learned hydrographer Fleurieu has very justly named this

[*] See *Transactions of the Cambridge Philosophical Society*, vol. vi., Part ii., 1837, p. 297. Other writers have given the ratio as $100 : 284$.

vast oceanic basin, which, under the tropics, extends over 145° of longitude, the *Great Ocean*, in contradistinction to all other seas. The southern and western hemispheres (reckoning the latter from the meridian of Teneriffe) are therefore more rich in water than any other region of the whole earth.

These are the main points involved in the consideration of the relative quantity of land and sea, a relation which exercises so important an influence on the distribution of temperature, the variations in atmospheric pressure, the direction of the winds, and the quantity of moisture contained in the air, with which the development of vegetation is so essentially connected. When we consider that nearly three fourths of the upper surface of our planet are covered with water,* we shall be less surprised at the imperfect condition of meteorology before the beginning of the present century, since it is only during the subsequent period that numerous accurate observations on the temperature of the sea at different latitudes and at different seasons have been made and numerically compared together.

The horizontal configuration of continents in their general relations of extension was already made a subject of intellectual contemplation by the ancient Greeks. Conjectures were advanced regarding the maximum of the extension from west to east, and Dicæarchus placed it, according to the testimony of Agathemerus, in the latitude of Rhodes, in the direction of a line passing from the Pillars of Hercules to Thine. This line, which has been termed *the parallel of the diaphragm of Dicæarchus*, is laid down with an astronomical accuracy of position, which, as I have stated in another work, is well worthy of exciting surprise and admiration.† Strabo, who was probably influenced by Eratosthenes, appears to have been so firmly convinced that this parallel of 36° was the maximum of the extension of the then existing world, that he supposed it had some intimate connection with the form of the earth, and therefore places under this line the continent whose existence

---

* In the Middle Ages, the opinion prevailed that the sea covered only one seventh of the surface of the globe, an opinion which Cardinal d'Ailly (*Imago Mundi*, cap. 8) founded on the fourth apocryphal book of Esdras. Columbus, who derived a great portion of his cosmographical knowledge from the cardinal's work, was much interested in upholding this idea of the smallness of the sea, to which the misunderstood expression of " the ocean stream" contributed not a little. See Humboldt, *Examen Critique de l'Hist. de la Géographie*, t. i., p. 186.

† Agathemerus, in Hudson, *Geographi Minores*, t. ii., p. 4. See Humboldt, *Asie Centr.*, t. i., p. 120–125.

he divined in the northern hemisphere, between Theria and
the coasts of Thine.*

As we have already remarked, one hemisphere of the earth
(whether we divide the sphere through the equator or through
the meridian of Teneriffe) has a much greater expansion of
elevated land than the opposite one : these two vast ocean-
girt tracts of land, which we term the eastern and western,
or the Old and New Continents, present, however, conjointly
with the most striking contrasts of configuration and position
of their axes, some similarities of form, especially with reference to the mutual relations of their opposite coasts.   In the
eastern continent, the predominating direction—the position
of the major axis—inclines from east to west (or, more correctly speaking, from southwest to northeast), while in the
western continent it inclines from south to north (or, rather,
from south-southeast to north-northwest).   Both terminate to
the north at a parallel coinciding nearly with that of 70°,
while they extend to the south in pyramidal points, having
submarine prolongations of islands and shoals.   Such, for instance, are the Archipelago of Tierra del Fuego, the Lagullas
Bank south of the Cape of Good Hope, and Van Diemen's
Land, separated from New Holland by Bass's Straits.   Northern Asia extends to the above parallel at Cape Taimura, which,
according to Krusenstern, is 78° 16', while it falls below it
from the mouth of the Great Tschukotschja River eastward
to Behring's Straits, in the eastern extremity of Asia—Cook's
East Cape—which, according to Beechey, is only 66° 3'.†
The northern shore of the New Continent follows with tolerable exactness the parallel of 70°, since the lands to the north
and south of Barrow's Strait, from Boothia Felix and Victoria
Land, are merely detached islands.

The pyramidal configuration of all the southern extremities
of continents belongs to the *similitudines physicæ in configu-
ratione mundi*, to which Bacon already called attention in his
Novum Organon, and with which Reinhold Foster, one of
Cook's companions in his second voyage of circumnavigation,
connected some ingenious considerations.   On looking eastward
from the meridian of Teneriffe, we perceive that the southern
extremities of the three continents, viz., Africa as the extreme

* Strabo, lib. i., p. 65, Casaub. See Humboldt, *Examen Crit.*, t. i.,
p. 152.
 † On the mean latitude of the Northern Asiatic shores, and the true
name of Cape Taimura (Cape Siewero-Wostotschnoi), and Cape Northeast (Schalagskoi Mys), see Humboldt, *Asie Centrale*, t. iii., p. 35, 37.

of the Old World, Australia, and South America, successively
approach nearer toward the south pole. New Zealand, whose
length extends fully 12° of latitude, forms an intermediate
link between Australia and South America, likewise terminating in an island, New Leinster. It is also a remarkable circumstance that the greatest extension toward the south falls
in the Old Continent, under the same meridian in which the
extremest projection toward the north pole is manifested. This
will be perceived on comparing the Cape of Good Hope and
the Lagullas Bank with the North Cape of Europe, and the
peninsula of Malacca with Cape Taimura in Siberia.* We
know not whether the poles of the earth are surrounded by
land or by a sea of ice. Toward the north pole the parallel
of 82° 55′ has been reached, but toward the south pole only
that of 78° 10′.

The pyramidal terminations of the great continents are variously repeated on a smaller scale, not only in the Indian Ocean,
and in the peninsulas of Arabia, Hindostan, and Malacca, but
also, as was remarked by Eratosthenes and Polybius, in the
Mediterranean, where these writers had ingeniously compared
together the forms of the Iberian, Italian, and Hellenic peninsulas.† Europe, whose area is five times smaller than that
of Asia, may almost be regarded as a multifariously articulated
western peninsula of the more compact mass of the continent
of Asia, the climatic relations of the former being to those of
the latter as the peninsula of Brittany is to the rest of France.‡
The influence exercised by the articulation and higher development of the form of a continent on the moral and intellectual condition of nations was remarked by Strabo,§ who extols

* Humboldt, *Asie Centrale*, t. i., p. 198–200. The southern point
of America, and the Archipelago which we call Terra del Fuego, lie in
the meridian of the northwestern part of Baffin's Bay, and of the great
polar land, whose limits have not as yet been ascertained, and which,
perhaps, belongs to West Greenland.

† Strabo, lib. ii., p. 92, 108, Casaub.

‡ Humboldt, *Asie Centrale*, t. iii., p. 25. As early as the year 1817,
in my work *De distributione Geographicâ Plantarum, secundum cœli
temperiem, et altitudinem Montium*, I directed attention to the import
ant influence of compact and of deeply-articulated continents on climate
and human civilization, "Regiones vel per sinus lunatos in longa cornua
porrectæ, angulosis littorum recessibus quasi membratim discerptæ, vel
spatia patentia in immensum, quorum littora nullis incisa augulis ambit
sine anfractu oceanus" (p. 81, 182). On the relations of the extent of
coast to the area of a continent (considered in some degree as a measure of the accessibility of the interior), see the inquiries in Berghaus,
Annalen der Erdkunde, bd. xii., 1835, s. 490, and *Physikal. Atlas*, 1839
No. iii , s. 69.   § Strabo, lib. ii., p. 92, 198, Casaub.

the varied form of our small continent as a special advantage. Africa* and South America, which manifest so great a resemblance in their configuration, are also the two continents that exhibit the simplest littoral outlines. It is only the eastern shores of Asia, which, broken as it were by the force of the currents of the ocean† (*fractas ex æquore terras*), exhibit a richly-variegated configuration, peninsulas and contiguous islands alternating from the equator to 60° north latitude.

Our Atlantic Ocean presents all the indications of a valley. It is as if a flow of eddying waters had been directed first toward the northeast, then toward the northwest, and back again to the northeast. The parallelism of the coasts north of 10° south latitude, the projecting and receding angles, the convexity of Brazil opposite to the Gulf of Guinea, that of Africa under the same parallel, with the Gulf of the Antilles, all favor this apparently speculative view.‡ In this Atlantic valley, as is almost every where the case in the configuration of large continental masses, coasts deeply indented, and rich in islands, are situated opposite to those possessing a different character. I long since drew attention to the geognostic importance of entering into a comparison of the western coast of Africa and of South America within the tropics. The deeply-curved indentation of the African continent at Fernando Po, 4° 30′ north latitude, is repeated on the coast of the Pacific at 18° 15′ south latitude, between the Valley of Arica and the Morro de Juan Diaz, where the Peruvian coast suddenly changes the direction from south to north which it had previously followed, and inclines to the northwest. This change

* Of Africa, Pliny says (v. 1), "Nec alia pars terrarum pauciores recipit sinus." The small Indian peninsula on this side the Ganges presents, in its triangular outline, a third analogous form. In ancient Greece there prevailed an opinion of the regular configuration of the dry land. There were four gulfs or bays, among which the Persian Gulf was placed in opposition to the Hyrcanian or Caspian Sea (Arrian, vii., 16; Plut., *in vita Alexandri*, cap. 44; Dionys. Perieg., v. 48 and 630, p. 11, 38, Bernh.). These four bays and the isthmuses were, according to the optical fancies of Agesianax, supposed to be reflected in the moon (Plut., *de Facie in Orbem Lunæ*, p. 921, 19). Respecting the *terra quadrifida*, or four divisions of the dry land, of which two lay north and two south of the equator, see Macrobius, *Comm. in Somnium Scipionis*, ii., 9. I have submitted this portion of the geography of the ancients, regarding which great confusion prevails, to a new and careful examination, in my *Examen Crit. de l'Hist. de la Géogr.*, t. i., p. 119, 145, 180–185, as also in *Asie Centr.*, t. ii., p. 172–178.

† Fleurieu, in *Voyage de Marchand autour du Monde*, t. iv., p. 38–42.

‡ Humboldt, in the *Journal de Physique*, liii., 1799, p. 33; and *Rel Hist.*, t. ii., p. 19; t. iii., p. 189, 198.

of direction extends in like manner to the chain of the Andes, which is divided into two parallel branches, affecting not only the littoral portions,* but even the eastern Cordilleras. In the latter, civilization had its earliest seat in the South American plateaux, where the small Alpine lake of Titicaca bathes the feet of the colossal mountains of Sorata and Illimani. Further to the south, from Valdivia and Chiloë (40° to 42° south latitude), through the Archipelago *de los Chonos* to *Terra del Fuego*, we find repeated that singular configuration of *fiords* (a blending of narrow and deeply-indented bays), which in the Northern hemisphere characterizes the western shores of Norway and Scotland.

These are the most general considerations suggested by the study of the upper surface of our planet with reference to the form of continents, and their expansion in a horizontal direction. We have collected facts and brought forward some analogies of configuration in distant parts of the earth, but we do not venture to regard them as fixed laws of form. When the traveler on the declivity of an active volcano, as, for instance, of Vesuvius, examines the frequent partial elevations by which portions of the soil are often permanently upheaved several feet above their former level, either immediately preceding or during the continuance of an eruption, thus forming roof-like or flattened summits, he is taught how accidental conditions in the expression of the force of subterranean vapors, and in the resistance to be overcome, may modify the form and direction of the elevated portions. In this manner, feeble perturbations in the equilibrium of the internal elastic forces of our planet may have inclined them more to its northern than to its southern direction, and caused the continent in the eastern part of the globe to present a broad mass, whose major axis is almost parallel with the equator, while in the western and more oceanic part the southern extremity is extremely narrow.

Very little can be empirically determined regarding the causal connection of the phenomena of the formation of continents, or of the analogies and contrasts presented by their

* Humboldt, in Poggendorf's *Annalen der Physik*, bd. xl., s. 171. On the remarkable fiord formation at the southeast end of America, see Darwin's Journal (*Narrative of the Voyages of the Adventure and Beagle*, vol. iii.), 1839, p. 266. The parallelism of the two mountain chains is maintained from 5° south to 5° north latitude. The change in the direction of the coast at Arica appears to be in consequence of the altered course of the fissure, above which the Cordillera of the Andes has been upheaved.

configuration. All that we know regarding this subject re-solves itself into this one point, that the active cause is sub-terranean; that continents did not arise at once in the form they now present, but were, as we have already observed, in-creased by degrees by means of numerous oscillatory elevations and depressions of the soil, or were formed by the fusion of separate smaller continental masses. Their present form is, therefore, the result of two causes, which have exercised a con-secutive action the one on the other: the first is the expression of subterranean force, whose direction we term accidental, owing to our inability to define it, from its removal from with-in the sphere of our comprehension, while the second is derived from forces acting on the surface, among which volcanic erup-tions, the elevation of mountains, and currents of sea water play the principal parts. How totally different would be the condition of the temperature of the earth, and, consequently, of the state of vegetation, husbandry, and human society, if the major axis of the New Continent had the same direction as that of the Old Continent; if, for instance, the Cordilleras, instead of having a southern direction, inclined from east to west; if there had been no radiating tropical continent, like Africa, to the south of Europe; and if the Mediterranean, which was once connected with the Caspian and Red Seas, and which has become so powerful a means of furthering the intercommunication of nations, had never existed, or if it had been elevated like the plains of Lombardy and Cyrene?

The changes of the reciprocal relations of height between the fluid and solid portions of the earth's surface (changes which, at the same time, determine the outlines of continents, and the greater or lesser submersion of low lands) are to be ascribed to numerous unequally working causes. The most powerful have incontestably been the force of elastic vapors inclosed in the interior of the earth, the sudden change of tem-perature of certain dense strata,* the unequal secular loss of

---

* De la Beche, *Sections and Views illustrative of Geological Phenome-na*, 1830, tab. 40; Charles Babbage, *Observations on the Temple of Serapis at Pozzuoli, near Naples, and on certain Causes which may produce Geological Cycles of great Extent*, 1834. "If a stratum of sand-stone five miles in thickness should have its temperature raised about 100°, its surface would rise twenty-five feet. Heated beds of clay would, on the contrary, occasion a sinking of the ground by their con-traction." See Bischof, *Wärmelehre des Innern unseres Erdkörpers*, s. 303, concerning the calculations for the secular elevation of Sweden, on the supposition of a rise by so small a quantity as 7° in a stratum of about 155,000 feet in thickness, and heated to a state of fusion.

heat experienced by the crust and nucleus of the earth, occasioning ridges in the solid surface, local modifications of gravitation,* and, as a consequence of these alterations, in the curvature of a portion of the liquid element. According to the views generally adopted by geognosists in the present day, and which are supported by the observation of a series of well-attested facts, no less than by analogy with the most important volcanic phenomena, it would appear that the elevation of continents is actual, and not merely apparent or owing to the configuration of the upper surface of the sea. The merit of having advanced this view belongs to Leopold von Buch, who first made his opinions known to the scientific world in the narrative of his memorable *Travels through Norway and Sweden* in 1806 and 1807.† While the whole coast of Sweden and Finland, from Sölvitzborg, on the limits of Northern Scania, past Gefle to Tornea, and from Tornea to Abo, experiences a gradual rise of four feet in a century, the southern part of Sweden is, according to Neilson, undergoing a simultaneous depression.‡ The maximum of this elevating

* The opinion so implicitly entertained regarding the invariability of the force of gravity at any given point of the earth's surface, has in some degree been controverted by the gradual rise of large portions of the earth's surface. See Bessel, *Ueber Maas und Gewicht,* in Schumacher's *Jahrbuch für* 1840, s. 134.

† Th. ii. (1810), s. 389. See Hallström, in *Kongl. Vetenskaps-Academiens Handlingar* (Stockh.). 1823, p. 30; Lyell, in the *Philos. Trans.* for 1835; Blom (Amtmann in Budskerud), *Stat. Beschr. von Norwegen,* 1843, s. 89–116. If not before Von Buch's travels through Scandinavia, at any rate before their publication, Playfair, in 1802, in his illustrations of the Huttonian theory, § 393, and, according to Keilhau (*Om Landjordens Stigning in Norge,* in the *Nyt Magazine für Naturvidenskaberne*), and the Dane Jessen, even before the time of Playfair, had expressed the opinion that it was not the sea which was sinking, but the solid land of Sweden which was rising. Their ideas, however, were wholly unknown to our great geologist, and exerted no influence on the progress of physical geography. Jessen, in his work, *Kongeriget Norge fremstillet efter dets naturlige og borgerlige Tilstand,* Kjobenh., 1763, sought to explain the causes of the changes in the relative levels of the land and sea, basing his views on the early calculations of Celsius, Kalm, and Dalin. He broaches some confused ideas regarding the possibility of an internal growth of rocks, but finally declares himself in favor of an upheaval of the land by earthquakes, "although," he observes, "no such rising was apparent immediately after the earthquake of Egersund, yet the earthquake may have opened the way for other causes producing such an effect."

‡ See Berzelius, *Jahrsbericht über die Fortschritte der Physischen Wiss.,* No. 18, s. 686. The islands of Saltholm, opposite to Copenhagen, and Björnholm, however, rise but very little—Björnholm scarcely one foot in a century. See Forchhammer, in *Philos. Magazine,* 3d Series, vol. ii., p 309.

force appears to be in the north of Lapland, and to diminish gradually to the south toward Calmar and Sölvitzborg. Lines marking the ancient level of the sea in pre-historic times are indicated throughout the whole of Norway,* from Cape Lindesnæs to the extremity of the North Cape, by banks of shells identical with those of the present seas, and which have lately been most accurately examined by Bravais during his long winter sojourn at Bosekop. These banks lie nearly 650 feet above the present mean level of the sea, and reappear, according to Keilhau and Eugene Robert, in a north-northwest direction on the coasts of Spitzbergen, opposite the North Cape. Leopold von Buch, who was the first to draw attention to the high banks of shells at Tromsoe (latitude 69° 40′), has, however, shown that the more ancient elevations on the North Sea appertain to a different class of phenomena, from the regular and gradual retrogressive elevations of the Swedish shores in the Gulf of Bothnia. This latter phenomenon, which is well attested by historical evidence, must not be confounded with the changes in the level of the soil occasioned by earthquakes, as on the shores of Chili and of Cutch, and which have recently given occasion to similar observations in other countries. It has been found that a perceptible sinking resulting from a disturbance of the strata of the upper surface sometimes occurs, corresponding with an elevation elsewhere, as, for instance, in West Greenland, according to Pingel and Graah, in Dalmatia and in Scania.

Since it is highly probable that the oscillatory movements of the soil, and the rising and sinking of the upper surface, were more strongly marked in the early periods of our planet than at present, we shall be less surprised to find in the interior of continents some few portions of the earth's surface lying below the general level of existing seas. Instances of this kind occur in the soda lakes described by General Andreossy, the small bitter lakes in the narrow Isthmus of Suez, the Caspian Sea, the Sea of Tiberias, and especially the Dead Sea.† The level of the water in the two last-named seas is

* Keilhau, in *Nyt Mag. für Naturvid.*, 1832, bd. i., p. 105–254; bd. ii., p. 57 ; Bravais, *Sur les Lignes d'ancien Niveau de la Mer*, 1843, p. 15–40. See, also, Darwin, "on the Parallel Roads of Glen-Roy and Lochaber," in *Philos. Trans. for* 1839, p. 60.

† Humboldt, *Asie Centrale*, t. ii., p. 319–324; t. iii., p. 549–551. The depression of the Dead Sea has been successively determined by the barometrical measurements of Count Bertou, by the more careful ones of Russegger, and by the trigonometrical survey of Lieutenant Symond, of the Royal Navy, who states that the difference of level be-

666 and 1312 feet below the level of the Mediterranean. If
we could suddenly remove the alluvial soil which covers the
rocky strata in many parts of the earth's surface, we should
discover how great a portion of the rocky crust of the earth
was then below the present level of the sea. The periodic,
although irregularly alternating rise and fall of the water of
the Caspian Sea, of which I have myself observed evident
traces in the northern portions of its basin, appears to prove,*
as do also the observations of Darwin on the coral seas,† that
without earthquakes, properly so called, the surface of the
earth is capable of the same gentle and progressive oscillations as those which must have prevailed so generally in the
earliest ages, when the surface of the hardening crust of the
earth was less compact than at present.

The phenomena to which we would here direct attention
remind us of the instability of the present order of things, and
of the changes to which the outlines and configuration of continents are probably still subject at long intervals of time.
That which may scarcely be perceptible in one generation,
accumulates during periods of time, whose duration is revealed
to us by the movement of remote heavenly bodies. The eastern coast of the Scandinavian peninsula has probably risen

tween the surface of the Dead Sea and the highest houses of Jaffa is
about 1605 feet. Mr. Alderson, who communicated this result to the
Geographical Society of London in a letter, of the contents of which I
was informed by my friend, Captain Washington, was of opinion (Nov.
28, 1841) that the Dead Sea lay about 1400 feet under the level of the
Mediterranean. A more recent communication of Lieutenant Symond
(Jameson's *Edinburgh New Philosophical Journal*, vol. xxxiv., 1843, p.
178) gives 1312 feet as the final result of two very accordant trigonometrical operations.

* *Sur la Mobilité du fond de la Mer Caspienne*, in my *Asie Centr.*, t.
ii., p. 283–294. The Imperial Academy of Sciences of St. Petersburgh,
in 1830, at my request, charged the learned physicist Lenz to place
marks indicating the mean level of the sea, for definite epochs, in different places near Baku, in the peninsula of Abscheron. In the same
manner, in an appendix to the instructions given to Captain (now Sir
James C.) Ross for his Antarctic expedition, I urged the necessity of
causing marks to be cut in the rocks of the southern hemisphere, as
had already been done in Sweden and on the shores of the Caspian
Sea. Had this measure been adopted in the early voyages of Bougain-
ville and Cook, we should now know whether the secular relative
changes in the level of the seas and land are to be considered as a general, or merely a local natural phenomenon, and whether a law of di
rection can be recognized in the points which have simultaneous elevation or depression.

† On the elevation and depression of the bottom of the South Sea,
and the different areas of alternate movements, see Darwin's *Journal*,
p. 557, 561–566.

about 320 feet in the space of 8000 years; and in 12,000 years, if the movement be regular, parts of the bottom of the sea which lie nearest the shores, and are in the present day covered by nearly fifty fathoms of water, will come to the surface and constitute dry land. But what are such intervals of time compared to the length of the geognostic periods revealed to us in the stratified series of formations, and in the world of extinct and varying organisms! We have hitherto only considered the phenomena of elevation; but the analogies of observed facts lead us with equal justice to assume the possibility of the depression of whole tracts of land. The mean elevation of the non-mountainous parts of France amounts to less than 480 feet. It would not, therefore, require any long period of time, compared with the old geognostic periods, in which such great changes were brought about in the interior of the earth, to effect the permanent submersion of the northwestern part of Europe, and induce essential alterations in its littoral relations.

The depression and elevation of the solid or fluid parts of the earth—phenomena which are so opposite in their action that the effect of elevation in one part is to produce an apparent depression in another—are the causes of all the changes which occur in the configuration of continents. In a work of this general character, and in an impartial exposition of the phenomena of nature, we must not overlook the *possibility* of a diminution of the quantity of water, and a constant depression of the level of seas. There can scarcely be a doubt that, at the period when the temperature of the surface of the earth was higher, when the waters were inclosed in larger and deeper fissures, and when the atmosphere possessed a totally different character from what it does at present, great changes must have occurred in the level of seas, depending upon the increase and decrease of the liquid parts of the earth's surface. But in the actual condition of our planet, there is no direct evidence of a real continuous increase or decrease of the sea, and we have no proof of any gradual change in its level at certain definite points of observation, as indicated by the mean range of the barometer. According to experiments made by Daussy and Antonio Nobile, an increase in the height of the barometer would in itself be attended by a depression in the level of the sea. But as the mean pressure of the atmosphere at the level of the sea is not the same at all latitudes, owing to meteorological causes depending upon the direction of the wind and varying degrees of moisture, the barometer alone can not afford a certain evidence of the general change of level in the ocean. The remarkable fact that some of the ports in the Mediterranean were repeatedly left dry during several hours at the beginning of this century, appears to show that currents may, by changes occurring in their direction and force, occasion a *local* retreat of the sea, and a permanent drying of a small portion of the shore, without being followed by any actual diminution of water, or any permanent depression of the ocean. We must, however, be very cautious in applying the knowledge which we have lately arrived at, regarding these involved phenomena, since we might otherwise be led to ascribe to water, as the elder element, what ought to be referred to the two other elements, earth and air.

As the *external* configuration of continents, which we have already described in their horizontal expansion, exercises, by their variously-indented littoral outlines, a favorable influence on climate, trade, and the progress of civilization, so likewise does their internal articulation, or the vertical elevation of the soil (chains of mountains and elevated plateaux), give rise to equally important results. Whatever produces a polymorphic diversity of forms on the surface of our planetary habitation — such as mountains, lakes, grassy savannas, or even deserts encircled by a band of forests—impresses some peculiar character on the social condition of the inhabitants. Ridges of high land covered by snow impede intercourse ; but a blending of low, discontinued mountain chains* and tracts of valleys, as we see so happily presented in the west and south of Europe, tends to the multiplication of meteorological processes and the products of vegetation, and, from the variety manifested in different kinds of cultivation in each district, even under the same degree of latitude, gives rise to wants that stimulate the activity of the inhabitants. Thus the awful revolutions, during which, by the action of the interior on the crust of the earth, great mountain chains have been elevated by the sudden upheaval of a portion of the oxydized exterior of our planet, have served, after the establishment of repose, and on the revival of organic life, to furnish a richer and more beautiful variety of individual forms, and in a great measure to remove from the earth that aspect of dreary

* Humboldt, *Rel. Hist.*, t. iii., p. 232–234. See, also, the able remarks on the configuration of the earth, and the position of its lines of elevation, in Albrechts von Roon, *Grundzügen der Erd Völker und Staatenkunde*, Abth. i., 1837, s. 158. 270, 276.

uniformity which exercises so impoverishing an influence on
the physical and intellectual powers of mankind.

According to the grand views of Elie de Beaumont, we
must ascribe a relative age to each system of mountain chains[*]
on the supposition that their elevation must necessarily have
occurred between the period of the deposition of the vertically elevated strata and that of the horizontally inclined strata
running at the base of the mountains. The ridges of the
Earth's crust—elevations of strata which are of the same geognostic age—appear, moreover, to follow one common direction. The line of strike of the horizontal strata is not always
parallel with the axis of the chain, but intersects it, so that,
according to my views,[†] the phenomenon of elevation of the
strata, which is even found to be repeated in the neighboring
plains, must be more ancient than the elevation of the chain.
The main direction of the whole continent of Europe (from
southwest to northeast) is opposite to that of the great fissures
which pass from northwest to southeast, from the mouths of
the Rhine and Elbe, through the Adriatic and Red Seas, and
through the mountain system of Putschi-Koh in Luristan, toward the Persian Gulf and the Indian Ocean. This almost
rectangular intersection of geodesic lines exercises an important influence on the commercial relations of Europe, Asia,
and the northwest of Africa, and on the progress of civilization
on the formerly more flourishing shores of the Mediterranean.[‡]

Since grand and lofty mountain chains so strongly excite
our imagination by the evidence they afford of great terrestrial revolutions, and when considered as the boundaries of
climates, as lines of separation for waters, or as the site of a
different form of vegetation, it is the more necessary to demonstrate, by a correct numerical estimation of their volume,
how small is the quantity of their elevated mass when compared with the area of the adjacent continents. The mass
of the Pyrenees, for instance, the mean elevation of whose
summits, and the areal quantity of whose base have been ascertained by accurate measurements, would, if scattered over

---

[*] Leop. von Buch, *Ueber die Geognostischen Systeme von Deutschland*,
in his *Geogn. Briefen an Alexander von Humboldt*, 1824, s. 265–271;
Elie de Beaumont, *Recherches sur les Révolutions de la Surface du Globe*,
1829, p. 297–307.

[†] Humboldt, *Asie Centrale*, t. i., p. 277–283. See, also, my *Essai
sur le Gisement des Roches*, 1822, p. 57, and *Relat. Hist.*, t. iii., p.
244–250.

[‡] *Asie Centrale*, t. i., p. 284, 286. The Adriatic Sea likewise follows
a direction from S.E. to N.W.

the surface of France, only raise its mean level about 115 feet. The mass of the eastern and western Alps would in like manner only increase the height of Europe about 21½ feet above its present level. I have found by a laborious investigation,* which, from its nature, can only give a maximum limit, that the center of gravity of the volume of the land raised above the present level of the sea in Europe and North America is respectively situated at an elevation of 671 and 748 feet, while it is at 1132 and 1152 feet in Asia and South America. These numbers show the low level of northern regions. In Asia the vast steppes of Siberia are compensated for by the great elevations of the land (between the Himalaya, the North Thibetian chain of Kuen-lun, and the Celestial Mountains), from 28° 30' to 40° north latitude. We may, to a certain extent, trace in these numbers the portions of the Earth in which the Plutonic forces were most intensely manifested in the interior by the upheaval of continental masses.

There are no reasons why these Plutonic forces may not, in future ages, add new mountain systems to those which Elie de Beaumont has shown to be of such different ages, and inclined in such different directions. Why should the crust of the Earth have lost its property of being elevated in ridges? The recently-elevated mountain systems of the Alps and the Cordilleras exhibit in Mont Blanc and Monte Rosa, in Sorata, Illimani, and Chimborazo, colossal elevations which do not favor the assumption of a decrease in the intensity of the subterranean forces. All geognostic phenomena indicate the periodic alternation of activity and repose ;† but the quiet we now enjoy is only apparent. The tremblings which still agitate the surface under all latitudes, and in every species of rock, the elevation of Sweden, the appearance of new islands of eruption, are all conclusive as to the unquiet condition of our planet.

* *De la hauteur Moyenne des Continents*, in my *Asie Centrale,* t. i., p 82–90, 165–189. The results which I have obtained are to be regarded as the extreme value (*nombres-limites*). Laplace's estimate of the mean height of continents at 3280 feet is at least three times too high. The immortal author of the *Mécanique Celeste* (t. v., p. 14) was led to this conclusion by hypothetical views as to the mean depth of the sea. I have shown (*Asie Centr.*, t. i., p. 93) that the old Alexandrian mathematicians, on the testimony of Plutarch (*in Æmilio Paulo,* cap. 15), believed this depth to depend on the height of the mountains. The height of the center of gravity of the volume of the continental masses is probably subject to slight variations in the course of many centuries.

† *Zweiter Geologischer Brief von Elie de Beaumont an Alexander von Humboldt*, in Poggendorf's *Annalen*, bd. xxv., s. 1–58.

The two envelopes of the solid surface of our planet—the liquid and the aëriform—exhibit, owing to the mobility of their particles, their currents, and their atmospheric relations, many analogies combined with the contrasts which arise from the great difference in the condition of their aggregation and elasticity. The depths of ocean and of air are alike unknown to us. At some few places under the tropics no bottom has been found with soundings of 276,000 feet (or more than four miles), while in the air, if, according to Wollaston, we may assume that it has a limit from which waves of sound may be reverberated, the phenomenon of twilight would incline us to assume a height at least nine times as great.* The aërial ocean rests partly on the solid earth, whose mountain chains and elevated plateaux rise, as we have already seen, like green wooded shoals, and partly on the sea, whose surface forms a moving base, on which rest the lower, denser, and more saturated strata of air.

Proceeding upward and downward from the common limit of the aërial and liquid oceans, we find that the strata of air and water are subject to determinate laws of decrease of temperature. This decrease is much less rapid in the air than in the sea, which has a tendency under all latitudes to maintain its temperature in the strata of water most contiguous to the atmosphere, owing to the sinking of the heavier and more cooled particles. A large series of the most carefully conducted observations on temperature shows us that in the ordinary and mean condition of its surface, the ocean from the equator to the forty-eighth degree of north and south latitude is somewhat warmer than the adjacent strata of air.† Owing to this decrease of temperature at increasing depths, fishes and other inhabitants of the sea, the nature of whose digestive and respiratory organs fits them for living in deep water, may even, under the tropics, find the low degree of temperature and the coolness of climate characteristic of more temperate and more northern latitudes. This circumstance, which is analogous to the prevalence of a mild and even cold air on the elevated plains of the torrid zone, exercises a special influence on the migration and geographical distribution of many marine animals. Moreover, the depths at which fishes live, modify, by the increase of pressure, their cutaneous respiration, and the

* [See Wilson's Paper, *On Wollaston's Argument from the Limitation of the Atmosphere as to the finite Divisibility of Matter.*— *Trans. of the Royal Society of Edinb.*, vol. xvi., p. 1, 1845.]— *Tr.*
† Humboldt, *Relation Hist.*, t. iii., chap. xxix., p. 514–530.

oxygenous and nitrogenous contents of their swimming blad-
ders.

As fresh and salt water do not attain the maximum of
their density at the same degree of temperature, and as the
saltness of the sea lowers the thermometrical degree corresponding to this point, we can understand how the water
drawn from great depths of the sea during the voyages of
Kotzebue and Dupetit-Thouars could have been found to have
only the temperature of 37° and 36°·5. This icy temperature
of sea water, which is likewise manifested at the depths of
tropical seas, first led to a study of the lower polar currents,
which move from both poles toward the equator. Without
these submarine currents, the tropical seas at those depths
could only have a temperature equal to the local maximum
of cold possessed by the falling particles of water at the radiating and cooled surface of the tropical sea. In the Mediterranean, the cause of the absence of such a refrigeration of the
lower strata is ingeniously explained by Arago, on the assumption that the entrance of the deeper polar currents into
the Straits of Gibraltar, where the water at the surface flows
in from the Atlantic Ocean from west to east, is hindered by
the submarine counter-currents which move from east to
west, from the Mediterranean into the Atlantic.

The ocean, which acts as a general equalizer and moder-
ator of climates, exhibits a most remarkable uniformity and
constancy of temperature, especially between 10° north and
10° south latitude,* over spaces of many thousands of square
miles, at a distance from land where it is not penetrated by
currents of cold and heated water. It has, therefore, been
justly observed, that an exact and long-continued investigation of these thermic relations of the tropical seas might most
easily afford a solution to the great and much-contested problem of the permanence of climates and terrestrial tempera
tures.† Great changes in the luminous disk of the sun would,

* See the series of observations made by me in the South Sea, from
0° 5' to 13° 16' N. lat., in my *Asie Centrale*, t. iii., p. 234.

† "We might (by means of the temperature of the ocean under the
tropics) enter into the consideration of a question which has hitherto
remained unanswered, namely, that of the constancy of terrestrial tem
peratures, without taking into account the very circumscribed local
influences arising from the diminution of wood in the plains and on
mountains, and the drying up of lakes and marshes. Each age might
easily transmit to the succeeding one some few data, which would per·
haps furnish the most simple, exact, and direct means of deciding whetn-
er the sun, which is almost the sole and exclusive source of the heat of

if they were of long duration, be reflected with more certainty in the mean temperature of the sea than in that of the solid land.

The zones, at which occur the maxima of the oceanic temperature and of the density (the saline contents) of its waters, do not correspond with the equator. The two maxima are separated from one another, and the waters of the highest temperature appear to form two nearly parallel lines north and south of the geographical equator. Lenz, in his voyage of circumnavigation, found in the Pacific the maxima of density in 22° north and 17° south latitude, while its minimum was situated a few degrees to the south of the equator. In the region of calms the solar heat can exercise but little influence on evaporation, because the stratum of air impregnated with saline aqueous vapor, which rests on the surface of the sea, remains still and unchanged.

The surface of all connected seas must be considered as having a general perfectly equal level with respect to their mean elevation. Local causes (probably prevailing winds and currents) may, however, produce permanent, although trifling changes in the level of some deeply-indented bays, as, for instance, the Red Sea. The highest level of the water at the Isthmus of Suez is at different hours of the day from 24 to 30 feet above that of the Mediterranean. The form of the Straits of Bab-el-Mandeb, through which the waters appear to find an easier ingress than egress, seems to contribute to this remarkable phenomenon, which was known to the ancients.* The admirable geodetic operations of Corabœuf and Delcrois show that no perceptible difference of level exists between the upper surfaces of the Atlantic and the Mediterranean, along the chain of the Pyrenees, or between the coasts of northern Holland and Marseilles.†

our planet, changes its physical constitution and splendor, like the great er number of the stars, or whether, on the contrary, that luminary has attained to a permanent condition."—Arago, in the *Comptes Rendus des Séances de l'Acad. des Sciences*, t. xi., Part ii., p. 309.

* Humboldt, *Asie Centrale*, t. ii., p. 321, 327.

† See the numerical results in p. 328–333 of the volume just named. From the geodesical levelings which, at my request, my friend General Bolivar caused to be taken by Lloyd and Falmarc, in the years 1828 and 1829, it was ascertained that the level of the Pacific is at the utmost 3½ feet higher than that of the Caribbean Sea; and even that at different hours of the day each of the seas is in turn the higher, according to their respective hours of flood and ebb. If we reflect that in a distance of 64 miles, comprising 933 stations of observation, an error of three feet would be very apt to occur, we may say that in these new

Disturbances of equilibrium and consequent movements of
the waters are partly irregular and transitory, dependent upon
winds, and producing waves which sometimes, at a distance
from the shore and during a storm, rise to a height of more
than 35 feet; partly regular and periodic, occasioned by the
position and attraction of the sun and moon, as the ebb and
flow of the tides; and partly permanent, although less in·
tense, occurring as oceanic currents. The phenomena of
tides, which prevail in all seas (with the exception of the
smaller ones that are completely closed in, and where the ebb-
ing and flowing waves are scarcely or not at all perceptible),
have been perfectly explained by the Newtonian doctrine,
and thus brought "within the domain of necessary facts."
Each of these periodically-recurring oscillations of the waters
of the sea has a duration of somewhat more than half a day.
Although in the open sea they scarcely attain an elevation of
a few feet, they often rise considerably higher where the waves
are opposed by the configuration of the shores, as, for instance,
at St. Malo and in Nova Scotia, where they reach the respective elevations of 50 feet, and of 65 to 70 feet. "It has
been shown by the analysis of the great geometrician Laplace, that, supposing the depth to be wholly inconsiderable
when compared with the radius of the earth, the stability of
the equilibrium of the sea requires that the density of its fluid
should be less than that of the earth; and, as we have already
seen, the earth's density is in fact five times greater than
that of water. The elevated parts of the land can not therefore be overflowed, nor can the remains of marine animals
found on the summits of mountains have been conveyed to
those localities by any previous high tides."* It is no slight

operations we have further confirmation of the equilibrium of the waters which communicate round Cape Horn. (Arago, in the *Annuaire
du Bureau des Longitudes pour* 1831, p. 319.) I had inferred, from
barometrical observations instituted in 1799 and 1804, that if there were
any difference between the level of the Pacific and the Atlantic (Caribbean Sea), it could not exceed three meters (nine feet three inches).
See my *Relat. Hist.*, t. iii., p. 555–557, and *Annales de Chimie*, t. i.,
p. 55–64. The measurements, which appear to establish an excess of
height for the waters of the Gulf of Mexico, and for those of the northern part of the Adriatic Sea, obtained by combining the trigonometrical
operations of Delcrois and Choppin with those of the Swiss and Austrian engineers, are open to many doubts. Notwithstanding the form
of the Adriatic, it is improbable that the level of its waters in its northern portion should be 28 feet higher than that of the Mediterranean at
Marseilles, and 25 feet higher than the level of the Atlantic Ocean.
See my *Asie Centrale*, t. ii., p. 332.
* Bessel, *Ueber Fluth und Ebbe*, in Schumacher's *Jahrbuch*, 1838, s. 225

evidence of the importance of analysis, which is too often regarded with contempt among the unscientific, that Laplace's
perfect theory of tides has enabled us, in our astronomical
ephemerides, to predict the height of spring-tides at the periods of new and full moon, and thus put the inhabitants of the
sea-shore on their guard against the increased danger attend-
ng these lunar revolutions.

Oceanic currents, which exercise so important an influence
on the intercourse of nations and on the climatic relations of
adjacent coasts, depend conjointly upon various causes, differing alike in nature and importance. Among these we may
reckon the periods at which tides occur in their progress round
the earth ; the duration and intensity of prevailing winds ;
the modifications of density and specific gravity which the particles of water undergo in consequence of differences in the
temperature and in the relative quantity of saline contents at
different latitudes and depths ;* and, lastly, the horary variations of the atmospheric pressure, successively propagated from
east to west, and occurring with such regularity in the tropics. These currents present a remarkable spectacle ; like rivers of uniform breadth, they cross the sea in different directions, while the adjacent strata of water, which remain undisturbed, form, as it were, the banks of these moving streams.
This difference between the moving waters and those at rest
is most strikingly manifested where long lines of sea-weed,
borne onward by the current, enable us to estimate its velocity. In the lower strata of the atmosphere, we may sometimes, during a storm, observe similar phenomena in the limited aërial current, which is indicated by a narrow line of
trees, which are often found to be overthrown in the midst of
a dense wood.

The general movement of the sea from east to west be-

* The relative density of the particles of water depends simultaneously on the temperature and on the amount of the saline contents—a
circumstance that is not sufficiently borne in mind in considering the
cause of currents. The submarine current, which brings the cold polar water to the equatorial regions, would follow an exactly opposite
course, that is to say, from the equator toward the poles, if the difference in saline contents were alone concerned. In this view, the geographical distribution of temperature and of density in the water of
the ocean, under the different zones of latitude and longitude, is of
great importance. The numerous observations of Lenz (Poggendorf's
Annalen, bd. xx., 1830, s. 129), and those of Captain Beechey, collected in his *Voyage to the   acific,* vol. ii., p. 727, deserve particular attention. See Humboldt, *Relat. Hist.,* t. i., p. 74, and *Asie Centrale,*
t. iii., p. 356.

tween the tropics (termed the equatorial or rotation current) is considered to be owing to the propagation of tides and to the trade winds. Its direction is changed by the resistance it experiences from the prominent eastern shores of continents. The results recently obtained by Daussy regarding the veloc ity of this current, estimated from observations made on the distances traversed by bottles that had purposely been thrown into the sea, agree within one eighteenth with the velocity of motion (10 French nautical miles, 952 toises each, in 24 hours) which I had found from a comparison with earlier experiments.* Christopher Columbus, during his third voyage, when he was seeking to enter the tropics in the meridian of Teneriffe, wrote in his journal as follows :† " I regard it as proved that the waters of the sea move from east to west, as do the heavens (*las aguas van con los cielos*), that is to say, like the apparent motion of the sun, moon, and stars."

The narrow currents, or true oceanic rivers which traverse the sea, bring warm water into higher and cold water into lower latitudes. To the first class belongs the celebrated Gulf Stream,‡ which was known to Anghiera,§ and more especially to Sir Humphrey Gilbert in the sixteenth century. Its first impulse and origin is to be sought to the south of the Cape of Good Hope ; after a long circuit it pours itself from the Caribbean Sea and the Mexican Gulf through the Straits of the Bahamas, and, following a course from south-southwest to north-northeast, continues to recede from the shores of the United States, until, further deflected to the eastward by the Banks of Newfoundland, it approaches the European coasts, frequently throwing a quantity of tropical seeds (*Mimosa scandens, Guilandina bonduc, Dolichos urens*) on the shores of Ireland, the Hebrides, and Norway. The northeastern prolongation tends to mitigate the cold of the ocean, and to ameliorate the climate on the most northern extremity of Scandinavia. At the point where the Gulf Stream

* Humboldt, *Relat. Hist.*, t. i., p. 64 ; *Nouvelles Annales des Voyages*, 1839, p. 255.

† Humboldt, *Examen Crit. de l'Hist. de la Géogr.*, t. iii., p. 100. Columbus adds shortly after (Navarrete, *Coleccion de los Viages y Descubrimientos de los Espanoles*, t. i., p. 260), that the movement is strongest in the Caribbean Sea. In fact, Rennell terms this region, " not a current, but a sea in motion" (*Investigation of Currents*, p. 23).

‡ Humboldt, *Examen Critique*, t. ii., p. 250; *Relat. Hist.*, t. i., p. 66–74.

§ Petrus Martyr de Anghiera, *De Rebus Oceanicis et Orbe Novo*, Bas., 1523, Dec. iii., lib. vi., p. 57. See Humboldt, *Examen Critique*, t. ii., p. 254–257, and t. iii., p. 108.

is deflected from the Banks of Newfoundland toward the east, it sends off branches to the south near the Azores.* This is the situation of the Sargasso Sea, or that great bank of weeds which so vividly occupied the imagination of Christopher Columbus, and which Oviedo calls the sea-weed meadows (*Praderias de yerva*). A host of small marine animals inhabits these gently-moved and evergreen masses of *Fucus natans*, one of the most generally distributed of the social plants of the sea.

The counterpart of this current (which in the Atlantic Ocean, between Africa, America, and Europe, belongs almost exclusively to the northern hemisphere) is to be found in the South Pacific, where a current prevails, the effect of whose low temperature on the climate of the adjacent shores I had an opportunity of observing in the autumn of 1802. It brings the cold waters of the high southern latitudes to the coast of Chili, follows the shores of this continent and of Peru, first from south to north, and is then deflected from the Bay of Arica onward from south-southeast to north-northwest. At certain seasons of the year the temperature of this cold oceanic current is, in the tropics, only 60°, while the undisturbed adjacent water exhibits a temperature of 81°·5 and 83°·7. On that part of the shore of South America south of Payta, which inclines furthest westward, the current is suddenly deflected in the same direction from the shore, turning so sharply to the west that a ship sailing northward passes suddenly from cold into warm water.

It is not known to what depth cold and warm oceanic currents propagate their motion; but the deflection experienced by the South African current, from the Lagullas Bank, which is fully from 70 to 80 fathoms deep, would seem to imply the existence of a far-extending propagation. Sand banks and shoals lying beyond the line of these currents may, as was first discovered by the admirable Benjamin Franklin, be recognized by the coldness of the water over them. This depression of the temperature appears to me to depend upon the fact that, by the propagation of the motion of the sea, deep waters rise to the margin of the banks and mix with the upper strata. My lamented friend, Sir Humphrey Davy, ascribed this phenomenon (the knowledge of which is often of great practical utility in securing the safety of the navigator) to the descent of the particles of water that had been cooled by nocturnal ra-

* Humboldt, *Examen Crit.*, t. iii., p. 64–109.

diation, and which remain nearer to the surface, owing to the hinderance placed in the way of their greater descent by the intervention of sand-banks. By his observations Franklin may be said to have converted the thermometer into a sounding line. Mists are frequently found to rest over these depths, owing to the condensation of the vapor of the atmosphere by the cooled waters. I have seen such mists in the south of Jamaica, and also in the Pacific, defining with sharpness and clearness the form of the shoals below them, appearing to the eye as the aërial reflection of the bottom of the sea. A still more striking effect of the cooling produced by shoals is manifested in the higher strata of air, in a somewhat analogous manner to that observed in the case of flat coral reefs, or sand islands. In the open sea, far from the land, and when the air is calm, clouds are often observed to rest over the spots where shoals are situated, and their bearing may then be taken by the compass in the same manner as that of a high mountain or isolated peak.

Although the surface of the ocean is less rich in living forms than that of continents, it is not improbable that, on a further investigation of its depths, its interior may be found to possess a greater richness of organic life than any other portion of our planet. Charles Darwin, in the agreeable narrative of his extensive voyages, justly remarks that our forests do not conceal so many animals as the low woody regions of the ocean, where the sea-weed, rooted to the bottom of the shoals, and the sev ered branches of fuci, loosened by the force of the waves and currents, and swimming free, unfold their delicate foliage, upborne by air-cells.* The application of the microscope increases, in the most striking manner, our impression of the rich luxuriance of animal life in the ocean, and reveals to the astonished senses a consciousness of the universality of life. In the oceanic depths, far exceeding the height of our loftiest mountain chains, every stratum of water is animated with polygastric sea-worms, Cyclidiæ, and Ophrydinæ. The waters swarm with countless hosts of small luminiferous animalcules, Mammaria (of the order of Acalephæ), Crustacea, Peridinea, and circling Nereides, which, when attracted to the surface by peculiar meteorological conditions, convert every wave into a foaming band of flashing light.

---

* [See *Structure and Distribution of Coral Reefs*, by Charles Darwin, London, 1842. Also, *Narrative of the Surveying Voyage of H.M.S. "Fly," in the Eastern Archipelago, during the Years* 1842–1846, by J. B. Jukes, Naturalist to the expedition, 1847.]—*Tr.*

The abundance of these marine animalcules, and the anima₄ matter yielded by their rapid decomposition, are so vast that the sea water itself becomes a nutrient fluid to many of the larger animals. However much this richness in animated forms, and this multitude of the most various and highly-developed microscopic organisms may agreeably excite the fancy, the imagination is even more seriously, and, I might say, more solemnly moved by the impression of boundlessness and immeasurability, which are presented to the mind by every sea voyage. All who possess an ordinary degree of mental activity, and delight to create to themselves an inner world of thought, must be penetrated with the sublime image of the infinite when gazing around them on the vast and boundless sea, when involuntarily the glance is attracted to the distant horizon, where air and water blend together, and the stars continually rise and set before the eyes of the mariner. This contemplation of the eternal play of the elements is clouded, like every human joy, by a touch of sadness and of longing.

A peculiar predilection for the sea, and a grateful remembrance of the impression which it has excited in my mind, when I have seen it in the tropics in the calm of nocturnal rest, or in the fury of the tempest, have alone induced me to speak of the individual enjoyment afforded by its aspect before I entered upon the consideration of the favorable influence which the proximity of the ocean has incontrovertibly exercised on the cultivation of the intellect and character of many nations, by the multiplication of those bands which ought to encircle the whole of humanity, by affording additional means of arriving at a knowledge of the configuration of the earth, and furthering the advancement of astronomy, and of all other mathematical and physical sciences. A portion of this influence was at first limited to the Mediterranean and the shores of southwestern Africa, but from the sixteenth century it has widely spread, extending to nations who live at a distance from the sea, in the interior of continents. Since Columbus was sent to " unchain the ocean"* (as the unknown voice whispered to him in a dream when he lay on a sick-bed near

* The voice addressed him in these words, " Maravillosamente Dios hizo sonar tu nombre en la tierra; de los atamientos de la mar Oceana, que estaban cerrados con cadenas tan fuertes, te dió las llaves"—" God will cause thy name to be wonderfully resounded through the earth, and give thee the keys of the gates of the ocean, which are closed with strong chains." The dream of Columbus is related in the letter to the Catholic monarchs of July the 7th, 1503. (Humboldt, *Examen Critique*, t. iii., p. 234.)

the River Belem), man has ever boldly ventured onward toward the discovery of unknown regions.

The second external and general covering of our planet, the aërial ocean, in the lower strata, and on the shoals of which we live, presents six classes of natural phenomena, which manifest the most intimate connection with one another. They are dependent on the chemical composition of the atmosphere, the variations in its transparency, polarization, and color, its density or pressure, its temperature and humidity, and its electricity. The air contains in oxygen the first element of physical animal life, and, besides this benefit, it possesses another, which may be said to be of a nearly equally high character, namely, that of conveying sound ; a faculty by which it likewise becomes the conveyer of speech and the means of communicating thought, and, consequently, of maintaining social intercourse. If the Earth were deprived of an atmosphere, as we suppose our moon to be, it would present itself to our imagination as a soundless desert.

The relative quantities of the substances composing the strata of air accessible to us have, since the beginning of the nineteenth century, become the object of investigations, in which Gay-Lussac and myself have taken an active part ; it is, however, only very recently that the admirable labors of Dumas and Boussingault have, by new and more accurate methods, brought the chemical analysis of the atmosphere to a high degree of perfection. According to this analysis, a volume of dry air contains 20·8 of oxygen and 79·2 of nitrogen, besides from two to five thousandth parts of carbonic acid gas, a still smaller quantity of carbureted hydrogen gas,* and, according to the important experiments of Saussure and Liebig, traces of ammoniacal vapors,† from which plants derive their nitrogenous contents. Some observations of Lewy render it probable that the quantity of oxygen varies percep-

---

* Boussingault, *Recherches sur la Composition de l'Atmosphère*, in the *Annales de Chimie et de Physique*, t. lvii., 1834, p. 171–173 ; and lxxi. 1839, p. 116. According to Boussingault and Lewy, the proportion of carbonic acid in the atmosphere at Audilly, at a distance, therefore, from the exhalations of a city, varied only between 0·00028 and 0·00031 in volume.

† Liebig, in his important work, entitled *Die Organische Chemie in ihrer Anwendung auf Agricultur und Physiologie*, 1840, s. 62–72. On the influence of atmospheric electricity in the production of nitrate of ammonia, which, coming into contact with carbonate of lime, is changed into carbonate of ammonia, see Boussingault's *Economie Rurale considérée dans ses Rapports avec la Chimie et la Météorologie*, 1844, t. ii., p. 247, 267, and t. i., p. 84.

tibly, although but slightly, over the sea and in the interior
of continents, according to local conditions or to the seasons of
the year. We may easily conceive that changes in the oxygen held in solution in the sea, produced by microscopic animal organisms, may be attended by alterations in the strata
of air in immediate contact with it.* The air which Martins
collected at Faulhorn at an elevation of 8767 feet, contained
as much oxygen as the air at Paris.†

The admixture of carbonate of ammonia in the atmosphere
may probably be considered as older than the existence of organic beings on the surface of the earth. The sources from
which carbonic acid‡ may be yielded to the atmosphere are
most numerous. In the first place we would mention the respiration of animals, who receive the carbon which they inhale
from vegetable food, while vegetables receive it from the atmosphere ; in the next place, carbon is supplied from the interior of the earth in the vicinity of exhausted volcanoes and
thermal springs, from the decomposition of a small quantity of
carbureted hydrogen gas in the atmosphere, and from the electric discharges of clouds, which are of such frequent occurrence
within the tropics. Besides these substances, which we have
considered as appertaining to the atmosphere at all heights
that are accessible to us, there are others accidentally mixed
with them, especially near the ground, which sometimes, in
the form of miasmatic and gaseous contagia, exercise a noxious
influence on animal organization. Their chemical nature has
not yet been ascertained by direct analysis ; but, from the consideration of the processes of decay which are perpetually going on in the animal and vegetable substances with which the
surface of our planet is covered, and judging from analogies
deduced from the domain of pathology, we are led to infer the
existence of such noxious local admixtures. Ammoniacal and
other nitrogenous vapors, sulphureted hydrogen gas, and compounds analogous to the polybasic ternary and quaternary combinations of the vegetable kingdom, may produce miasmata,§

* Lewy, in the *Comptes Rendus de l'Acad. des Sciences*, t. xvii., Part
ii., p. 235–248.

† Dumas, in the *Annales de Chimie*, 3e *Série*, t. iii., 1841, p. 257.

‡ In this enumeration, the exhalation of carbonic acid by plants during the night, while they inhale oxygen, is not taken into account, because the increase of carbonic acid from this source is amply counterbalanced by the respiratory process of plants during the day. See Boussingault's *Econ. Rurale*, t. i., p. 53–68, and Liebig's *Organische Chemie*,
s. 16, 21.

§ Gay-Lussac, in *Annales de Chimie*, t. liii., p. 120 ; Payen, *Mém. sur*

which, under various forms, may generate ague and typhus fever (not by any means exclusively on wet, marshy ground, or on coasts covered by putrescent mollusca, and low bushes of *Rhizophora mangle* and Avicennia). Fogs, which have a peculiar smell at some seasons of the year, remind us of these accidental admixtures in the lower strata of the atmosphere. Winds and currents of air caused by the heating of the ground even carry up to a considerable elevation solid substances reduced to a fine powder. The dust which darkens the air for an extended area, and falls on the Cape Verd Islands, to which Darwin has drawn attention, contains, according to Ehrenberg's discovery, a host of silicious-shelled infusoria.

As principal features of a general descriptive picture of the atmosphere, we may enumerate :

1. *Variations of atmospheric pressure :* to which belong the horary oscillations, occurring with such regularity in the tropics, where they produce a kind of ebb and flow in the atmosphere, which can not be ascribed to the attraction of the moon,* and which differs so considerably according to geographical latitude, the seasons of the year, and the elevation above the level of the sea.

2. *Climatic distribution of heat,* which depends on the relative position of the transparent and opaque masses (the fluid and solid parts of the surface of the earth), and on the hypsometrical configuration of continents ; relations which determine the geographical position and curvature of the isothermal lines (or curves of equal mean annual temperature) both in a horizontal and vertical direction, or on a uniform plane, or in different superposed strata of air.

3. *The distribution of the humidity of the atmosphere.* The quantitative relations of the humidity depend on the differences in the solid and oceanic surfaces ; on the distance from the equator and the level of the sea ; on the form in which the

la Composition Chimique des Végétaux, p. 36, 42 ; Liebig, *Org. Chemie.* s. 229–345; Boussingault, *Econ. Rurale,* t. i., p. 142–153.

* Bouvard, by the application of the formulæ, in 1827, which Laplace had deposited with the Board of Longitude shortly before his death, found that the portion of the horary oscillations of the pressure of the atmosphere, which depends on the attraction of the moon, can not raise the mercury in the barometer at Paris more than the 0·018 of a millimeter, while eleven years' observations at the same place show the mean barometric oscillation, from 9 A.M. to 3 P.M., to be 0·756 millim., and from 3 P.M. to 9 P.M., 0·373 millim. See *Mémoires de l'Acad. des Sciences,* t. vii., 1827, p. 267.

aqueous vapor is precipitated, and on the connection existing between these deposits and the changes of temperature, and the direction and succession of winds.

4. *The electric condition of the atmosphere.* The primary cause of this condition, when the heavens are serene, is still much contested. Under this head we must consider the relation of ascending vapors to the electric charge and the form of the clouds, according to the different periods of the day and year; the difference between the cold and warm zones of the earth, or low and high lands; the frequency or rarity of thunder storms, their periodicity and formation in summer and winter; the causal connection of electricity, with the infrequent occurrence of hail in the night, and with the phenomena of water and sand spouts, so ably investigated by Peltier.

The horary oscillations of the barometer, which in the tropics present two maxima (viz., at 9 or $9\frac{1}{4}$ A.M., and $10\frac{1}{2}$ or $10\frac{3}{4}$ P.M., and two minima, at 4 or $4\frac{1}{4}$ P.M., and 4 A.M., occurring, therefore, in almost the hottest and coldest hours), have long been the object of my most careful diurnal and nocturnal observations.* Their regularity is so great, that, in the daytime especially, the hour may be ascertained from the height of the mercurial column without an error, on the average, of more than fifteen or seventeen minutes. In the torrid zones of the New Continent, on the coasts as well as at elevations of nearly 13,000 feet above the level of the sea, where the mean temperature falls to $44°·6$, I have found the regularity of the ebb and flow of the aërial ocean undisturbed by storms, hurricanes, rain, and earthquakes. The amount of the daily oscillations diminishes from $1·32$ to $0·18$ French lines from the equator to 70° north latitude, where Bravais made very accurate observations at Bosekop.† The supposition that, much nearer the pole, the height of the barometer is really less at 10 A.M. than at 4 P.M., and, consequently, that the maximum and minimum influences of these hours

* *Observations faites pour constater la Marche des Variations Horaires du Baromètre sous les Tropiques*, in my *Relation Historique du Voyage aux Régions Equinoxiales*, t. iii., p. 270–313.

† Bravais, in Kaemtz and Martins, *Météorologie*, p. 263. At Halle (51° 29′ N. lat.), the oscillation still amounts to $0·28$ lines. It would seem that a great many observations will be required in order to obtain results that can be trusted in regard to the hours of the maximum and minimum on mountains in the temperate zone. See the observations of horary variations, collected on the Faulhorn in 1832, 1841, and 1842 (Martins, *Météorologie*, p. 254.)

are inverted, is not confirmed by Parry's observations at Port Bowen (73° 14′).

The mean height of the barometer is somewhat less under the equator and in the tropics, owing to the effect of the rising current,[*] than in the temperate zones, and it appears to attain its maximum in Western Europe between the parallels of 40° and 45°. If with Kämtz we connect together by *isobarometric* lines those places which present the same mean difference between the monthly extremes of the barometer, we shall have curves whose geographical position and inflections yield important conclusions regarding the influence exercised by the form of the land and the distribution of seas on the oscillations of the atmosphere. Hindostan, with its high mountain chains and triangular peninsulas, and the eastern coasts of the New Continent, where the warm Gulf Stream turns to the east at the Newfoundland Banks, exhibit greater isobarometric oscillations than do the group of the Antilles and Western Europe. The prevailing winds exercise a principal influence on the diminution of the pressure of the atmosphere, and this, as we have already mentioned, is accompanied, according to Daussy, by an elevation of the mean level of the sea.[†]

As the most important fluctuations of the pressure of the atmosphere, whether occurring with horary or annual regularity, or accidentally, and then often attended by violence and danger,[‡] are, like all the other phenomena of the weather, mainly owing to the heating force of the sun's rays, it has long been suggested (partly according to the idea of Lambert) that the direction of the wind should be compared with the height of the barometer, alternations of temperature, and the increase and decrease of humidity. Tables of atmospheric pressure during different winds, termed *barometric windroses*, afford a deeper insight into the connection of meteorological phenomena.[§] Dove has, with admirable sagacity, recognized, in the "law of rotation" in both hemispheres, which he himself established, the cause of many important processes in the aërial ocean.[‖] The difference of temperature between the

[*] Humboldt, *Essai sur la Géographie des Plantes*, 1807, p. 90; and in *Rel. Hist.*, t. iii., p. 313; and on the diminution of atmospheric pressure in the tropical portions of the Atlantic, in Poggend., *Annalen der Physik*, bd. xxxvii., s. 245–258, and s. 468–486.

[†] Daussy, in the *Comptes Rendus*, t. iii., p. 136.

[‡] Dove, *Ueber die Stürme*, in Poggend., *Annalen*, bd. lii., s. 1.

[§] Leopold von Buch, *Barometrische Windrose*, in *Abhandl. der Akad. der Wiss. zu Berlin aus den Jahren* 1818–1819, s. 187.

[‖] See Dove, *Meteorologische Untersuchungen*, 1837, s. ^^–313; and

equatorial and polar regions engenders two opposite currents
in the upper strata of the atmosphere and on the Earth's sur·
face. Owing to the difference between the rotatory velocity
at the poles and at the equator, the polar current is deflected
eastward, and the equatorial current westward. The great
phenomena of atmospheric pressure, the warming and cooling
of the strata of air, the aqueous deposits, and even, as Dove
has correctly represented, the formation and appearance of
clouds, alike depend on the opposition of these two currents,
on the place where the upper one descends, and on the displacement of the one by the other. Thus the figures of the
clouds, which form an animated part of the charms of a landscape, announce the processes at work in the upper regions of
the atmosphere, and, when the air is calm, the clouds will
often present, on a bright summer sky, the " projected image"
of the radiating soil below.

Where this influence of radiation is modified by the relative
position of large continental and oceanic surfaces, as between
the eastern shore of Africa and the western part of the Indian
peninsula, its effects are manifested in the Indian monsoons,
which change with the periodic variations in the sun's decli·
nation,* and which were known to the Greek navigators under the name of *Hippalos.* In the knowledge of the monsoons, which undoubtedly dates back thousands of years among
the inhabitants of Hindostan and China, of the eastern parts
of the Arabian Gulf and of the western shores of the Malayan

the excellent observations of Kämtz on the descent of the west wind
of the upper current in high latitudes, and the general phenomena of
the direction of the wind, in his *Vorlesungen über Meterologie,* 1840, s.
58–66, 196–200, 327–336, 353–364; and in Schumacher's *Jahrbuch für*
1838, s. 291–302. A very satisfactory and vivid representation of meteorological phenomena is given by Dove, in his small work entitled
Witterungsverhältnisse von Berlin, 1842. On the knowledge of the
earlier navigators of the rotation of the wind, see Churruca, *Viage al
Magellanes,* 1793, p. 15; and on a remarkable expression of Columbus,
which his son Don Fernando Colon has presented to us in his *Vida del
Almirante,* cap. 55, see Humboldt, *Examen Critique de l'Hist. de Géographie,* t. iv., p. 253.

* *Monsun* (Malayan *musim,* the *hippalos* of the Greeks) is derived
from the Arabic word *mausim,* a set time or season of the year, the time
of the assemblage of pilgrims at Mecca. The word has been applied
to the seasons at which certain winds prevail, which are, besides, named
from places lying in the direction from whence they come; thus, for
instance, there is the *mausim* of Aden, of Guzerat, Malabar, &c. (Lassen, *Indische Alterthumskunde,* bd. i., 1843, s. 211). On the contrasts
between the solid or fluid substrata of the atmosphere, see Dove, in *Der
Abhandl. der Akad. der Wiss. zu Berlin aus dem Jahr* 1842, s. 239

Sea, and in the still more ancient and more general acquaintance with land and sea winds, lies concealed, as it were, the germ of that meteorological science which is now making such rapid progress. The long chain of *magnetic stations* extending from Moscow to Pekin, across the whole of Northern Asia, will prove of immense importance in determining the *law of the winds*, since these stations have also for their object the investigation of general meteorological relations. The comparison of observations made at places lying so many hundred miles apart, will decide, for instance, whether the same east wind blows from the elevated desert of Gobi to the interior of Russia, or whether the direction of the aërial current first began in the middle of the series of the stations, by the descent of the air from the higher regions. By means of such observations, we may learn, in the strictest sense, *whence* the wind cometh. If we only take the results on which we may depend from those places in which the observations on the direction of the winds have been continued more than twenty years, we shall find (from the most recent and careful calculations of Wilhelm Mahlmann) that in the middle latitudes of the temperate zone, in both continents, the prevailing aërial current has a west-southwest direction.

Our insight into the *distribution of heat* in the atmosphere has been rendered more clear since the attempt has been made to connect together by lines those places where the mean annual summer and winter temperatures have been ascertained by correct observations. The system of *isothermal, isotheral,* and *isochimenal* lines, which I first brought into use in 1817, may, perhaps, if it be gradually perfected by the united efforts of investigators, serve as one of the main foundations of *comparative climatology.* Terrestrial magnetism did not acquire a right to be regarded as a science until partial results were graphically connected in a system of lines of *equal declination, equal inclination,* and *equal intensity.*

The term *climate,* taken in its most general sense, indicates all the changes in the atmosphere which sensibly affect our organs, as temperature, humidity, variations in the barometrical pressure, the calm state of the air or the action of opposite winds, the amount of electric tension, the purity of the atmosphere or its admixture with more or less noxious gaseous exhalations, and, finally, the degree of ordinary transparency and clearness of the sky, which is not only important with respect to the increased radiation from the Earth, the organic development of plants, and the ripening of fruits, but also with reference to its influence on the feelings and mental condition of men.

If the surface of the Earth consisted of one and the same homogeneous fluid mass, or of strata of rock having the same color, density, smoothness, and power of absorbing heat from the solar rays, and of radiating it in a similar manner through the atmosphere, the isothermal, isotheral, and isochimenal lines would all be parallel to the equator. In this hypothetical condition of the Earth's surface, the power of absorbing and emitting light and heat would every where be the same under the same latitudes. The mathematical consideration of climate, which does not exclude the supposition of the existence of currents of heat in the interior, or in the external crust of the earth, nor of the propagation of heat by atmospheric currents, proceeds from this mean, and, as it were, primitive condition. Whatever alters the capacity for absorption and radiation, at places lying under the same parallel of latitude, gives rise to inflections in the isothermal lines. The nature of these inflections, the angles at which the isothermal, isotheral, or isochimenal lines intersect the parallels of latitude, their convexity or concavity with respect to the pole of the same hemisphere, are dependent on causes which more or less modify the temperature under different degrees of longitude.

The progress of *Climatology* has been remarkably favored by the extension of European civilization to two opposite coasts, by its transmission from our western shores to a continent which is bounded on the east by the Atlantic Ocean. When, after the ephemeral colonization from Iceland and Greenland, the British laid the foundation of the first permanent settlements on the shores of the United States of America, the emigrants (whose numbers were rapidly increased in consequence either of religious persecution, fanaticism, or love of freedom, and who soon spread over the vast extent of territory lying between the Carolinas, Virginia, and the St. Lawrence) were astonished to find themselves exposed to an intensity of winter cold far exceeding that which prevailed in Italy, France, and Scotland, situated in corresponding parallels of latitude. But, however much a consideration of these climatic relations may have awakened attention, it was not attended by any practical results until it could be based on the numerical data of *mean annual temperature*. If, between 58° and 30° north latitude, we compare Nain, on the coast of Labrador, with Gottenburg ; Halifax with Bordeaux ; New

York with Naples ; St. Augustine, in Florida, with Cairo, we find that, under the same degrees of latitude, the differences of the mean annual temperature between Eastern America and Western Europe, proceeding from north to south, are successively $20^{\circ}\cdot7$, $13^{\circ}\cdot9$, $6^{\circ}\cdot8$, and almost 0°. The gradual decrease of the differences in this series extending over 28° of latitude is very striking. Further to the south, under the tropics, the isothermal lines are every where parallel to the equator in both hemispheres. We see, from the above examples, that the questions often asked in society, how many degrees America (without distinguishing between the eastern and western shores) is colder than Europe ? and how much the mean annual temperature of Canada and the United States is lower than that of corresponding latitudes in Europe ? are, when thus *generally expressed*, devoid of meaning. There is a separate difference for each parallel of latitude, and without a special comparison of the winter and summer temperatures of the opposite coasts, it will be impossible to arrive at a correct idea of climatic relations, in their influence on agriculture and other industrial pursuits, or on the individual comfort or discomfort of mankind in general.

In enumerating the causes which produce disturbances in the form of the isothermal lines, I would distinguish between those which *raise* and those which *lower* the temperature. To the first class belong the proximity of a western coast in the temperate zone ; the divided configuration of a continent into peninsulas, with deeply-indented bays and inland seas ; the aspect or the position of a portion of the land with reference either to a sea of ice spreading far into the polar circle, or to a mass of continental land of considerable extent, lying in the same meridian, either under the equator, or, at least, within a portion of the tropical zone ; the prevalence of southerly or westerly winds on the western shore of a continent in the temperate northern zone ; chains of mountains acting as protecting walls against winds coming from colder regions ; the infrequency of swamps, which, in the spring and beginning of summer, long remain covered with ice, and the absence of woods in a dry, sandy soil ; finally, the constant serenity of the sky in the summer months, and the vicinity of an oceanic current, bringing water which is of a higher temperature than that of the surrounding sea.

Among the causes which tend to *lower* the mean annual temperature I include the following : elevation above the level of the sea, when not forming part of an extended plain ; the

vicinity of an eastern coast in high and middle latitudes ; the
compact configuration of a continent having no littoral curvatures or bays ; the extension of land toward the poles into
the region of perpetual ice, without the intervention of a sea
remaining open in the winter ; a geographical position, in
which the equatorial and tropical regions are occupied by the
sea, and, consequently, the absence, under the same meridian,
of a continental tropical land having a strong capacity for the
absorption and radiation of heat ; mountain chains, whose
mural form and direction impede the access of warm winds ,
the vicinity of isolated peaks, occasioning the descent of cold
currents of air down their declivities ; extensive woods, which
hinder the insolation of the soil by the vital activity of their
foliage, which produces great evaporation, owing to the extension of these organs, and increases the surface that is cooled by radiation, acting consequently in a three-fold manner,
by shade, evaporation, and radiation ; the frequency of swamps
or marshes, which in the north form a kind of subterranean
glacier in the plains, lasting till the middle of the summer ; a
cloudy summer sky, which weakens the action of the solar
rays ; and, finally, a very clear winter sky, favoring the radiation of heat.*

The simultaneous action of these disturbing causes, whether productive of an increase or decrease of heat, determines,
as the total effect, the inflection of the isothermal lines, especially with relation to the expansion and configuration of solid
continental masses, as compared with the liquid oceanic.
These perturbations give rise to convex and concave summits
of the isothermal curves.   There are, however, different orders of disturbing causes, and each one must, therefore, be
considered separately, in order that their total effect may afterward be investigated with reference to the motion (direction, local curvature) of the isothermal lines, and the actions
by which they are connected together, modified, destroyed, or
increased in intensity, as manifested in the contact and intersection of small oscillatory movements.   Such is the method
by which, I hope, it may some day be possible to connect together, by empirical and numerically expressed laws, vast series of apparently isolated facts, and to exhibit the mutual dependence which must necessarily exist among them.

The trade winds—easterly winds blowing within the tropics—give rise, in both temperate zones, to the west, or west-

* Humboldt, *Recherches sur les Causes des Inflexions des Lignes Isothermes*, in *Asie Centr.*, t. iii., p. 103–114, 118, 122, 188.

southwest winds which prevail in those regions, and which are land winds to eastern coasts, and sea winds to western coasts, extending over a space which, from the great mass and the sinking of its cooled particles, is not capable of any considerable degree of cooling, and hence it follows that the east winds of the Continent must be cooler than the west winds, where their temperature is not affected by the occurrence of oceanic currents near the shore. Cook's young companion on his second voyage of circumnavigation, the intelligent George Forster, to whom I am indebted for the lively interest which prompted me to undertake distant travels, was the first who drew attention, in a definite manner, to the climatic differences of temperature existing in the eastern and western coasts of both continents, and to the similarity of temperature of the western coast of North America in the middle latitudes, with that of Western Europe.* Even in northern latitudes exact observations show a striking difference between the *mean annual temperature* of the east and west coasts of America. The mean annual temperature of Nain, in Labrador (lat. 57° 10'), is fully 6°·8 *below* the freezing point, while on the northwest coast, at New Archangel, in Russian America (lat. 57° 3'), it is 12°·4 *above* this point. At the first-named place, the mean summer temperature hardly amounts to 43°, while at the latter place it is 57°. Pekin (39° 54'), on the eastern coast of Asia, has a mean annual temperature of 52°·3, which is 9° below that of Naples, situated somewhat further to the north. The mean winter temperature of Pekin is at least 5°·4 below the freezing point, while in Western Europe, even at Paris (48° 50'), it is nearly 6° above the freezing point. Pekin has also a mean winter cold which is 4°·5 lower than that of Copenhagen, lying 17° further to the north.

We have already seen the slowness with which the great mass of the ocean follows the variations of temperature in the atmosphere, and how the sea acts in equalizing temperatures, moderating simultaneously the severity of winter and the heat of summer. Hence arises a second more important contrast —that, namely, between insular and littoral climates enjoyed by all articulated continents having deeply-indented bays and peninsulas, and between the climate of the interior of great masses of solid land. This remarkable contrast has been fully

* George Forster, *Kleine Schriften*, th. iii., 1794, s. 87 ; Dove, in Schumacher's *Jahrbuch für* 1841, s. 289 ; Kämtz, *Meteorologie*, bd. ii., s. 41, 43, 67, and 96 ; Arago, in the *Comptes Rendus*, t. i., p. 268.

developed by Leopold von Buch in all its various phenomena, both with respect to its influence on vegetation and agriculture, on the transparency of the atmosphere, the radiation of the soil, and the elevation of the line of perpetual snow. In the interior of the Asiatic Continent, Tobolsk, Barnaul on the Oby, and Irkutsk, have the same mean summer heat as Berlin, Munster, and Cherbourg in Normandy, the thermometer sometimes remaining for weeks together at 86° or 88°, while the mean winter temperature is, during the coldest month, as low as —0°·4 to —4°. These continental climates have therefore justly been termed *excessive* by the great mathematician and physicist Buffon; and the inhabitants who live in countries having such *excessive* climates seem almost condemned, as Dante expresses himself,

" A sofferir tormenti caldi e geli."*

In no portion of the earth, neither in the Canary Islands, in Spain, nor in the south of France, have I ever seen more luxuriant fruit, especially grapes, than in Astrachan, near the shores of the Caspian Sea (46° 21'). Although the mean annual temperature is about 48°, the mean summer heat rises to 70°, as at Bordeaux, while not only there, but also further to the south, as at Kislar on the mouth of the Terek (in the latitude of Avignon and Rimini), the thermometer sinks in the winter to —13° or —22°.

Ireland, Guernsey, and Jersey, the peninsula of Brittany, the coasts of Normandy, and of the south of England, present, by the mildness of their winters, and by the low temperature and clouded sky of their summers, the most striking contrast to the continental climate of the interior of Eastern Europe. In the northeast of Ireland (54° 56'), lying under the same parallel of latitude as Königsberg in Prussia, the myrtle blooms as luxuriantly as in Portugal. The mean temperature of the month of August, which in Hungary rises to 70°, scarcely reaches 61° at Dublin, which is situated on the same isothermal line of 49°; the mean winter temperature, which falls to about 28° at Pesth, is 40° at Dublin (whose mean annual temperature is not more than 49°); 3°·6 higher than that of Milan, Pavia, Padua, and the whole of Lombardy, where the mean annual temperature is upward of 55°. At Stromness, in the Orkneys, scarcely half a degree further south than Stockholm, the winter temperature is 39°, and consequently higher than that of Paris, and nearly as high as that of London.

* Dante, *Divina Commedia, Purgatorio*, canto iii.

Even in the Färoë Islands, at 62° latitude, the inland waters never freeze, owing to the favoring influence of the west winds and of the sea. On the charming coasts of Devonshire, near Salcombe Bay, which has been termed, on account of the mildness of its climate, the *Montpellier of the North*, the Agave Mexicana has been seen to blossom in the open air, while orange-trees trained against espaliers, and only slightly protected by matting, are found to bear fruit. There, as well as at Penzance and Gosport, and at Cherbourg on the coast of Normandy, the mean winter temperature exceeds 42°, falling short by only 2°·4 of the mean winter temperature of Montpellier and Florence.* These observations will suffice to show the important influence exercised on vegetation and agriculture, on the cultivation of fruit, and on the comfort of mankind, by differences in the distribution of the same mean annual temperature, through the different seasons of the year.

The lines which I have termed *isochimenal* and *isotheral* (lines of equal winter and equal summer temperature) are by no means parallel with the *isothermal* lines (lines of equal annual temperature). If, for instance, in countries where myrtles grow wild, and the earth does not remain covered with snow in the winter, the temperature of the summer and autumn is barely sufficient to bring apples to perfect ripeness, and if, again, we observe that the grape rarely attains the ripeness necessary to convert it into wine, either in islands or in the vicinity of the sea, even when cultivated on a western coast, the reason must not be sought only in the low degree of summer heat, indicated, in littoral situations, by the thermometer when suspended in the shade, but likewise in another cause that has not hitherto been sufficiently considered, although it exercises an active influence on many other phenomena (as, for instance, in the inflammation of a mixture of chlorine and hydrogen), namely, the difference between direct and diffused light, or that which prevails when the sky is clear and when it is overcast by mist. I long since endeavored to attract the attention of physicists and physiologists† to this

* Humboldt, *Sur les Lignes Isothermes*, in the *Mémoires de Physique et de Chimie de la Société d'Arcueil*, t. iii., Paris, 1817, p. 143–165 ; Knight, in the *Transactions of the Horticultural Society of London*, vol. , p. 32 ; Watson, *Remarks on the Geographical Distribution of British Plants*, 1835, p. 60 ; Trevelyan, in Jamieson's *Edinburgh New Phil. Journal*, No. 18, p. 154 ; Mahlmann, in his admirable German transla tion of my *Asie Centrale*, th. ii., s. 60.

† " Hæc de temperie aeris, qui terram late circumfundit, ac in quo, longe a solo, instrumenta nostra meteorologica suspensa habemus. Sed

difference, and to the *unmeasured* heat which is locally developed in the living vegetable cell by the action of direct light.

If, in forming a thermic scale of different kinds of cultivation,* we begin with those plants which require the hottest climate, as the vanilla, the cacao, banana, and cocoa-nut, and proceed to pine-apples, the sugar-cane, coffee, fruit-bearing date-trees, the cotton-tree, citrons, olives, edible chestnuts, and vines producing potable wine, an exact geographical consideration of the limits of cultivation, both on plains and on the declivities of mountains, will teach us that other climatic relations besides those of mean annual temperature are involved in these phenomena. Taking an example, for instance, from the cultivation of the vine, we find that, in order to procure *potable* wine,† it is requisite that the mean annual heat should exceed 49°, that the winter temperature should be upward of 33°, and the mean summer temperature upward of 64°. At Bordeaux, in the valley of the Garonne (41° 50′ lat.), the mean annual, winter, summer, and autumn temperatures are respectively 57°, 43°, 71°, and 58°. In the plains near the

alia est caloris vis, quem radii solis nullis nubibus velati, in foliis ipsis et fructibus maturescentibus, magis minusve coloratis, gignunt, quemque, ut egregia demonstrant experimenta amicissimorum Gay-Lussacii et Thenardi de combustione chlori et hydrogenis, ope thermometri metiri nequis. Etenim locis planis et montanis, vento libe spirante, circumfusi aeris temperies eadem esse potest cœlo sudo vel nebuloso ; ideoque ex observationibus solis thermometricis, nullo adhibito Photometro, haud cognosces, quam ob causam Galliæ septentrionalis tractus Armoricanus et Nervicus, versus littora, cœlo temperato sed sole raro utentia, Vitem fere non tolerant. Egent enim stirpes non solum caloris stimulo, sed et lucis, quæ magis intensa locis excelsis quam planis, duplici modo plantas movet, vi sua tum propria, tum calorem in superficie earum excitante."—Humboldt, *De Distributione Geographica Plantarum*, 1817, p. 163–164.

* Humboldt, op. cit., p. 156–161 ; Meyen, in his *Grundriss der Pflanzengeographie*, 1836, s. 379–467 ; Boussingault, *Economie Rurale*, t. ii., p. 675.

† The following table illustrates the cultivation of the vine in Europe, and also the depreciation of its produce according to climatic relations. See my *Asie Centrale*, t. iii., p. 159. The examples quoted in the text for Bordeaux and Potsdam are, in respect of numerical relation, alike applicable to the countries of the Rhine and Maine (48° 35′ to 50° 7′ N. lat.). Cherbourg in Normandy, and Ireland, show in the most remarkable manner how, with thermal relations very nearly similar to those prevailing in the interior of the Continent (as estimated by the thermometer in the shade), the results are nevertheless extremely different as regards the ripeness or the unripeness of the fruit of the vine, this difference undoubtedly depending on the circumstance whether the vegetation of the plant proceeds under a bright sunny sky, or under a sky that is habitually obscured by clouds:

Baltic (52° 30′ lat.), where a wine is produced that can
scarcely be considered potable, these numbers are as follows :
47°·5, 31°, 63°·7, and 47°·5. If it should appear strange
that the great differences indicated by the influence of climate
on the production of wine should not be more clearly manifested by our thermometers, the circumstance will appear less
singular when we remember that a thermometer standing in
the shade, and protected from the effect of direct insolation
and nocturnal radiation can not, at all seasons of the year, and
during all periodic changes of heat, indicate the true superficial
temperature of the ground exposed to the whole effect of the
sun's rays.

The same relations which exist between the equable littoral
climate of the peninsula of Brittany, and the lower winter and

Places.	Latitude.	Elevation.	Mean of the Year.	Winter.	Spring.	Summer.	Autumn.	Number of the Years of the Observation.
	o ′	Eng. ft.	Fahr.					
Bordeaux ...	44 50	25·6	57·0	43·0	56·0	71·0	58·0	10
Strasbourg...	48 35	479·0	49·6	34·5	50·0	64·6	50·0	35
Heidelberg..	49 24	333·5	49·5	34·0	50·0	64·3	49·7	20
Manheim ...	49 29	300·5	50·6	34·6	50·8	67·1	49·5	12
Würzburg...	49 48	562·5	50·2	35·5	50·5	65·7	49·4	27
Frankfort on Maine....	50 7	388·5	49·5	33·3	50·0	64·4	49·4	19
Berlin	52 31	102·3	47·5	31·0	46·6	63·6	47·5	23
Cherbourg (no wine)	49 39	....	52·1	41·5	50·8	61·7	54·3	3
Dublin (ditto)	53 23	....	49·1	40·2	47·1	59·6	49·7	13

The great accordance in the distribution of the annual temperature
through the different seasons, as presented by the results obtained for
the valleys of the Rhine and Maine, tends to confirm the accuracy of
these meteorological observations. The months of December, January,
and February are reckoned as winter months. When the different
qualities of the wines produced in Franconia, and in the countries
around the Baltic, are compared with the mean summer and autumn
temperature of Würzburg and Berlin, we are almost surprised to find
a difference of only about two degrees. The difference in the spring
is about four degrees. The influence of late May frosts on the flowering season, and after a correspondingly cold winter, is almost as im
portant an element as the time of the subsequent ripening of the grape,
and the influence of direct, not diffused, light of the unclouded sun
The difference alluded to in the text between the true temperature of
the surface of the ground and the indications of a thermometer sus
pended in the shade and protected from extraneous influences, is in
ferred by Dove from a consideration of the results of fifteen years' ob
servations made at the Chiswick Gardens. See Dove, in *Bericht über
die Verhandl. der Berl. Akad. der Wiss.*, August, 1844, s. 285.

higher summer temperature of the remainder of the continent
of France, are likewise manifested, in some degree, between
Europe and the great continent of Asia, of which the former
may be considered to constitute the western peninsula. Europe owes its milder climate, in the first place, to its position
with respect to Africa, whose wide extent of tropical land is
favorable to the ascending current, while the equatorial region
to the south of Asia is almost wholly oceanic; and next to its
deeply-articulated configuration, to the vicinity of the ocean
on its western shores; and, lastly, to the existence of an open
sea, which bounds its northern confines. Europe would therefore become colder* if Africa were to be overflowed by the
ocean; or if the mythical Atlantis were to arise and connect
Europe with North America; or if the Gulf Stream were no
longer to diffuse the warming influence of its waters into the
North Sea; or if, finally, another mass of solid land should be
upheaved by volcanic action, and interposed between the
Scandinavian peninsula and Spitzbergen. If we observe that
in Europe the mean annual temperature falls as we proceed,
from west to east, under the same parallel of latitude, from
the Atlantic shores of France through Germany, Poland, and
Russia, toward the Uralian Mountains, the main cause of this
phenomenon of increasing cold must be sought in the form of
the continent (which becomes less indented, and wider, and
more compact as we advance), in the increasing distance from
seas, and in the diminished influence of westerly winds. Beyond the Uralian Mountains these winds are converted into
cool land-winds, blowing over extended tracts covered with
ice and snow.   The cold of western Siberia is to be ascribed
to these relations of configuration and atmospheric currents,
and not—as Hippocrates and Trogus Pompeius, and even celebrated travelers of the eighteenth century conjectured—to the
great elevation of the soil above the level of the sea.†

If we pass from the differences of temperature manifested in
the plains to the inequalities of the polyhedric form of the surface of our planet, we shall have to consider mountains either
in relation to their influence on the climate of neighboring

* See my memoir, *Ueber die Haupt-Ursachen der Temperaturver-
schiedenheit auf der Erdoberfläche*, in the *Abhandl. der Akad. der Wissensch. zu Berlin von dem Jahr* 1827, s. 311.

† The general level of Siberia, from Tobolsk, Tomsk, and Barnaul,
from the Altai Mountains to the Polar Sea, is not so high as that of
Mauheim and Dresden; indeed, Irkutsk, far to the east of the Jenisei,
is only 1330 feet above the level of the sea, or about one third lower
than Munich.

valleys, or according to the effects of the hypsometrical relations on their own summits, which often spread into elevated
plateaux. The division of mountains into chains separates
the earth's surface into different basins, which are often nar
row and walled in, forming caldron-like valleys, and (as in
Greece and in part of Asia Minor) constitute an individual
local climate with respect to heat, moisture, transparency of
atmosphere, and frequency of winds and storms. These circumstances have at all times exercised a powerful influence
on the character and cultivation of natural products, and on
the manners and institutions of neighboring nations, and even
on the feelings with which they regard one another. This
character of *geographical individuality* attains its maximum,
if we may be allowed so to speak, in countries where the differences in the configuration of the soil are the greatest possible, either in a vertical or horizontal direction, both in relief
and in the articulation of the continent. The greatest contrast to these varieties in the relations of the surface of the
earth are manifested in the Steppes of Northern Asia, the
grassy plains (savannahs, llanos, and pampas) of the New
Continent, the heaths (*Ericeta*) of Europe, and the sandy and
stony deserts of Africa.

The law of the decrease of heat with the increase of elevation at different latitudes is one of the most important subjects
involved in the study of meteorological processes, of the geography of plants, of the theory of terrestrial refraction, and of
the various hypotheses that relate to the determination of the
height of the atmosphere. In the many mountain journeys
which I have undertaken, both within and without the tropics, the investigation of this law has always formed a special
object of my researches.*

Since we have acquired a more accurate knowledge of the
true relations of the distribution of heat on the surface of the
earth, that is to say, of the inflections of isothermal and isotheral lines, and their unequal distance apart in the different
eastern and western systems of temperature in Asia, Central
Europe, and North America, we can no longer ask the general question, what fraction of the mean annual or summer
temperature corresponds to the difference of one degree of
geographical latitude, taken in the same meridian? In each
system of *isothermal* lines of equal curvature there reigns a

---

* Humboldt, *Recueil d'Observations Astronomiques*, t. i., p. 126–140;
Rélation Historique, t. i., p. 119, 141, 227; Biot, in *Connaissance des
Temps pour l'an* 1841, p. 90–109.

close and necessary connection between three elements, namely, the decrease of heat in a vertical direction from below upward, the difference of temperature for every one degree of geographical latitude, and the uniformity in the mean temperature of a mountain station, and the latitude of a point situated at the level of the sea.

In the system of Eastern America, the mean annual temperature from the coast of Labrador to Boston changes $1°·6$ for every degree of latitude; from Boston to Charleston about $1°·7$; from Charleston to the tropic of Cancer, in Cuba, the variation is less rapid, being only $1°·2$. In the tropics this diminution is so much greater, that from the Havana to Cumana the variation is less than $0°·4$ for every degree of latitude.

The case is quite different in the isothermal system of Central Europe. Between the parallels of $38°$ and $71°$ I found that the decrease of temperature was very regularly $0°·9$ for every degree of latitude. But as, on the other hand, in Central Europe the decrease of heat is $1°·8$ for about every 534 feet of vertical elevation, it follows that a difference of elevation of about 267 feet corresponds to the difference of one degree of latitude. The same mean annual temperature as that occurring at the Convent of St. Bernard, at an elevation of 8173 feet, in lat. $45° 50'$, should therefore be met with at the level of the sea in lat. $75° 50'$.

In that part of the Cordilleras which falls within the tropics, the observations I made at various heights, at an elevation of upward of 19,000 feet, gave a decrease of $1°$ for every 341 feet; and my friend Boussingault found, thirty years afterward, as a mean result, 319 feet. By a comparison of places in the Cordilleras, lying at an equal elevation above the level of the sea, either on the declivities of the mountains or even on extensive elevated plateaux, I observed that in the latter there was an increase in the annual temperature varying from $2°·7$ to $4°·1$. This difference would be still greater if it were not for the cooling effect of nocturnal radiation. As the different climates are arranged in successive strata, the one above the other, from the cacao woods of the valleys to the region of perpetual snow, and as the temperature in the tropics varies but little throughout the year, we may form to ourselves a tolerably correct representation of the climatic relations to which the inhabitants of the large cities in the Andes are subjected, by comparing these climates with the temperatures of particular months in the plains of France and Italy. While

the heat which prevails daily on the woody shores of the
Orinoco exceeds by 7°·2 that of the month of August at Palermo, we find, on ascending the chain of the Andes, at Popayan, at an elevation of 5826 feet, the temperature of the
three summer months of Marseilles; at Quito, at an elevation of 9541 feet, that of the close of May at Paris; and on
the Paramos, at a height of 11,510 feet, where only stunted
Alpine shrubs grow, though flowers still bloom in abundance, that of the beginning of April at Paris. The intelligent
observer, Peter Martyr de Anghiera, one of the friends of
Christopher Columbus, seems to have been the first who recognized (in the expedition undertaken by Rodrigo Enrique
Colmenares, in October, 1510) that the limit of perpetual
snow continues to ascend as we approach the equator. We
read, in the fine work *De Rebus Oceanicis*,* "the River Gaira
comes from a mountain in the Sierra Nevada de Santa Marta,
which, according to the testimony of the companions of Colmenares, is higher than any other mountain hitherto discovered. It must undoubtedly be so if *it retain snow perpetually* in a zone which is not more than 10° from the equinoctial line." The lower limit of perpetual snow, in a given
latitude, is the lowest line at which snow continues during
summer, or, in other words, it is the maximum of height to
which the snow-line recedes in the course of the year. But
this elevation must be distinguished from three other phenomena, namely, the annual fluctuation of the snow-line, the
occurrence of sporadic falls of snow, and the existence of glaciers, which appear to be peculiar to the temperate and cold
zones. This last phenomenon, since Saussure's immortal
work on the Alps, has received much light, in recent times,
from the labors of Venetz, Charpentier, and the intrepid and
persevering observer Agassiz.

We know only the *lower*, and not the *upper* limit of perpetual snow; for the mountains of the earth do not attain to
those ethereal regions of the rarefied and dry strata of air, in
which we may suppose, with Bouguer, that the vesicles of
aqueous vapor are converted into crystals of ice, and thus rendered perceptible to our organs of sight. The lower limit of
snow is not, however, a mere function of geographical latitude
or of mean annual temperature; nor is it at the equator, or

---

* Anglerius, *De Rebus Oceanicis*, Dec. xi., lib. ii., p. 140 (ed. Col.,
1574). In the Sierra de Santa Marta, the highest point of which appears to exceed 19,000 feet (see my *Rélat. Hist.*, t. ii., p. 214), there is
a peak that is still called Pico de Gaira.

even in the region of the tropics, that this limit attains its greatest elevation above the level of the sea. The phenomenon of which we are treating is extremely complicated, depending on the general relations of temperature and humidity, and on the form of mountains. On submitting these relations to the test of special analysis, as we may be permitted to do from the number of determinations that have recently been made,[*] we shall find that the controlling causes are the differences in the temperature of different seasons of the year; the direction of the prevailing winds and their relations to the land and sea; the degree of dryness or humidity in the upper strata of the air; the absolute thickness of the accumulated masses of fallen snow; the relation of the snow-line to the total height of the mountain; the relative position of the latter in the chain to which it belongs, and the steepness of its declivity; the vicinity of other summits likewise perpetually covered with snow; the expansion, position, and elevation of the plains from which the snow-mountain rises as an isolated peak or as a portion of a chain; whether this plain be part of the sea-coast or of the interior of a continent; whether it be covered with wood or waving grass; and whether, finally, it consist of a dry and rocky soil, or of a wet and marshy bottom.

The snow-line which, under the equator in South America, attains an elevation equal to that of the summit of Mont Blanc in the Alps, and descends, according to recent measurements, about 1023 feet lower toward the northern tropic in the elevated plateaux of Mexico (in 19° north latitude), rises, according to Pentland, in the southern tropical zone (14° 30′ to 18° south latitude), being more than 2665 feet higher in the maritime and western branch of the Cordilleras of Chili than under the equator near Quito on Chimborazo, Cotopaxi, and Antisana. Dr. Gillies even asserts that much further to the south, on the declivity of the volcano of Peuquenes (latitude 33°), he found the snow-line at an elevation of between 14,520 and 15,030 feet. The evaporation of the snow in the extremely dry air of the summer, and under a cloudless sky, is so powerful, that the volcano of Aconcagua, northeast of Valparaiso (latitude 32° 30′), which was found in the expedition of the Beagle to be more than 1400 feet higher than Chimborazo, was on one occasion seen free from snow.[†] In

---

[*] See my table of the height of the line of perpetual snow, in both hemispheres, from 71° 15′ north lat. to 53° 54′ south lat., in my *Asie Centrale*, t. iii., p. 360.

[†] Darwin, *Journal of the Voyages of the Adventure and Beagle*, p. 297.

an almost equal northern latitude (from 30° 45′ to 31°), the snow-line on the southern declivity of the Himalaya lies at an elevation of 12,982 feet, which is about the same as the height which we might have assigned to it from a comparison with other mountain chains; on the northern declivity, however, under the influence of the high lands of Thibet (whose mean elevation appears to be about 11,510 feet), the snow-line is situated at a height of 16,630 feet. This phenomenon, which has long been contested both in Europe and in India, and whose causes I have attempted to develop in various works, published since 1820,* possesses other grounds of interest than

As the volcano of Aconcagua was not at that time in a state of eruption, we must not ascribe the remarkable phenomenon of the absence of snow to the internal heat of the mountain (to the escape of heated air through fissures), as is sometimes the case with Cotopaxi. Gillies, in the *Journal of Natural Science*, 1830, p. 316.

* See my *Second Mémoire sur les Montagnes de l'Inde*, in the *Annales de Chimie et de Physique*, t. xiv., p. 5–55; and *Asie Centrale*, t. iii., p. 281–327. While the most learned and experienced travelers in India, Colebrooke, Webb, and Hodgson, Victor Jacquemont, Forbes Royle, Carl von Hügel, and Vigne, who have all personally examined the Himalaya range, are agreed regarding the greater elevation of the snow-line on the Thibetian side, the accuracy of this statement is called in question by John Gerard, by the geognosist MacClelland, the editor of the *Calcutta Journal*, and by Captain Thomas Hutton, assistant surveyor of the Agra Division. The appearance of my work on Central Asia gave rise to a rediscussion of this question. A recent number (vol. iv., January, 1844) of MacClelland and Griffith's *Calcutta Journal of Natural History* contains, however, a very remarkable and decisive notice of the determination of the snow-line in the Himalayas. Mr. Batten, of the Bengal service, writes as follows from Camp Semulka, on the Cosillah River, Kumaon: "In the July, 1843, No. 14 of your valuable Journal of Natural History, which I have only lately had the opportunity of seeing, I read Captain Hutton's paper on the snow of the Himalayas, and as I differed almost entirely from the conclusions so confidently drawn by that gentleman, I thought it right, for the interest of scientific truth, to prepare some kind of answer; as, however, on a more attentive perusal, I find that you yourself appear implicitly to adopt Captain Hutton's views, and actually use these words, ' We have long been conscious of the error here so well pointed out by Captain Hutton, *in common with every one who has visited the Himalayas*,' I feel more inclined to address *you*, in the first instance, and to ask whether you will publish a short reply which I meditate; and whether your note to Captain Hutton's paper was written after your own full and careful examination of the subject, or merely on a general kind of acquiescence with the fact and opinions of your able contributor, who is so well known and esteemed as a collector of scientific data? Now I am one who have visited the Himalaya on the western side; I have crossed the Borendo or Boorin Pass into the Buspa Valley, in Lower Kanawar, returning into the Rewaien Mountains of Ghurwal by the Koopin Pass; I have visited the source of the Jumna at Jumnootree;

those of a purely physical nature, since it exercises no incon‑ siderable degree of influence on the mode of life of numerous tribes—the meteorological processes of the atmosphere being the controlling causes on which depend the agricultural or pastoral pursuits of the inhabitants of extensive tracts of con‑ tinents.

As the quantity of moisture in the atmosphere increases with the temperature, this element, which is so important for the whole organic creation, must vary with the hours of the day, the seasons of the year, and the differences in latitude and elevation. Our knowledge of the hygrometric relations of the Earth's surface has been very materially augmented of late years by the general application of August's psychrom‑ eter, framed in accordance with the views of Dalton and Daniell, for determining the relative quantity of vapor, or the

and, moving eastward, the sources of the Kalee or Mundaknee branch of the Ganges at Kadarnath; of the Vishnoo Gunga, or Aluknunda, at Buddrinath and Mana; of the Pindur at the foot of the Great Peak Nundidevi; of the Dhoulee branch of the Ganges, beyond Neetee, cross‑ ing and recrossing the pass of that name into Thibet; of the Goree or great branch of the Sardah, or Kalee, near Oonta Dhoora, beyond Me‑ lum. I have also, in my official capacity, made the settlement of the Bhote Mehals of this province. My residence of more than six years in the hills has thrown me constantly in the way of European and na‑ tive travelers, nor have I neglected to acquire information from the re‑ corded labors of others. Yet, with all this experience, I am prepared to affirm that *the perpetual snow-line is at a higher elevation* on the north‑ ern slope of ' the Himalaya' than on the southern slope.

"The facts mentioned by Captain Hutton appear to me only to refer to the northern sides of all mountains in these regions, and not to affect, in any way, the reports of Captain Webb and others, on which Hum‑ boldt formed his theory. Indeed, how can any facts of one observer in one place falsify the facts of another observer in another place? I will‑ ingly allow that the north side of a hill retains the snow longer and deeper than the south side, and this observation applies equally to heights in Bhote; but Humboldt's theory is on the question of the per‑ petual snow-line, and Captain Hutton's references to Simla and Mus‑ sooree, and other mountain sites, are out of place in this question, or else he fights against a shadow, or an objection of his own creation. In no part of his paper does he quote accurately the dictum which he wishes to oppose."

If the mean altitude of the Thibetian highlands be 11,510 feet, they admit of comparison with the lovely and fruitful plateau of Caxamarca in Peru. But at this estimate they would still be 1300 feet lower than the plateau of Bolivia at the Lake of Titicaca, and the causeway of the town of Potosi. Ladak, as appears from Vigne's measurement, by de‑ termining the boiling-point, is 9994 feet high. This is probably also the altitude of H'Lassa (Yul-sung), a monastic city, which Chinese writers describe as the *realm of pleasure,* and which is surrounded by vineyards. Must not these lie in deep valleys?

condition of moisture of the atmosphere, by means of the difference of the *dew point* and of the temperature of the air.
Temperature, atmospheric pressure, and the direction of the
wind, are all intimately connected with the vivifying action
of atmospheric moisture.   This influence is not, however, so
much a consequence of the quantity of moisture held in solution in different zones, as of the nature and frequency of the
precipitation which moistens the ground, whether in the form
of dew, mist, rain, or snow. According to the exposition made
by Dove of the law of rotation, and to the general views of
this distinguished physicist,* it would appear that, in our
northern zone, " the elastic force of the vapor is greatest with
a southwest, and least with a northeast wind. On the western side of the windrose this elasticity diminishes, while it increases on the eastern side ; on the former side, for instance,
the cold, dense, and dry current of air repels the warmer,
lighter current containing an abundance of aqueous vapor,
while on the eastern side it is the former current which is
repulsed by the latter.   The southwest is the equatorial current, while the northeast is the sole prevailing polar current."

The agreeable and fresh verdure which is observed in many
trees in districts within the tropics, where, for five or seven
months of the year, not a cloud is seen on the vault of heaven,
and where no perceptible dew or rain falls, proves that the
leaves are capable of extracting water from the atmosphere
by a peculiar vital process of their own, which perhaps is not
alone that of producing cold by radiation.   The absence of
rain in the arid plains of Cumana, Coro, and Ceara in North
Brazil, forms a striking contrast to the quantity of rain which
falls in some tropical regions, as, for instance, in the Havana,
where it would appear, from the average of six years' observation by Ramon de la Sagra, the mean annual quantity of
rain is 109 inches, equal to four or five times that which falls
at Paris or at Geneva.†   On the declivity of the Cordilleras,

* See Dove, *Meteorologische Vergleichung von Nordamerika und Eu-
ropa*, in Schumacher's *Jahrbuch für* 1841, s. 311 ; and his *Meteorologische
Untersuchungen*, s. 140.
† The mean annual quantity of rain that fell in Paris between 1805
and 1822 was found by Arago to be 20 inches; in London, between
1812 and 1827, it was determined by Howard at 25 inches; while at
Geneva the mean of thirty-two years' observation was 30·5 inches.   In
Hindostan, near the coast, the quantity of rain is from 115 to 128 inches;
and in the island of Cuba, fully 142 inches fell in the year 1821.   With
regard to the distribution of the quantity of rain in Central Europe, at
different periods of the year, see the admirable researches of Gasparin,
Schouw, and Bravais, in the *Bibliothèque Universelle*, t. xxxviii., p. 54

the quantity of rain, as well as the temperature, diminishes with the increase in the elevation.* My South American fellow-traveler, Caldas, found that, at Santa Fé de Bogota, at an elevation of almost 8700 feet, it did not exceed 37 inches, being consequently little more than on some parts of the western shore of Europe. Boussingault occasionally observed at Quito that Saussure's hygrometer receded to 26° with a temperature of from 53°·6 to 55°·4. Gay-Lussac saw the same hygrometer standing at 25°·3 in his great aërostatic ascent in a stratum of air 7034 feet high, and with a temperature of 39°·2. The greatest dryness that has yet been observed on the surface of the globe in low lands is probably that which Gustav Rose, Ehrenberg, and myself found in Northern Asia, between the valleys of the Irtisch and the Oby. In the Steppe of Platowskaja, after southwest winds had blown for a long time from the interior of the Continent, with a temperature of 74°·7, we found the dew point at 24°. The air contained only $\frac{16}{100}$ths of aqueous vapor.† The accurate observers Kämtz, Bravais, and Martins have raised doubts during the last few years regarding the greater dryness of the mountain air, which appeared to be proved by the hygrometric measurements made by Saussure and myself in the higher regions of the Alps and the Cordilleras. The strata of air at Zurich and on the Faulhorn, which can not be considered as an elevated mountain when compared with non-European elevations, furnished the data employed in the comparisons made by these observers.‡ In the tropical region of the Paramos (near the region where snow begins to fall, at an elevation of between 12,000 and 14,000 feet), some species of large flowering myrtle-leaved alpine shrubs are almost constantly bathed in moisture; but this fact does not actually prove the existence of any great and absolute quantity of aqueous vapor at such an elevation, merely affording

and 264; *Tableau du Climat de l'Italie*, p. 76; and Martins's notes to his excellent French translation of Kämtz's *Vorlesungen über Meteorologie*, p. 142.

* According to Boussingault (*Economie Rurale*, t. ii., p. 693), the mean quantity of rain that fell at Marmato (latitude 5° 27′, altitude 4675 feet, and mean temperature 69°) in the years 1833 and 1834 was 64 inches, while at Santa Fé de Bogota (latitude 4° 36′, altitude 8685 feet, and mean temperature 58°) it only amounted to 39½ inches.

† For the particulars of this observation, see my *Asie Centrale*, t. iii., p. 85–89 and 567; and regarding the amount of vapor in the atmosphere in the lowlands of tropical South America, consult my *Rélat. Hist.*, t. i., p. 242–248; t. ii., p. 45, 164.

‡ Kämtz, *Vorlesungen über Meteorologie*, s. 117.

an evidence of the frequency of aqueous precipitation, in like manner as do the frequent mists with which the lovely plateau of Bogota is covered. Mists arise and disappear several times in the course of an hour in such elevations as these, and with a calm state of the atmosphere. These rapid alterna tions characterize the Paramos and the elevated plains of the chain of the Andes.

The electricity of the atmosphere, whether considered in the lower or in the upper strata of the clouds, in its silent problematical diurnal course, or in the explosion of the lightning and thunder of the tempest, appears to stand in a manifold relation to all phenomena of the distribution of heat, of the pressure of the atmosphere and its disturbances, of hydrometeoric exhibitions, and probably, also, of the magnetism of the external crust of the earth. It exercises a powerful influence on the whole animal and vegetable world; not merely by meteorological processes, as precipitations of aqueous vapor, and of the acids and ammoniacal compounds to which it gives rise, but also directly as an electric force acting on the nerves, and promoting the circulation of the organic juices. This is not a place in which to renew the discussion that has been started regarding the actual source of atmospheric electricity when the sky is clear, a phenomenon that has alternately been ascribed to the evaporation of impure fluids impregnated with earths and salts,* to the growth of plants,† or to some other chemical decompositions on the surface of the earth, to the unequal distribution of heat in the strata of the air,‡ and, finally, according to Peltier's intelligent researches,§ to the agency of a constant charge of negative electricity in the terrestrial globe. Limiting itself to results yielded by electrometric observations, such, for instance, as are furnished by the ingenious electro-magnetic apparatus first proposed by Colladon, the physical description of the universe should merely notice the incontestable increase of intensity in the general positive electricity of the atmosphere,‖ accompanying an increase of altitude and the absence of trees, its daily variations (which, according to Clark's experiments at Dublin,

---

* Regarding the conditions of electricity from evaporation at high temperatures, see Peltier, in the *Annales de Chimie*, t. lxxv., p. 330

† Pouillet, in the *Annales de Chimie*, t. xxxv., p. 405.

‡ De la Rive, in his admirable *Essai Historique sur l'Electricité*, p. 140.

§ Peltier, in the *Comptes Rendus de l'Acad. des Sciences*, t. xii., p 307 ; Becquerel, *Traité de l'Electricité et du Magnétisme*, t. iv., p. 107

‖ Duprez, *Sur l'Electricité de l'Air* (Bruxelles, 1844), p. 56–61

take place at more complicated periods than those found by Saussure and myself), and its variations in the different seasons of the year, at different distances from the equator, and in the different relations of continental or oceanic surface.

The electric equilibrium is less frequently disturbed where the aërial ocean rests on a liquid base than where it impends over the land ; and it is very striking to observe how, in extensive seas, small insular groups affect the condition of the atmosphere, and occasion the formation of storms.   In fogs, and in the commencement of falls of snow, I have seen, in a long series of observations, the previously permanent positive electricity rapidly pass into the negative condition, both on the plains of the colder zones, and in the Paramos of the Cordilleras, at elevations varying from 11,000 to 15,000 feet. The alternate transition was precisely similar to that indicated by the electrometer shortly before and during a storm.* When the vesicles of vapor have become condensed into clouds, having definite outlines, the electric tension of the external surface will be increased in proportion to the amount of electricity which passes over to it from the separate vesicles of vapor.† Slate-gray clouds are charged, according to Peltier's experiments at Paris, with negative, and white, red, and orange-colored clouds with positive electricity.   Thunder clouds not only envelop the highest summits of the chain of the Andes (I have myself seen the electric effect of lightning on one of the rocky pinnacles which project upward of 15,000 feet above the crater of the volcano of Toluca), but they have also been observed at a vertical height of 26,650 feet over the low

* Humboldt, *Rélation Historique*, t. iii., p. 318.   I here only refer to those of my experiments in which the three-foot metallic conductor of Saussure's electrometer was neither moved upward nor downward, nor, according to Volta's proposal, armed with burning sponge.   Those of my readers who are well acquainted with the *quæstiones vexatæ* of atmospheric electricity will understand the grounds for this limitation. Respecting the formation of storms in the tropics, see my *Rél. Hist.*, t. ii., p. 45 and 202–209.

† Gay-Lussac, in the *Annales de Chimie et de Physique*, t. viii., p. 167. In consequence of the discordant views of Lamé, Becquerel, and Peltier, it is difficult to come to a conclusion regarding the cause of the specific distribution of electricity in clouds, some of which have a positive, and others a negative tension.   The negative electricity of the air, which near high water-falls is caused by a disintegration of the drops of water—a fact originally noticed by Tralles, and confirmed by myself in various latitudes—is very remarkable, and is sufficiently intense to produce an appreciable effect on a delicate electrometer at a distance of 300 or 400 feet.

lands in the temperate zone.* Sometimes, however, the stratum of cloud from which the thunder proceeds sinks to a distance of 5000, or, indeed, only 3000 feet above the plain.

According to Arago's investigations—the most comprehensive that we possess on this difficult branch of meteorology—the evolution of light (lightning) is of three kinds—zigzag, and sharply defined at the edges; in sheets of light, illuminating a whole cloud, which seems to open and reveal the light within it; and in the form of fire-balls.† The duration of the two first kinds scarcely continues the thousandth part of a second; but the globular lightning moves much more slowly remaining visible for several seconds. Occasionally (as is proved by the recent observations, which have confirmed the description given by Nicholson and Beccaria of this phenomenon), isolated clouds, standing high above the horizon, continue uninterruptedly for some time to emit a luminous radiance from their interior and from their margins, although there is no thunder to be heard, and no indication of a storm; in some cases even hail-stones, drops of rain, and flakes of snow have been seen to fall in a luminous condition, when the phenomenon was not preceded by thunder. In the geographical distribution of storms, the Peruvian coast, which is not visited by thunder or lightning, presents the most striking contrast to the rest of the tropical zone, in which, at certain seasons of the year, thunder-storms occur almost daily, about four or five hours after the sun has reached the meridian. According to the abundant evidence collected by Arago‡ from the testimony of navigators (Scoresby, Parry, Ross, and Franklin), there can be no doubt that, in general, electric explosions are extremely rare in high northern regions (between 70° and 75° latitude).

The meteorological portion of the descriptive history of na ture which we are now concluding shows that the processes of the absorption of light, the liberation of heat, and the variations in the elastic and electric tension, and in the hygrometric condition of the vast aërial ocean, are all so intimately connected together, that each individual meteorological process is modified by the action of all the others. The com-

---

* Arago, in the *Annuaire du Bureau des Longitudes pour* 1838, p. 246.

† Arago, op. cit., p. 249–266. (See, also, p. 268–279.)

‡ Arago, op. cit., p. 388–391. The learned academician Von Baer, who has done so much for the meteorology of Northern Asia, has not taken into consideration the extreme rarity of storms in Iceland and Greenland; he has only remarked (*Bulletin de l'Academie de St. Péters·bourg*, 1839, Mai) that in Nova Zembla and Spitzbergen it is sometimes heard to thunder.

plicated nature of these disturbing causes (which involuntarily remind us of those which the near and especially the smallest cosmical bodies, the satellites, comets, and shooting stars, are subjected to in their course) increases the difficulty of giving a full explanation of these involved meteorological phenomena, and likewise limits, or wholly precludes, the possibility of that predetermination of atmospheric changes which would be so important for horticulture, agriculture, and navigation, no less than for the comfort and enjoyment of life. Those who place the value of meteorology in this problematic species of prediction rather than in the knowledge of the phenomena themselves, are firmly convinced that this branch of science, on account of which so many expeditions to distant mountainous regions have been undertaken, has not made any very considerable progress for centuries past. The confidence which they refuse to the physicist they yield to changes of the moon, and to certain days marked in the calendar by the superstition of a by-gone age.

" Great local deviations from the distribution of the mean temperature are of rare occurrence, the variations being in general uniformly distributed over extensive tracts of land. The deviation, after attaining its maximum at a certain point, gradually decreases to its limits ; when these are passed, however, decided deviations are observed in the *opposite direction.* Similar relations of weather extend more frequently from south to north than from west to east. At the close of the year 1829 (when I had just completed my Siberian journey), the maximum of cold was at Berlin, while North America enjoyed an unusually high temperature. It is an entirely arbitrary assumption to believe that a hot summer succeeds a severe winter, and that a cool summer is preceded by a mild winter." Opposite relations of weather in contiguous countries, or in two corn-growing continents, give rise to a beneficent equalization in the prices of the products of the vine, and of agricultural and horticultural cultivation. It has been justly remarked, that it is the barometer alone which indicates to us the changes that occur in the pressure of the air throughout all the aërial strata from the place of observation to the extremest confines of the atmosphere, while* the thermometer and psychrometer only acquaint us with all the variations occurring in the local heat and moisture of the lower strata of

* Kämtz, in Schumacher's *Jahrbuch für* 1838, s. 285. Regarding the opposite distribution of heat in the east and the west of Europe and North America, see Dove, *Repertorium der Physik,* bd. iii., s. 392–395

air in contact with the ground. The simultaneous thermic and hygrometric modifications of the upper regions of the air can only be learned (when direct observations on mountain stations or aerostatic ascents are impracticable) from hypothetical combinations, by making the barometer serve both as a thermometer and an hygrometer. Important changes of weather are not owing to merely local causes, situated at the place of observation, but are the consequence of a disturbance in the equilibrium of the aërial currents at a great distance from the surface of the Earth, in the higher strata of the atmosphere, bringing cold or warm, dry or moist air, rendering the sky cloudy or serene, and converting the accumulated masses of clouds into light feathery *cirri*. As, therefore, the inaccessibility of the phenomenon is added to the manifold nature and complication of the disturbances, it has always appeared to me that meteorology must first seek its foundation and progress in the torrid zone, where the variations of the atmospheric pressure, the course of hydro-meteors, and the phenomena of electric explosion, are all of periodic occurrence.

As we have now passed in review the whole sphere of inorganic terrestrial life, and have briefly considered our planet with reference to its form, its internal heat, its electro-magnetic tension, its phenomena of polar light, the volcanic reaction of its interior on its variously composed solid crust, and, lastly, the phenomena of its two-fold envelopes—the aërial and liquid ocean—we might, in accordance with the older method of treating physical geography, consider that we had completed our descriptive history of the globe. But the nobler aim I have proposed to myself, of raising the contemplation of nature to a more elevated point of view, would be defeated, and this delineation of nature would appear to lose its most attractive charm, if it did not also include the sphere of organic life in the many stages of its typical development. The idea of vitality is so intimately associated with the idea of the existence of the active, ever-blending natural forces which animate the terrestrial sphere, that the creation of plants and animals is ascribed in the most ancient mythical representations of many nations to these forces, while the condition of the surface of our planet, before it was animated by vital forms, is regarded as coeval with the epoch of a chaotic conflict of the struggling elements. But the empirical domain of objective contemplation, and the delineation of our planet in its present condition, do not include a consideration

of the mysterious and insoluble problems of origin and existence.

A cosmical history of the universe, resting upon facts as its
basis, has, from the nature and limitations of its sphere, necessarily no connection with the obscure domain embraced by a
history of organisms,* if we understand the word *history* in
its broadest sense. It must, however, be remembered, that
the inorganic crust of the Earth contains within it the same
elements that enter into the structure of animal and vegetable organs. A physical cosmography would therefore be in

* The *history of plants*, which Endlicher and Unger have described
in a most masterly manner (*Grundzüge der Botanik*, 1843, s. 449–468),
I myself separated from the *geography of plants* half a century ago
In the aphorisms appended to my *Subterranean Flora*, the following
passage occurs : "Geognosia naturam animantem et inanimam vel, ut
vocabulo minus apto, ex antiquitate saltem haud petito, utar, corpora
organica æque ac inorganica considerat. Sunt enim tria quibus absol
vitur capita : Geographia oryctologica quam simpliciter Geognosiam vel
Geologiam dicunt, virque acutissimus Wernerus egregie digessit ; Geographia zoologica, cujus doctrinæ fundamenta Zimmermannus et Tre-
viranus jecerunt ; et Geographia plantarum quam æquales nostri diu in-
tactam reliquerunt. Geographia plantarum vincula et cognationem
tradit, quibus omnia vegetabilia inter se connexa sint, terræ tractus
quos teneant, in aerem atmosphæricum quæ sit eorum vis ostendit, saxa
atque rupes quibus potissimum algarum primordiis radicibusque destru-
antur docet, et quo pacto in telluris superficie humus nascatur, commemorat. Est itaque quod differat inter Geognosiam et Physiographiam,
historia naturalis perperam nuncupatam quum Zoognosia, Phytognosia,
et Oryctognosia, quæ quidem omnes in naturæ investigatione versantur,
non nisi singulorum animalium, plantarum, rerum metallicarum vel
(venia sit verbo) fossilium formas, anatomen, vires scrutantur. Historia
Telluris, Geognosiæ magis quam Physiographiæ affinis, nemini adhuc
tentata, plantarum animaliumque genera orbem inhabitantia primævum,
migrationes eorum compluriumque interitum, ortum quem montes,
valles, saxorum strata et venæ metalliferæ ducunt, aerem, mutatis tem-
porum vicibus, modo purum, modo vitiatum, terræ superficiem humo
plantisque paulatim obtectam, fluminum inundantium impetu denuo
nudatam, iterumque siccatam et gramine vestitam commemorat. Igi-
tur Historia zoologica, Historia plantarum et Historia oryctologica, quæ
non nisi pristinum orbis terræ statum indicant, a Geognosia probe dis-
tinguendæ."—Humboldt, *Flora Friburgensis Subterranea, cui accedunt
Aphorismi ex Physiologia Chemica Plantarum*, 1793, p. ix.-x. Respecting the "spontaneous motion," which is referred to in a subsequent
part of the text, see the remarkable passage in Aristotle, *De Cœlo*, ii.,
2, p. 284, Bekker, where the distinction between animate and inanimate
bodies is made to depend on the internal or external position of the
seat of the determining motion. "No movement," says the Stagirite,
"proceeds from the vegetable spirit, because plants are buried in a
still sleep, from which nothing can arouse them" (Aristotle, *De Generat.
Animal.*, v. i., p. 778, Bekker); and again, "because plants have no
desires which incite them to spontaneous motion." (Arist., *De Somno
et Vigil.*, cap. i., p. 455, Bekker.)

complete if it were to omit a consideration of these forces, and of the substances which enter into solid and fluid combina- tions in organic tissues, under conditions which, from our igno- rance of their actual nature, we designate by the vague term of *vital forces*, and group into various systems, in accordance with more or less perfectly conceived analogies. The nat- ural tendency of the human mind involuntarily prompts us to follow the physical phenomena of the Earth, through all their varied series, until we reach the final stage of the mor- phological evolution of vegetable forms, and the self-determin- ing powers of motion in animal organisms. And it is by these links that *the geography of organic beings—of plants and animals*—is connected with the delineation of the inorganic phenomena of our terrestrial globe.

Without entering on the difficult question of *spontaneous motion*, or, in other words, on the difference between vegeta- ble and animal life, we would remark, that if nature had en- dowed us with microscopic powers of vision, and the integu- ments of plants had been rendered perfectly transparent to our eyes, the vegetable world would present a very different aspect from the apparent immobility and repose in which it is now manifested to our senses. The interior portion of the cellular structure of their organs is incessantly animated by the most varied currents, either rotating, ascending and de- scending, ramifying, and ever changing their direction, as manifested in the motion of the granular mucus of marine plants (Naiades, Characeæ, Hydrocharidæ), and in the hairs of phanerogamic land plants ; in the molecular motion first dis- covered by the illustrious botanist Robert Brown, and which may be traced in the ultimate portions of every molecule of matter, even when separated from the organ ; in the gyratorv currents of the globules of cambium (*cyclosis*) circulating in their peculiar vessels ; and, finally, in the singularly articula- ted self-unrolling filamentous vessels in the antheridia of the chara, and in the reproductive organs of liverworts and algæ, in the structural conditions of which Meyen, unhappily too early lost to science, believed that he recognized an analogy with the spermatozoa of the animal kingdom.* If to these

* [" In certain parts, probably, of all plants, are found peculiar spiral filaments, having a striking resemblance to the spermatozoa of animals. They have been long known in the organs called the antheridia of mosses, Hepaticæ, and Characeæ, and have more recently been dis- covered in peculiar cells on the germinal frond of ferns, and on the very young leaves of the buds of Phanerogamia. They are found in peculiar cells, and when these are placed in water they are torn by the

manifold currents and gyratory movements we add the phenomena of endosmosis, nutrition, and growth, we shall have
some idea of those forces which are ever active amid the apparent repose of vegetable life.

Since I attempted in a former work, *Ansichten der Natur*
(Views of Nature), to delineate the universal diffusion of life
over the whole surface of the Earth, in the distribution of
organic forms, both with respect to elevation and depth, our
knowledge of this branch of science has been most remarkably
increased by Ehrenberg's brilliant discovery " on microscopic
life in the ocean, and in the ice of the polar regions"—a discovery based, not on deductive conclusions, but on direct observation.   The sphere of vitality, we might almost say, the
horizon of life, has been expanded before our eyes.   " Not
only in the polar regions is there an uninterrupted development of active microscopic life, where larger animals can no
longer exist, but we find that the microscopic animals collected in the Antarctic expedition of Captain James Ross exhibit
a remarkable abundance of unknown and often most beautiful
forms.   Even in the residuum obtained from the melted ice,
swimming about in round fragments in the latitude of 70° 10',
there were found upward of fifty species of silicious-shelled
Polygastria and Coscinodiscæ with their green ovaries, and
therefore living and able to resist the extreme severity of the
cold.   In the Gulf of Erebus, sixty-eight silicious-shelled Polygastria and Phytolitharia, and only one calcareous-shelled Polythalamia, were brought up by lead sunk to a depth of from
1242 to 1620 feet."

The greater number of the oceanic microscopic forms hitherto discovered have been silicious-shelled, although the analysis of sea water does not yield silica as the main constituent,
and it can only be imagined to exist in it in a state of suspension.   It is not only at particular points in inland seas, or in
the vicinity of the land, that the ocean is densely inhabited
by living atoms, invisible to the naked eye, but samples of

filament, which commences an active spiral motion.   The signification
of these organs is at present quite unknown; they appear, from the
researches of Nägeli, to resemble the cell mucilage, or proto-plasma,
in composition, and are developed from it.   Schleiden regards them as
mere mucilaginous deposits, similar to those connected with the circulation in cells, and he contends that the movement of these bodies in
water is analogous to the molecular motion of small particles of organic
and inorganic substances, and depends on mechanical causes."—*Outlines
of Structural and Physiological Botany*, by A. Henfrey, F.L.S., &c.,
1846, p. 23.]—*Tr*

water taken up by Schayer on his return from Van Diemen's Land (south of the Cape of Good Hope, in 57° latitude, and under the tropics in the Atlantic) show that the ocean in its ordinary condition, without any apparent discoloration, contains numerous microscopic moving organisms, which bear no resemblance to the swimming fragmentary silicious filaments of the genus Chætoceros, similar to the Oscillatoriæ so common in our fresh waters.   Some few Polygastria, which have been found mixed with sand and excrements of penguins in Cockburn Island, appear to be spread over the whole earth, while others seem to be peculiar to the polar regions.*

We thus find from the most recent observations that animal life predominates amid the eternal night of the depths of ocean, while vegetable life, which is so dependent on the periodic action of the solar rays, is most prevalent on continents. The mass of vegetation on the Earth very far exceeds that of animal organisms ; for what is the volume of all the large living Cetacea and Pachydermata when compared with the thickly-crowded colossal trunks of trees, of from eight to twelve feet in diameter, which fill the vast forests covering the tropical region of South America, between the Orinoco, the Amazon, and the Rio da Madeira ?   And although the character of different portions of the earth depends on the combination of external phenomena, as the outlines of mountains—the physiognomy of plants and animals—the azure of the sky— the forms of the clouds—and the transparency of the atmosphere—it must still be admitted that the vegetable mantle with which the earth is decked constitutes the main feature of the picture.   Animal forms are inferior in mass, and their powers of motion often withdraw them from our sight.   The

* See Ehrenberg's treatise *Ueber das kleinste Leben im Ocean*, read before the Academy of Science at Berlin on the 9th of May, 1844.

[Dr. J. Hooker found Diatomaceæ in countless numbers between the parallels of 60° and 80° south, where they gave a color to the sea, and also to the icebergs floating in it.   The death of these bodies in the South Arctic Ocean is producing a submarine deposit, consisting entirely of the silicious particles of which the skeletons of these vegetables are composed.   This deposit exists on the shores of Victoria Land and at the base of the volcanic mountain Erebus.   Dr. Hooker accounted for the fact that the skeletons of Diatomaceæ had been found in the lava of volcanic mountains, by referring to these deposits at Mount Erebus, which lie in such a position as to render it quite possible that the skeletons of these vegetables should pass into the lower fissures of the mountain, and then passing into the stream of lava, be thrown out, unacted upon by the heat to which they have been exposed.   See Dr. Hooker's Paper, read before the British Association at Oxford, July, 1847.]—*Tr.*

vegetable kingdom, on the contrary, acts upon our imagination by its continued presence and by the magnitude of its forms; for the size of a tree indicates its age, and here alone age is associated with the expression of a constantly renewed vigor.[*] In the animal kingdom (and this knowledge is also the result of Ehrenberg's discoveries), the forms which we term microscopic occupy the largest space, in consequence of their rapid propagation.[†] The minutest of the Infusoria, the Monadidæ, have a diameter which does not exceed $\frac{1}{3000}$th of a line, and yet these silicious-shelled organisms form in humid districts subterranean strata of many fathoms in depth.

The strong and beneficial influence exercised on the feelings of mankind by the consideration of the diffusion of life throughout the realms of nature is common to every zone, but the impression thus produced is most powerful in the equatorial regions, in the land of palms, bamboos, and arborescent ferns, where the ground rises from the shore of seas rich in mollusca and corals to the limits of perpetual snow. The local distribution of plants embraces almost all heights and all depths. Organic forms not only descend into the interior of the earth, where the industry of the miner has laid open extensive excavations and sprung deep shafts, but I have also found snow-white stalactitic columns encircled by the delicate web of an Usnea, in caves where meteoric water could alone penetrate through fissures. Podurellæ penetrate into the icy crevices of the glaciers on Mount Rosa, the Grindelwald, and the Upper Aar ; the Chionæa araneoides described by Dalman, and the microscopic Discerea nivalis (formerly known as Protococcus), exist in the polar snow as well as in that of our high mountains. The redness assumed by the snow after lying on the ground for some time was known to Aristotle, and was probably observed by him on the mountains of Macedonia.[‡]

<hr />

* Humboldt, *Ansichten der Natur* (2te Ausgabe, 1826), bd. ii., s. 21.

† On multiplication by spontaneous division of the mother-corpuscle and intercalation of new substance, see Ehrenberg, *Von den jetzt lebenden Thierarten der Kreidebildung*, in the *Abhandl. der Berliner Akad. der Wiss.*, 1839, s. 94. The most powerful productive faculty in nature is that manifested in the Vorticellæ. Estimations of the greatest possible development of masses will be found in Ehrenberg's great work, *Die Infusionsthierchen als vollkommne Organismen*, 1838, s. xiii., xix., and 244. "The Milky Way of these organisms comprises the genera Monas, Vibrio, Bacterium, and Bodo." The universality of life is so profusely distributed throughout the whole of nature, that the smaller Infusoria live as parasites on the larger, and are themselves inhabited by others, s. 194, 211, and 512.

‡ Aristot., *Hist. Animal.*, v. xix., p. 552, Bekk.

While, on the loftiest summits of the Alps, only Lecideæ, Parmeliæ, and Umbilicariæ cast their colored but scanty covering over the rocks, exposed by the melted snow, beautiful phanerogamic plants, as the Culcitium rufescens, Sida pinchinchensis, and Saxifraga Boussingaulti, are still found to flourish in the tropical region of the chain of the Andes, at an elevation of more than 15,000 feet. Thermal springs contain small insects (Hydroporus thermalis), Gallionellæ, Oscillatoria, and Confervæ, while their waters bathe the root-fibers of phanerogamic plants. As air and water are animated at different temperatures by the presence of vital organisms, so likewise is the interior of the different portions of animal bodies. Animalcules have been found in the blood of the frog and the salmon ; according to Nordmann, the fluids in the eyes of fishes are often filled with a worm that lives by suction (Diplostomum), while in the gills of the bleak the same observer has discovered a remarkable double animalcule (Diplozoon paradoxum), having a cross-shaped form with two heads and two caudal extremities.

Although the existence of meteoric Infusoria is more than doubtful, it can not be denied that, in the same manner as the pollen of the flowers of the pine is observed every year to fall from the atmosphere, minute infusorial animalcules may likewise be retained for a time in the strata of the air, after having been passively borne up by currents of aqueous vapor.* This circumstance merits serious attention in reconsidering the old discussion respecting *spontaneous generation*,† and the

* Ehrenberg, op. cit., s. xiv., p. 122 and 493. This rapid multiplication of microscopic organisms is, in the case of some (as, for instance, in wheat-eels, wheel-animals, and water-bears or tardigrade animalcules), accompanied by a remarkable tenacity of life. They have been seen to come to life from a state of apparent death after being dried for twenty-eight days in a vacuum with chloride of lime and sulphuric acid, and after being exposed to a heat of 248°. See the beautiful experiments of Doyère, in *Mém. sur les Tardigrades et sur leur propriété de revenir à la vie*, 1842, p. 119, 129, 131, 133. Compare, also, Ehrenberg, s. 492–496, on the revival of animalcules that had been dried during a space of many years.

† On the supposed " primitive transformation" of organized or unorganized matter into plants and animals, see Ehrenberg, in Poggendorf's *Annalen der Physik*, bd. xxiv., s. 1–48, and also his *Infusionsthierchen*, s. 121, 525, and Joh. Müller, *Physiologie des Menschen* (4te Aufl., 1844), bd. i., s. 8–17. It appears to me worthy of notice that one of the early fathers of the Church, St. Augustine, in treating of the question how islands may have been covered with new animals and plants after the flood, shows himself in no way disinclined to adopt the view of the so-called " spontaneous generation" (*generatio æquivoca*,

more so, as Ehrenberg, as I have already remarked, has discovered that the nebulous dust or sand which mariners often
encounter in the vicinity of the Cape Verd Islands, and even
at a distance of 380 geographical miles from the African shore,
contains the remains of eighteen species of silicious-shelled polygastric animalcules.

Vital organisms, whose relations in space are comprised under the head of the geography of plants and animals, may be
considered either according to the difference and relative numbers of the types (their arrangement into genera and species),
or according to the number of individuals of each species on a
given area. In the mode of life of plants as in that of animals, an important difference is noticed ; they either exist in
an isolated state, or live in a social condition. Those species
of plants which I have termed *social** uniformly cover vast
extents of land. Among these we may reckon many of the
marine Algæ—Cladoniæ and mosses, which extend over the
desert steppes of Northern Asia—grasses, and cacti growing

spontanea aut primaria). " If," says he, " animals have not been
brought to remote islands by angels, or perhaps by inhabitants of con
tinents addicted to the chase, they must have been spontaneously produced upon the earth ; although here the question certainly arises, to
what purpose, then, were animals of all kinds assembled in the ark ?"
" Si e terra exortæ sunt (bestiæ) secundum originem primam, quando
dixit Deus: *Producat terra animam vivam !* multo clarius apparet, non
tam reparandorum animalium causa, quam figurandarum variarum gen-
tium (?) propter ecclesiæ sacramentum in arca fuisse omnia genera, si in
insulis quo transire non possent, multa animalia terra produxit." Augus-
tinus, *De Civitate Dei*, lib. xvi., cap. 7 ; *Opera, ed. Monach. Ordinis S.
Benedicti*, t. vii., Venet., 1732, p. 422. Two centuries before the time of
the Bishop of Hippo, we find, by extracts from Trogus Pompeius, that
the *generatio primaria* was brought forward in connection with the
earliest drying up of the ancient world, and of the high table-land of
Asia, precisely in the same manner as the terraces of Paradise, in the
theory of the great Linnæus, and in the visionary hypotheses entertained in the eighteenth century regarding the fabled Atlantis : " Quod si
omnes quondam terræ submersæ profundo fuerunt, profecto editissi-
mam quamque partem decurrentibus aquis primum detectam ; humil-
limo autem solo eandem aquam diutissime immoratam, et quanto prior
quæque pars terrarum siccata sit, tanto prius animalia generare cœpisse.
Porro Scythiam adeo editiorem omnibus terris esse ut cuncta flumina
ibi nata in Mæotium, tum deinde in Ponticum et Ægyptium mare de-
currant."—Justinus, lib. ii., cap. 1. The erroneous supposition that the
land of Scythia is an elevated table-land, is so ancient that we meet
with it most clearly expressed in Hippocrates, *De Ære et Aquis*, cap.
6, § 96, Coray. " Scythia," says he, " consists of high and naked
plains, which, without being crowned with mountains, ascend higher
and higher toward the north."

* Humboldt, *Aphorismi ex Physiologia Chemica Plantarum*, in the
Flora Fribergensis Subterranea, 1793, p. 178.

together like the pipes of an organ—Avicenniæ and mangroves in the tropics—and forests of Coniferæ and of birches in the plains of the Baltic and in Siberia. This mode of geographical distribution determines, together with the individual form of the vegetable world, the size and type of leaves and flowers, in fact, the principal physiognomy of the district ;* its character being but little, if at all, influenced by the ever-moving forms of animal life, which, by their beauty and diversity, so powerfully affect the feelings of man, whether by exciting the sensations of admiration or horror. Agricultural nations increase artificially the predominance of social plants, and thus augment, in many parts of the temperate and northern zones, the natural aspect of uniformity ; and while their labors tend to the extirpation of some wild plants, they likewise lead to the cultivation of others, which follow the colonist in his most distant migration. The luxuriant zone of the tropics offers the strongest resistance to these changes in the natural distribution of vegetable forms.

Observers who in short periods of time have passed over vast tracts of land, and ascended lofty mountains, in which climates were ranged, as it were, in strata one above another, must have been early impressed by the regularity with which vegetable forms are distributed. The results yielded by their observations furnished the rough materials for a science, to which no name had as yet been given. The same zones or regions of vegetation which, in the sixteenth century, Cardinal Bembo, when a youth,† described on the declivity of Ætna, were observed on Mount Ararat by Tournefort. He ingeniously compared the Alpine flora with the flora of plains situated in different latitudes, and was the first to observe the influence exercised in mountainous regions, on the distribution of plants by the elevation of the ground above the level of the sea, and by the distance from the poles in flat countries. Menzel, in an inedited work on the flora of Japan, accidentally made use of the term *geography of plants ;* and the same expression occurs in the fanciful but graceful work of Bernardin de St. Pierre, *Etudes de la Nature.* A scientific treatment of the subject began, however, only when the geography of plants was intimately associated with the study of the dis-

* On the physiognomy of plants, see Humboldt, *Ansichten der Natur*, bd. ii., s. 1–125.

† *Ætna Dialogus. Opuscula*, Basil., 1556, p. 53, 54. A very beautiful geography of the plants of Mount Ætna has recently been published by Philippi. See *Linnæa*, 1832, s. 733.

tribution of heat over the surface of the earth, and when the arrangement of vegetable forms in natural families admitted of a numerical estimate being made of the different forms which increase or decrease as we recede from the equator toward the poles, and of the relations in which, in different parts of the earth, each family stood with reference to the whole mass of phanerogamic indigenous plants of the same region. I consider it a happy circumstance that, at the time during which I devoted my attention almost exclusively to botanical pursuits, I was led by the aspect of the grand and strongly characterized features of tropical scenery to direct my investigations toward these subjects.

The study of the geographical distribution of animals, regarding which Buffon first advanced general, and, in most instances, very correct views, has been considerably aided in its advance by the progress made in modern times in the geography of plants. The curves of the isothermal lines, and more especially those of the isochimenal lines, correspond with the limits which are seldom passed by certain species of plants, and of animals which do not wander far from their fixed habitation, either with respect to elevation or latitude.* The

* [The following valuable remarks by Professor Forbes, on the correspondence existing between the distribution of existing faunas and floras of the British Islands, and the geological changes that have affected their area, will be read with much interest; they have been copied, by the author's permission, from the *Survey Report,* p. 16:

"If the view I have put forward respecting the origin of the flora of the British mountains be true—and every geological and botanical probability, so far as the area is concerned, favors it—then must we endeavor to find some more plausible cause than any yet shown for the presence of numerous species of plants, and of some animals, on the higher parts of Alpine ranges in Europe and Asia, specifically identical with animals and plants indigenous in regions very far north, and not found in the intermediate lowlands. Tournefort first remarked, and Humboldt, the great organizer of the science of natural history geography, demonstrated, that zones of elevation on mountains correspond to parallels of latitude, the higher with the more northern or southern, as the case might be. It is well known that this correspondence is recognized in the general *facies* of the flora and fauna, dependent on generic correspondences, specific representatives, and, in some cases, specific identities. But when announcing and illustrating the law that climatal zones of animal and vegetable life are mutually repeated or represented by elevation and latitude, naturalists have not hitherto sufficiently (if at all) distinguished between the evidence of that law, as exhibited by *representative species* and by *identical.* In reality, the former essentially depend on the law, the latter being an *accident* not necessarily dependent upon it, and which has hitherto not been accounted for. In the case of the Alpine flora of Britain, the evidence of the activity of the law, and the influence of the accident, are inseparable, the law be-

elk, for instance, lives in the Scandinavian peninsula, almost ten degrees further north than in the interior of Siberia, where the line of equal winter temperature is so remarkably concave. Plants migrate in the germ ; and, in the case of many species, the seeds are furnished with organs adapting them to be conveyed to a distance through the air. When once they have taken root, they become dependent on the soil and on the strata of air surrounding them. Animals, on the contrary, can at pleasure migrate from the equator toward the poles ; and this they can more especially do where the isothermal lines are much inflected, and where hot summers succeed a great degree of winter cold. The royal tiger, which in no respect differs from the Bengal species, penetrates every summer into

ing maintained by a transported flora, for the transmission of which I have shown we can not account by an appeal to unquestionable geological events. In the case of the Alps and Carpathians, and some other mountain ranges, we find the law maintained partly by a representative flora, special in its region, *i. e.*, by specific centers of their own, and partly by an assemblage more or less limited in the several ranges of identical species, these latter in several cases so numerous that ordinary modes of transportation now in action can no more account for their presence than they can for the presence of a Norwegian flora on the British mountains. Now I am prepared to maintain that the same means which introduced a sub-Arctic (now mountain) flora into Britain, acting at the same epoch, originated the identity, as far as it goes, of the Alpine floras of Middle Europe and Central Asia ; for, now that we know the vast area swept by the glacial sea, including almost the whole of Central and Northern Europe, and belted by land, since greatly uplifted, which then presented to the water's edge those climatal conditions for which a sub-Arctic flora—destined to become Alpine—was specially organized, the difficulty of deriving such a flora from its parent north, and of diffusing it over the snowy hills bounding this glacial ocean, vanishes, and the presence of identical species at such distant points remain no longer a mystery. Moreover, when we consider that the greater part of the northern hemisphere was under such climatal conditions during the epoch referred to, the undoubted evidences of which have been made known in Europe by numerous British and Continental observers, on the bounds of Asia by Sir Roderick Murchison, in America by Mr. Lyell, Mr. Logan, Captain Bayfield, and others, and that the botanical (and zoological as well) region, essentially northern and Alpine, designated by Professor Schouw that ' of saxifrages and mosses,' and first in his classification, exists now only on the flanks of the great area which suffered such conditions ; and that, though similar conditions reappear, the relationship of Alpine and Arctic vegetation in the southern hemisphere, with that in the northern, is entirely maintained by *representative*, and not by identical species (the representative, too, being in great part generic, and not specific), the general truth of my explanation of Alpine floras, including identical species, becomes so strong, that the view proposed acquires fair claims to be ranked as a theory, and not considered merely a convenient or bold hypothesis."]— *Tr.*

the north of Asia as far as the latitudes of Berlin and Hamburg, a fact of which Ehrenberg and myself have spoken in other works.*

The grouping or association of different vegetable species, to which we are accustomed to apply the term *Floras*, do not appear to me, from what I have observed in different portions of the earth's surface, to manifest such a predominance of individual families as to justify us in marking the geographical distinctions between the regions of the Umbellatæ, of the Solidaginæ, of the Labiatæ, or the Scitamineæ. With reference to this subject, my views differ from those of several of my friends, who rank among the most distinguished of the botanists of Germany. The character of the floras of the elevated plateaux of Mexico, New Granada, and Quito, of European Russia, and of Northern Asia, consists, in my opinion, not so much in the relatively larger number of the species presented by one or two natural families, as in the more complicated relations of the coexistence of many families, and in the relative numerical value of their species. The Gramineæ and the Cyperaceæ undoubtedly predominate in meadow lands and steppes, as do Coniferæ, Cupuliferæ, and Betulineæ in our northern woods; but this predominance of certain forms is only apparent, and owing to the aspect imparted by the social plants. The north of Europe, and that portion of Siberia which is situated to the north of the Altai Mountains, have no greater right to the appellation of a region of Gramineæ and Coniferæ than have the boundless llanos between the Orinoco and the mountain chain of Caraccas, or the pine forests of Mexico. It is the coexistence of forms which may partially replace each other, and their relative numbers and association, which give rise either to the general impression of luxuriance and diversity, or of poverty and uniformity in the contemplation of the vegetable world.

In this fragmentary sketch of the phenomena of organization, I have ascended from the simplest cell†—the first manifestation of life—progressively to higher structures. " The

* Ehrenberg, in the *Annales des Sciences Naturelles*, t. **xxi.**, p. 387 412; Humboldt, *Asie Centrale*, t. i., p. 339–342, and t. iii., p. 96–101

† Schleiden, *Ueber die Entwicklungsweise der Pflanzenzellen*, in Müller's *Archiv für Anatomie und Physiologie*, 1838, s. 137–176; also his *Grundzüge der wissenschaftlichen Botanik*, th. i., s. 191, and th. ii., s 11. Schwann, *Mikroscopische Untersuchungen über die Uebereinstimmung in der Struktur und dem Wachsthum der Thiere und Pflanzen*, 1839, s. 45, 220. Compare also, on similar propagation, Joh. Müller *Physiologie des Menschen*, 1840 th. ii., s. 614.

association of mucous granules constitutes a definitely-formed cytoblast, around which a vesicular membrane forms a closed cell," this cell being either produced from another pre-existing cell,* or being due to a cellular formation, which, as in the case of the fermentation-fungus, is concealed in the obscurity of some unknown chemical process.† But in a work like the present we can venture on no more than an allusion to the mysteries that involve the question of modes of origin ; the geography of animal and vegetable organisms must limit itself to the consideration of germs already developed, of their habitation and transplantation, either by voluntary or involuntary migrations, their numerical relation, and their distribution over the surface of the earth.

The general picture of nature which I have endeavored to delineate would be incomplete if I did not venture to trace a few of the most marked features of the human race, considered with reference to physical gradations—to the geographical distribution of cotemporaneous types—to the influence exercised upon man by the forces of nature, and the reciprocal, although weaker action which he in his turn exercises on these natural forces. Dependent, although in a lesser degree than plants and animals, on the soil, and on the meteorological processes of the atmosphere with which he is surrounded —escaping more readily from the control of natural forces, by activity of mind and the advance of intellectual cultivation, no less than by his wonderful capacity of adapting himself to all climates—man every where becomes most essentially associated with terrestrial life. It is by these relations that the obscure and much-contested problem of the possibility of one common descent enters into the sphere embraced by a general physical cosmography. The investigation of this problem will impart a nobler, and, if I may so express myself, more purely human interest to the closing pages of this section of my work.

The vast domain of language, in whose varied structure we see mysteriously reflected the destinies of nations, is most intimately associated with the affinity of races ; and what even slight differences of races may effect is strikingly manifested in the history of the Hellenic nations in the zenith of their intellectual cultivation. The most important questions of the civilization of mankind are connected with the ideas of races,

* Schleiden, *Grundzüge der wissenschaftlichen Botanik*, 1842, th. i., s. 192–197.
† [On cellular formation, see Henfrey's *Outlines of Structural and Physiological Botany*, op. cit., p. 16–22.]—*Tr.*

community of language, and adherence to one original direction of the intellectual and moral faculties.

As long as attention was directed solely to the extremes in
varieties of color and of form, and to the vividness of the first
impression of the senses, the observer was naturally disposed
to regard races rather as originally different species than as
mere varieties. The permanence of certain types* in the midst
of the most hostile influences, especially of climate, appeared
to favor such a view, notwithstanding the shortness of the interval of time from which the historical evidence was derived.
In my opinion, however, more powerful reasons can be advanced in support of the theory of the unity of the human
race, as, for instance, in the many intermediate gradations†
in the color of the skin and in the form of the skull, which
have been made known to us in recent times by the rapid progress of geographical knowledge—the analogies presented by
the varieties in the species of many wild and domesticated animals—and the more correct observations collected regarding
the limits of fecundity in hybrids.‡　The greater number of
the contrasts which were formerly supposed to exist, have disappeared before the laborious researches of Tiedemann on the
brain of negroes and of Europeans, and the anatomical inves-

* Tacitus, in his speculations on the inhabitants of Britain (*Agricola*,
cap. ii.), distinguishes with much judgment between that which may
be owing to the local climatic relations, and that which, in the immigrating races, may be owing to the unchangeable influence of a hered-
itary and transmitted type.　"Britanniam qui mortales initio coluerunt,
indigenæ an advecti, ut inter barbaros, parum compertum.　Habitus
corporis varii, atque ex eo argumenta ; namque rutilæ Caledoniam hab-
itantium comæ, magni artus Germanicam originem adseverant.　Silu
rum colorati vultus et torti plerumque crines, et posita contra Hispania,
Iberos veteres trajecisse, easque cedes occupasse fidem faciunt : proxi-
mi Gallis, et similes sunt : seu durante originis vi ; seu procurrentibus
in diversa terris, positio cœli corporibus habitum dedit."　Regarding
the persistency of types of conformation in the hot and cold regions of
the earth, and in the mountainous districts of the New Continent, see
my *Relation Historique*, t. i., p. 498, 503, and t. ii., p. 572, 574.

† On the American races generally, see the magnificent work of
Samuel George Morton, entitled *Crania Americana*, 1839, p. 62, 86 ;
and on the skulls brought by Pentland from the highlands of Titicaca,
see the *Dublin Journal of Medical and Chemical Science*, vol. v., 1834,
p. 475 ; also Alcide d'Orbigny, *L'homme Américain considéré sous ses
rapports Physiol. et Mor.*, 1839, p. 221 ; and the work by Prince Maxi-
milian of Wied, which is well worthy of notice for the admirable ethno
graphical remarks in which it abounds, entitled *Reise in das Innere von
Nordamerika* (1839).

‡ Rudolph Wagner, *Ueber Blendlinge und Bastarderzeugung*, in his
notes to the German translation of Prichard's *Physical History of Mankind*, vol. i., p. 138–150.

tigations of Vrolik and Weber on the form of the pelvis. On comparing the dark-colored African nations, on whose physical history the admirable work of Prichard has thrown so much light, with the races inhabiting the islands of the South-Indian and West-Australian archipelago, and with the Papuas and Alfourous (Haroforas, Endamenes), we see that a black skin, woolly hair, and a negro-like cast of countenance are not necessarily connected together.* So long as only a small portion of the earth was known to the Western nations, partial views necessarily predominated, and tropical heat and a black skin consequently appeared inseparable. " The Ethiopians," said the ancient tragic poet Theodectes of Phaselis,† "are colored by the near sun-god in his course with a sooty luster, and their hair is dried and crisped with the heat of his rays." The campaigns of Alexander, which gave rise to so many new ideas regarding physical geography, likewise first excited a discussion on the problematical influence of climate on races. " Families of animals and plants," writes one of the greatest anatomists of the day, Johannes Müller, in his noble and comprehensive work, *Physiologie des Menschen*, " undergo, within certain limitations peculiar to the different races and species, various modifications in their distribution over the surface of the earth, propagating these variations as organic types of species.‡ The present races of animals have been produced by

* Prichard, op. cit., vol. ii., p. 324.

† Onesicritus, in Strabo, xv., p. 690, 695, Casaub. Welcker, *Griechische Tragödien*, abth. iii., s. 1078, conjectures that the verses of Theodectes, cited by Strabo, are taken from a lost tragedy, which probably bore the title of " Memnon."

‡ [In illustration of this, the conclusions of Professor Edward Forbes respecting the origin and diffusion of the British flora may be cited. See the *Survey Memoir* already quoted, *On the Connection between the Distribution of the existing Fauna and Flora of the British Islands*, &c., p. 65. " 1. The flora and fauna, terrestrial and marine, of the British islands and seas, have originated, so far as that area is concerned, since the meiocene epoch. 2. The assemblages of animals and plants composing that fauna and flora did not appear in the area they now inhabit simultaneously, but at several distinct points in time. 3. Both the fauna and flora of the British islands and seas are composed partly of species which, either permanently or for a time, appeared in that area before the glacial epoch ; partly of such as inhabited it during that epoch ; and in great part of those which did not appear there until afterward, and whose appearance on the earth was coeval with the elevation of the bed of the glacial sea and the consequent climatal changes. 4. The greater part of the terrestrial animals and flowering plants now inhabiting the British islands are members of specific centers beyond their area, and have migrated to it over continuous land before, during, or after the glacial epoch. 5. The climatal conditions of the area under

the combined action of many different internal as well as external conditions, the nature of which can not in all cases be
defined, the most striking varieties being found in those families which are capable of the greatest distribution over the surface of the earth. The different races of mankind are forms
of one sole species, by the union of two of whose members
descendants are propagated. They are not different species
of a genus, since in that case their hybrid descendants would
remain unfruitful. But whether the human races have descended from several primitive races of men, or from one alone,
is a question that can not be determined from experience."*

Geographical investigations regarding the ancient *seat*, the
so-called *cradle of the human race*, are not devoid of a myth-

discussion, and north, east, and west of it, were severer during the gla
cial epoch, when a great part of the space now occupied by the British
isles was under water, than they are now or were before; but there is
good reason to believe that, so far from those conditions having continued severe, or having gradually diminished in severity southward of
Britain, the cold region of the glacial epoch came directly into contact
with a region of more southern and thermal character than that in which
the most southern beds of glacial drift are now to be met with. 6. This
state of things did not materially differ from that now existing, under
corresponding latitudes, in the North American, Atlantic, and Arctic
seas, and on their bounding shores. 7. The Alpine floras of Europe
and Asia, so far as they are identical with the flora of the Arctic and
sub-Arctic zones of the Old World, are fragments of a flora which was
diffused from the north, either by means of transport not now in action
on the temperate coasts of Europe, or over continuous land which no
.onger exists. The deep sea fauna is in like manner a fragment of the
general glacial fauna. 8. The floras of the islands of the Atlantic region, between the Gulf-weed Bank and the Old World, are fragments
of the great Mediterranean flora, anciently diffused over a land constituted out of the upheaved and never again submerged bed of the (shallow) Meiocene Sea. This great flora, in the epoch anterior to, and
probably, in part, during the glacial period, had a greater extension
northward than it now presents. 9. The termination of the glacial
epoch in Europe was marked by a recession of an Arctic fauna and flora
northward, and of a fauna and flora of the Mediterranean type southward; and in the interspace thus produced there appeared on land the
Germanic fauna and flora, and in the sea that fauna termed Celtic.
10. The causes which thus preceded the appearance of a new assemblage of organized beings were the destruction of many species of animals, and probably also of plants, either forms of extremely local distribution, or such as were not capable of enduring many changes of conditions—species, in short, with very limited capacity for horizontal or
vertical diffusion. 11. All the changes before, during, and after the
glacial epoch appear to have been gradual, and not sudden, so that no
marked line of demarkation can be drawn between the creatures inhabiting the same element and the same locality during two proximate
periods."]—*Tr.*

* Joh. Müller, *Physiologie des Menschen*, bd. ii., s. 768.

ical character. "We do not know," says Wilhelm von Humboldt, in an unpublished work *On the Varieties of Languages and Nations*, " either from history or from authentic tradition, any period of time in which the human race has not been divided into social groups. Whether the gregarious condition was original, or of subsequent occurrence, we have no historic evidence to show. The separate mythical relations found to exist independently of one another in different parts of the earth, appear to refute the first hypothesis, and concur in ascribing the generation of the whole human race to the union of one pair. The general prevalence of this myth has caused it to be regarded as a traditionary record transmitted from the primitive man to his descendants. But this very circumstance seems rather to prove that it has no historical foundation, but has simply arisen from an identity in the mode of intellectual conception, which has every where led man to adopt the same conclusion regarding identical phenomena ; in the same manner as many myths have doubtlessly arisen, not from any historical connection existing between them, but rather from an identity in human thought and imagination. Another evidence in favor of the purely mythical nature of this belief is afforded by the fact that the first origin of mankind—a phenomenon which is wholly beyond the sphere of experience—is explained in perfect conformity with existing views, being considered on the principle of the colonization of some desert island or remote mountainous valley at a period when mankind had already existed for thousands of years. It is in vain that we direct our thoughts to the solution of the great problem of the first origin, since man is too intimately associated with his own race and with the relations of time to conceive of the existence of an individual independently of a preceding generation and age. A solution of those difficult questions, which can not be determined by inductive reasoning or by experience—whether the belief in this presumed traditional condition be actually based on historical evidence, or whether mankind inhabited the earth in gregarious associations from the origin of the race—can not, therefore, be determined from philological data, and yet its elucidation ought not to be sought from other sources."

The distribution of mankind is therefore only a distribution into *varieties*, which are commonly designated by the somewhat indefinite term *races*. As in the vegetable kingdom, and in the natural history of birds and fishes, a classification into many small families is based on a surer foundation than

where large sections are separated into a few but large divisions ; so it also appears to me, that in the determination of
races a preference should be given to the establishment of
small families of nations.   Whether we adopt the old classification of my master, Blumenbach, and admit *five* races (the
Caucasian, Mongolian, American, Ethiopian, and Malayan),
or that of Prichard, into *seven* races* (the Iranian, Turanian,
American, Hottentots and Bushmen, Negroes, Papuas, and
Alfourous), we fail to recognize any typical sharpness of definition, or any general or well-established principle in the division of these groups.   The extremes of form and color are
certainly separated, but without regard to the races, which
can not be included in any of these classes, and which have
been alternately termed Scythian and Allophyllic.   Iranian is
certainly a less objectionable term for the European nations
than Caucasian ; but it may be maintained generally that
geographical denominations are very vague when used to express the points of departure of races, more especially where
the country which has given its name to the race, as, for instance, Turan (Mawerannahr), has been inhabited at different periods† by Indo-Germanic and Finnish, and not by Mongolian tribes.

* Prichard, op. cit., vol. i., p. 247.
† The late arrival of the Turkish and Mongolian tribes on the Oxus
and on the Kirghis Steppes is opposed to the hypothesis of Niebuhr,
according to which the Scythians of Herodotus and Hippocrates were
Mongolians.   It seems far more probable that the Scythians (Scoloti)
should be referred to the Indo-Germanic Massagetæ (Alani).   The
Mongolian, true Tartars (the latter term was afterward falsely given to
purely Turkish tribes in Russia and Siberia), were settled, at that period, far in the eastern part of Asia.   See my *Asie Centrale*, t. i., p. 239,
400 ; *Examen Critique de l'Histoire de la Géogr.*, th. ii., p. 320.   A distinguished philologist, Professor Buschmann, calls attention to the circumstance that the poet Firdousi, in his half-mythical prefatory remarks
in the *Schahnameh*, mentions " a fortress of the Alani" on the sea-shore,
in which Selm took refuge, this prince being the eldest son of the
King Feridun, who in all probability lived two hundred years before
Cyrus.   The Kirghis of the Scythian steppe were originally a Finnish
tribe; their three hordes probably constitute in the present day the
most numerous nomadic nation, and their tribe dwelt, in the sixteenth
century, in the same steppe in which I have myself seen them.   The
Byzantine Menander (p. 380–382, ed. Nieb.) expressly states that the
Chacan of the Turks (Thu-Khiu), in 569, made a present of a Kirghis
slave to Zemarchus, the embassador of Justinian II.; he terms her a
χερχίς ; and we find in Abulgasi (*Historia Mongolorum et Tatarorum*)
that the Kirghis are called Kirkiz.   Similarity of manners, where the
nature of the country determines the principal characteristics, is a very
uncertain evidence of identity of race.   The life of the steppes produces among the Turks (Ti Tukiu), the Baschkirs (Fins), the Kirghis,

Languages, as intellectual creations of man, and as closely interwoven with the development of mind, are, independently of the *national* form which they exhibit, of the greatest importance in the recognition of similarities or differences in races. This importance is especially owing to the clew which a community of descent affords in treading that mysterious labyrinth in which the connection of physical powers and intellectual forces manifests itself in a thousand different forms. The brilliant progress made within the last half century, in Germany, in philosophical philology, has greatly facilitated our investigations into the *national* character* of languages and the influence exercised by descent. But here, as in all domains of ideal speculation, the dangers of deception are closely linked to the rich and certain profit to be derived.

Positive ethnographical studies, based on a thorough knowledge of history, teach us that much caution should be applied in entering into these comparisons of nations, and of the languages employed by them at certain epochs. Subjection, long association, the influence of a foreign religion, the blending of races, even when only including a small number of the more influential and cultivated of the immigrating tribes, have produced, in both continents, similarly recurring phenomena ; as, for instance, in introducing totally different families of languages among one and the same race, and idioms, having one common root, among nations of the most different origin. Great Asiatic conquerors have exercised the most powerful influence on phenomena of this kind.

But language is a part and parcel of the history of the development of mind ; and, however happily the human intellect, under the most dissimilar physical conditions, may unfettered pursue a self-chosen track, and strive to free itself from the dominion of terrestrial influences, this emancipation is never perfect. There ever remains, in the natural capacities of the mind, a trace of something that has been derived from the influences of race or of climate, whether they be associated with a land gladdened by cloudless azure skies, or with the vapory atmosphere of an insular region. As, therefore, richness and grace of language are unfolded from the most luxu-

the Torgodi and Dsungari (Mongolians), the same habits of nomadic life, and the same use of felt tents, carried on wagons and pitched among herds of cattle.

* Wilhelm von Humboldt, *Ueber die Verschiedenheit der menschlichen Sprachbaues*, in his great work *Ueber die Kawi-Sprache auf der Insel Java*, bd. i., s. xxi., xlviii., and ccxiv.

riant depths of thought, we have been unwilling wholly to disregard the bond which so closely links together the physical world with the sphere of intellect and of the feelings by depriving this general picture of nature of those brighter lights and tints which may be borrowed from considerations, however slightly indicated, of the relations existing between races and languages.

While we maintain the unity of the human species, we at the same time repel the depressing assumption of superior and inferior races of men.* There are nations more susceptible of cultivation, more highly civilized, more ennobled by mental cultivation than others, but none in themselves nobler than others. All are in like degree designed for freedom ; a freedom which, in the ruder conditions of society, belongs only to the individual, but which, in social states enjoying political institutions, appertains as a right to the whole body of the community. "If we would indicate an idea which, throughout the whole course of history, has ever more and more widely extended its empire, or which, more than any other, testifies to the much-contested and still more decidedly misunderstood perfectibility of the whole human race, it is that of establishing our common humanity—of striving to remove the barriers which prejudice and limited views of every kind have erected among men, and to treat all mankind, without reference to religion, nation, or color, as one fraternity, one great community, fitted for the attainment of one object, the unrestrained development of the physical powers. This is the ultimate and highest aim of society, identical with the direction implanted by nature in the mind of man toward the indefinite extension of his existence. He regards the earth in all its limits, and the heavens as far as his eye can scan their bright and starry depths, as inwardly his own, given to him as the objects of his contemplation, and as a field for the development of his energies. Even the child longs to pass the hills or the seas which inclose his narrow home ; yet, when his eager steps have borne him beyond those limits, he pines, like the plant, for his native soil ; and it is by this touching and beautiful attribute of man—this longing for that which is unknown, and this fond remembrance of that which is lost —that he is spared from an exclusive attachment to the pres-

* The very cheerless, and, in recent times, too often discussed doctrine of the unequal rights of men to freedom, and of slavery as an institution in conformity with nature, is unhappily found most systematically developed in Aristotle's *Politica*, i., 3, 5, 6.

ent. Thus deeply rooted in the innermost nature of man, and even enjoined upon him by his highest tendencies, the recognition of the bond of humanity becomes one of the noblest leading principles in the history of mankind."*

With these words, which draw their charm from the depths of feeling, let a brother be permitted to close this general description of the natural phenomena of the universe. From the remotest nebulæ and from the revolving double stars, we have descended to the minutest organisms of animal creation, whether manifested in the depths of ocean or on the surface of our globe, and to the delicate vegetable germs which clothe the naked declivity of the ice-crowned mountain summit; and here we have been able to arrange these phenomena according to partially known laws; but other laws of a more mysterious nature rule the higher spheres of the organic world, in which is comprised the human species in all its varied conformation, its creative intellectual power, and the languages to which it has given existence. A physical delineation of nature terminates at the point where the sphere of intellect begins, and a new world of mind is opened to our view. It marks the limit, but does not pass it.

---

* Wilhelm von Humboldt, *Ueber die Kawi-Sprache*, bd. iii., s. 426. I subjoin the following extract from this work: " The impetuous conquests of Alexander, the more politic and premeditated extension of territory made by the Romans, the wild and cruel incursions of the Mexicans, and the despotic acquisitions of the incas, have in both hemispheres contributed to put an end to the separate existence of many tribes as independent nations, and tended at the same time to establish more extended international amalgamation. Men of great and strong minds, as well as whole nations, acted under the influence of one idea, the purity of which was, however, utterly unknown to them. It was Christianity which first promulgated the truth of its exalted charity, although the seed sown yielded but a slow and scanty harvest. Before the religion of Christ manifested its form, its existence was only revealed by a faint foreshadowing presentiment. In recent times, the idea of civilization has acquired additional intensity, and has given rise to a desire of extending more widely the relations of national intercourse and of intellectual cultivation; even selfishness begins to learn that by such a course its interests will be better served than by violent and forced isolation. Language, more than any other attribute of mankind, binds together the whole human race. By its idiomatic proper ties it certainly seems to separate nations, but the reciprocal under-standing of foreign languages connects men together, on the other hand, without injuring individual national characteristics."

ADDITIONAL NOTES

TO THE PRESENT EDITION. MARCH, 1849.

GIGANTIC BIRDS OF NEW ZEALAND.—Vol. i., p. 287.

An extensive and highly interesting collection of bones, referrible to several species of the *Moa* (Dinornis of Owen), and to three or four other genera of birds, formed by Mr. Walter Mantell, of Wellington, New Zealand, has recently arrived in England, and is now deposited in the British Museum. This series consists of between 700 and 800 specimens, belonging to different parts of the skeletons of many individuals of various sizes and ages. Some of the largest vertebræ, tibiæ, and femora equal in magnitude the most gigantic previously known, while others are not larger than the corresponding bones of the living apteryx. Among these relics are the *skulls* and *mandibles* of two genera, the *Dinornis* and *Palapteryx;* and of an extinct genus, *Notornis*, allied to the *Rallidæ;* and the mandibles of a species of *Nestor*, a genus of nocturnal owl-like parrots, of which only two living species are known.*

These osseous remains are in a very different state of preservation from any previously received from New Zealand; they are light and porous, and of a light fawn-color; the most delicate processes are entire, and the articulating surfaces smooth and uninjured; *fragments of egg-shells*, and *even the bony rings of the trachea and air tubes, are preserved.*

The bones were dug up by Mr. Walter Mantell from a bed of marly sand, containing magnetic iron, crystals of hornblende and augite, and the detritus of augitic rocks and earthy volcanic tuff. This sand had filled up all the cavities and cancelli, but was in no instance consolidated or aggregated together; it was, therefore, easily removed by a soft brush, and the bones perfectly cleared without injury.

The spot whence these precious relics of the colossal birds that once inhabited the islands of New Zealand were obtained, is a flat tract of land, near the embouchure of a river, named Waingongoro, not far from Wanganui, which has its rise in the volcanic regions of Mount Egmont. The natives affirm that this level tract was one of the places first dwelt upon by their remote ancestors; and this tradition is corroborated by the existence of numerous heaps and pits of ashes and charred bones indicating ancient fires, long burning on the same spot. In these fire-heaps Mr. Mantell found burned bones of *men, moas,* and *dogs.*

The fragments of egg-shells, imbedded in the ossiferous deposits, had escaped the notice of all previous naturalists. They are, unfortunately, very small portions, the largest being only four inches long, but they afford a chord by which to estimate the size of the original. Mr. Mantell observes that the egg of the Moa must have been so large that a hat would form a good egg-cup for it. These relics evidently belong to two or more species, perhaps genera. In some examples the ex-

* See Professor Owen's Memoir on these fossil remains, in *Zoological Transactions,* 1848

ternal surface is smooth; in others it is marked with short intercepted linear grooves, resembling the eggs of some of the Struthionidæ, but distinct from all known recent types.   In this valuable collection only one bone of a mammal has been detected, namely, *the femur of a dog*.

An interesting memoir on the probable geological position and age of the ornithic bone deposits of New Zealand, by Dr. Mantell, based on the observations of his enterprising son, is published in the Quarter-ly Journal of the Geological Society of London (1848).   It appears that in many instances the bones are imbedded in sand and clay, which lie beneath a thick deposit of volcanic detritus, and rest on an argillaceous stratum abounding in marine shells.   The specimens found in the rivers and streams have been washed out of their banks by the currents which now flow through channels from ten to thirty feet deep, formed in the more ancient alluvial soil.   Dr. Mantell concludes that the islands of New Zealand were densely peopled at a period geologically recent, though historically remote, by tribes of gigantic brevi-pennate birds allied to the ostrich tribe, all, or almost all, of species and genera now extinct; and that, subsequently to the formation of the most ancient ornithic deposit, the sea-coast has been elevated from fifty to one hund-red feet above its original level; hence the terraces of shingle and loam which now skirt the maritime districts.   The existing rivers and mountain torrents flow in deep gulleys which they have eroded in the course of centuries in these pleistocene strata, in like manner as the river courses of Auvergne, in Central France, are excavated in the mammiferous tertiary deposits of that country.   The last of the gigantic birds were probably exterminated, like the dodo, by human agency: some small species allied to the apteryx may possibly be met with in the unexplored parts of the middle island.

THE DODO.—A most valuable and highly interesting history of the dodo and its kindred[*] has recently appeared, in which the history, affinities, and osteology of the *Dodo, Solitaire,* and other extinct birds of the islands Mauritius, Rodriguez, and Bourbon are admirably eluci-dated by H. G. Strickland (of Oxford), and Dr. G. A. Melville.   The historical part is by the former, the osteological and physiological por-tion by the latter eminent anatomist.   We would earnestly recommend the reader interested in the most perfect history that has ever appear-ed, of the extinction of a race of large animals, of which thousands ex-isted but three centuries ago, to refer to the original work.   We have only space enough to state that the authors have proved, upon the most incontrovertible evidence, that the dodo was neither a vulture, ostrich, nor galline, as previous anatomists supposed, but a *frugiverous pigeon.*

* *The Dodo and its Kindred.* By Messrs. Strickland and Melville.   1 vol. 4to, with numerous plates.   Reeves, London, 1848.

INDEX TO VOL. I.

END OF VOL. I.

Library of Congress-Cataloging-in-Publication Data

Humboldt, Alexander von. 1769–1859.
   [Kosmos. English]
   Cosmos : a sketch of a physical description of the universe /
  by Alexander von Humboldt ; translated by E. C. Otté.
      p.   cm. — (Foundations of natural history)
  Previously published: New York : Harper & Brothers, 1858.
  Includes index.
  ISBN 0-8018-5502-0 (vol. 1, pbk. : alk. paper)
  ISBN 0-8018-5503-9 (vol. 2, pbk. : alk. paper)
  1. Cosmology.   I. Series
QB981.H8613   1997
508—dc20   96-36421
                                        CIP

Ingram Content Group UK Ltd.
Milton Keynes UK
UKHW012112170423
420317UK00002B/198

9 780801 855023